Student Solutions Manual
to Accompany

Keller/Gettys/Skove

Physics SECOND EDITION

and

Physics: Classical and Modern SECOND EDITION

Volume I (Chapters 1-19)

Student Solutions Manual
to Accompany

Keller/Gettys/Skove

Physics SECOND EDITION

and

Physics: Classical and Modern SECOND EDITION

Volume I (Chapters 1-19)

Jai N. Dahiya and Giulio V. Venezian
Southeast Missouri State University

McGraw-Hill, Inc.
NEW YORK ST. LOUIS SAN FRANCISCO AUCKLAND BOGOTÁ CARACAS
LISBON LONDON MADRID MEXICO CITY MILAN MONTREAL
NEW DELHI SAN JUAN SINGAPORE SYDNEY TOKYO TORONTO

Student Solutions Manual to Accompany
Keller/Gettys/Skove: PHYSICS, Second Edition, and
PHYSICS: CLASSICAL AND MODERN, Second Edition
Volume I (Chapters 1-19)

This book is printed on recycled paper containing a minimum of 50% total recycled fiber with 10% postconsumer de-inked fiber.

2 3 4 5 6 7 8 9 0 MAL MAL 9 0 9 8 7 6 5 4 3

ISBN 0-07-033910-4

The editor was Jack Shira; the production supervisor was Diane Ficarra.
Malloy Lithographing, Inc., was printer and binder.

Contents

Preface

This manual contains complete, worked-out solutions to the odd-numbered questions, exercises, and problems that appear in Keller, Gettys, and Skove's *Physics*, second edition, and *Physics: Classical and Modern,* second edition. We hope that in observing the approach used in solving these problems, you will be able to understand the logic and reasoning behind the concepts presented in this text.

Every effort has been made to avoid errors in the solutions. Problems have been checked and rechecked to ensure accuracy. We sincerely apologize for any errors we may have missed and encourage you, the reader, to inform us of any inaccurate or confusing solutions you may find.

We wish to thank all who worked on this project, for authors alone do not make a book. We are grateful to Eric Sumner for his careful and detailed accuracy checking, copy editing, and proofreading; Anne R. Costa and Jeanne Wallace for their precise typesetting; John Bullock for his help in solving problems in Chapters 38-44; and Beth H. Morel for pulling it all together. Without their time, effort, and patience, this book would never have become a reality.

Jai N. Dahiya
Giulio V. Venezian

CHAPTER 1 INTRODUCTION

Questions:

1. The dimensions of the volume of a cube are $(\text{length})^3$.
The dimensions of the volume of a sphere are also $(\text{length})^3$.
The ratio of the volume of the sphere to that of the cube has no dimensions.

3. The dimensions of 60 mi/h are length/time and those of 2.5 kg/m^3 are $\text{mass}/(\text{length})^3$.

5. If we made contact with a civilization in another galaxy, we could explain our standard of time in terms of the period of oscillation of light emitted by given atoms, our standard of length by the distance travelled by light in a given time and our standard of mass in terms of the mass of a proton.

7. The correction is put into civil time because the rotation of the earth is not uniform.

9. We keep a standard kilogram because mass cannot be measured accurately in terms of atomic properties. Time and length can be measured to high accuracy in terms of atomic properties.

11. If an equation is consistent in one set of units, it is not necessarily dimensionally consistent because it may contain dimensional constants.

13. In a sum, the precision of the answer is determined by the number with the largest magnitude.

15. If 1 blink = 30 μs, light travels 3×10^8 m/s $\times 30 \times 10^{-6}$ s = 9000 m, so the speed of light is 9000 m per blink.

Exercises:

1. For the equation to be dimensionally consistent, all the terms must have the same dimensions. The first two terms have dimensions L^2/T^2. Thus the dimension of b must be L/T^2. These are the dimensions of acceleration.

3. The dimension of K is ML^2/T^2.

5. The dimension of $a^b v^c t^d$ is

$$\left(\frac{L}{T^2}\right)^b \left(\frac{L}{T}\right)^c T^d = L^{b+c} T^{d-2b-c}$$

If this is dimensionless,

$$b + c = 0$$
$$d - 2b - c = 0$$

Thus

$$c = -b, \quad d = b$$

Taking (arbitrarily) b = 1, then c = −1 and d = 1. The dimensionless variable is

$$at/v$$

Of course any power of this quantity is also dimensionless.

7. To convert km/hr to m/s, multiply by

$$1000\,\frac{\text{m}}{\text{km}} \times \frac{\text{hr}}{3600\ \text{s}} = 0.278\,\frac{\text{m/s}}{\text{km/hr}}$$

9. To convert mi/hr/s to m/s^2, multiply by

$$1609\,\frac{\text{m}}{\text{mi}} \times \frac{\text{hr}}{3600\ \text{s}} = 0.447\,\frac{\text{m/s}}{\text{mi/hr}}$$

So 12 mi/hr/s = 5.36 m/s^2.

11. To find the speed of light in miles per second, multiply

$$3.00 \times 10^8\,\frac{\text{m}}{\text{s}} \times \frac{\text{mi}}{1609\ \text{m}} = 186 \times 10^3\,\frac{\text{mi}}{\text{s}}$$

13.

$$\begin{array}{r} 84.626 \\ +\ 923.1 \\ \hline 1007.7 \end{array}$$

The second number has only one significant figure to the right of the decimal point. Thus the result cannot have more.

15.

$$\frac{5.47 \times 10^4\ \text{m/s}^2}{(26.67 \times 10^{-8}\ \text{s}^{-1})^2} - (3.63 \times 10^{11}\ \text{m}) \cos 56^\circ$$

$$= 7.69 \times 10^{17}\ \text{m} - 2.0 \times 10^{11}\ \text{m}$$

$$= 7.69 \times 10^{17}\ \text{m}$$

Note: Since 56° is not given with any more precision, cos 56° has only two significant figures. This does not affect the final result since the second term is insignificant in comparison with the first.

17. $M_e \sim \frac{4}{3}\pi R^3 \rho \sim \frac{4(3)}{3}(6.4 \times 10^6)^3(3000) = 3 \times 10^{24}$ kg

19. $(6.00 \times 10^2)(315) = 1.89 \times 10^5$. This result has 3 significant figures. The result of multiplying 600×315 can have 1, 2, or 3 significant figures depending on whether the 600 has 1, 2, or 3 significant figures. The answer would be 2×10^5, 1.9×10^5, or 1.89×10^5, respectively.

21. To estimate the mass of Mt. Everest, take it to be a cone 10^4 m high and 10^4 m in diameter at its base with a density of 3000 kg/m^3. Then

$$M \sim \left(\frac{1}{12}\pi 10^{12}\right)(3000) = 8 \times 10^{14}$$

or $\sim 10^{15}$ kg

For an elephant, estimate its volume as a cylinder 10 m long with a radius of 2 m and a density of 1000 kg/m^3. Then

$$M \sim 10^5 \text{ kg}$$

23. The radius of the earth is 6×10^6 m, so the surface area is 4×10^{14} m^2. Assuming the surface to be covered by a layer of water 1.6×10^3 m thick, the total volume of water would be 7×10^{17} m^3 and its mass 7×10^{20} kg.

25. One hectare = 10^4 m^2. Assuming 1 tree for every 10 m^2, there would be 1000 trees. To estimate the mass of each tree, consider it as a cylinder 1 meter in diameter and 30 meters tall, with a density of 700 kg/m^3. Doubling this amount to account for the limbs, this gives 3×10^5 kg per tree or 3×10^8 kg in one hectare.

CHAPTER 2 VECTORS

Questions:

1. A scalar can be positive or negative. The magnitude of a vector cannot be a negative quantity, but its components can be positive or negative. As an example, a vector in the second quadrant has a negative x–component and a positive y–component.

3. A position vector is a vector that is drawn from the origin of coordinates to any other point. On the other hand, a displacement vector can be drawn between any two points.

5. The maximum magnitude of the resultant vector will be the sum of two vectors in the same direction that will be twice as much as one of the vectors. The minimum magnitude of the resultant vector is zero when both vectors are in opposite direction.

7. $\mathbf{A} + \mathbf{B} \neq 0$ if $\mathbf{A} \neq -\mathbf{B}$

For example,

$$\mathbf{A} + \mathbf{B} \neq -\mathbf{B} + \mathbf{B}$$

$$\mathbf{A} + \mathbf{B} \neq 0$$

9. Your classmate describing the 400 m race is correct. The displacement is zero for a displacement vector PQ in which the two points P and Q coincide with each other. That does not mean the distance travelled during the race is zero.

11. The displacement of one meter is the magnitude of the vector. A unit vector is a mathematical device that specifies a direction only and it is a dimensionless quantity.

13. If $\mathbf{U} = -\mathbf{i} - \mathbf{j} - \mathbf{k}$, then the magnitude of this vector is given by

$$U = \sqrt{(-1)^2 + (-1)^2 + (-1)^2} = \sqrt{3},$$

and similarly the magnitude of **W** is given by

$$W = \sqrt{(1)^2 + (1)^2 + (-1)^2} = \sqrt{3}$$

15. For a vector $\mathbf{F} = F_x\mathbf{i} + F_y\mathbf{j} + F_z\mathbf{k}$ that lies in the yz plane, the x–component is obviously zero.

17. The unit vector **j** has a component of unity value in the j–direction only. The x– and y–components of this vector are zero.

19. Generally, calculators operate in the degree mode. Some calculators do have a degree/radian key. The easiest way to figure this out will be to press for, say, cosine of 0 equal to 1 or cosine of 60° equal to 0.5 that indicates the degree mode of the calculator.

21.

Symbol	Represents	Type	SI unit
$\|A\|$	magnitude of a vector	scalar	•
C_x	x–component of vector C	component	•
r	position vector	vector	m
i	unit vector	vector	none

- The SI unit for this entry depends on the quantity. If A or C_x**i** is a displacement, then the SI unit is a meter.

Exercises:

1. Use $r = \sqrt{x^2 + y^2 + z^2}$ for the magnitude of position vector **r**.

(a) r = 2.236 m; (b) r = 2.236 m; (c) r = 741 m

3. (a) A

(b) PQ

(c) MN

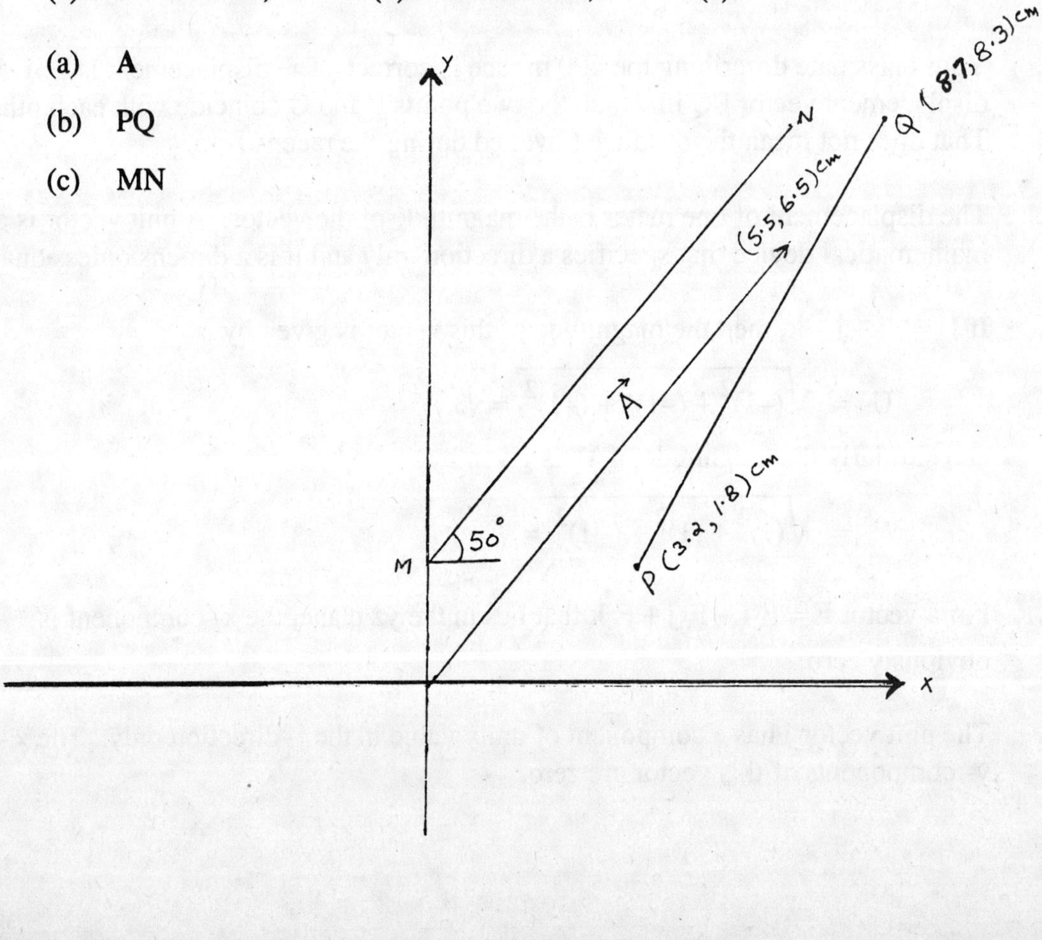

5.

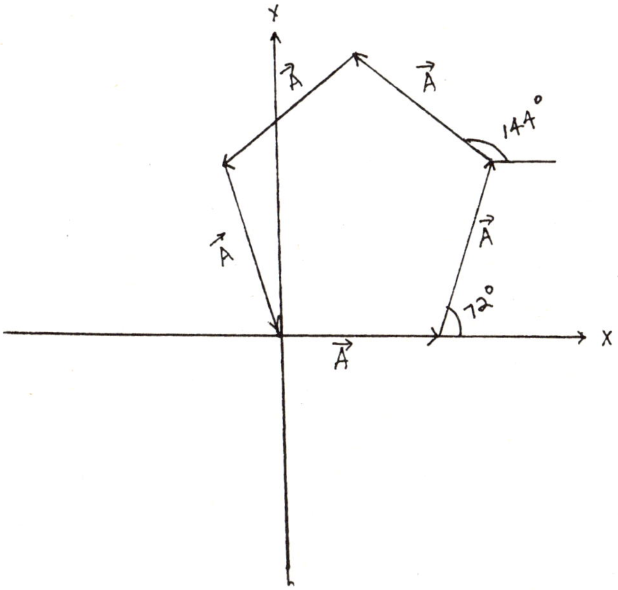

The sum of the five vectors with magnitude A as shown above using the head-to-tail method is zero.

7. (a)

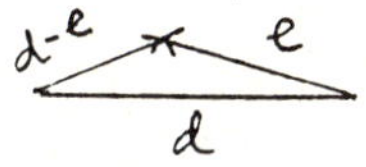

(b)

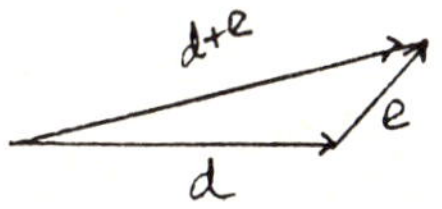

(c) Same as (b) except vector e is negative.

(d)

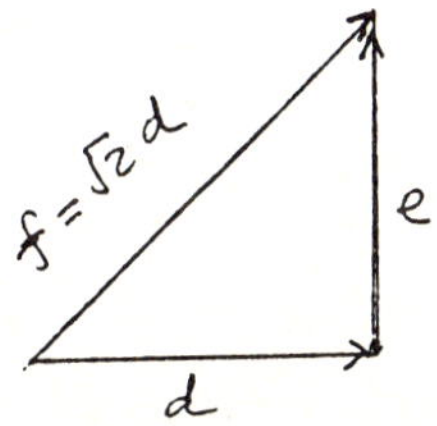

9. (a) Position vector **A** as shown.

(b) Vector **A** has a magnitude A = 6 cm, and it makes an angle of 22° with the positive x–axis.

(c) $A_x = A \cos 22^\circ = 5.56$ cm
$A_y = A \sin 22^\circ = 2.25$ cm

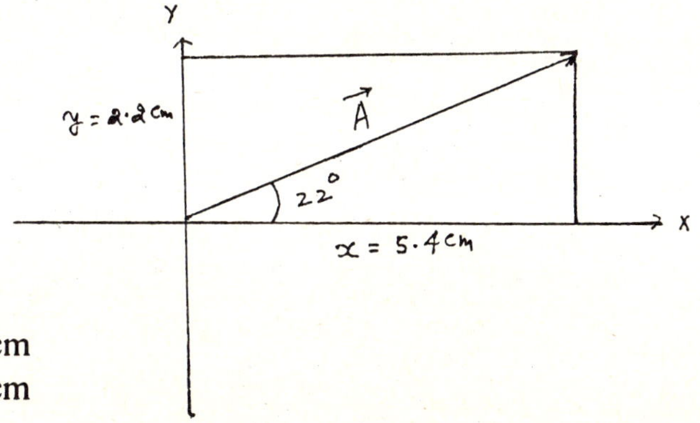

11.

L = Library

P = Physics Hall

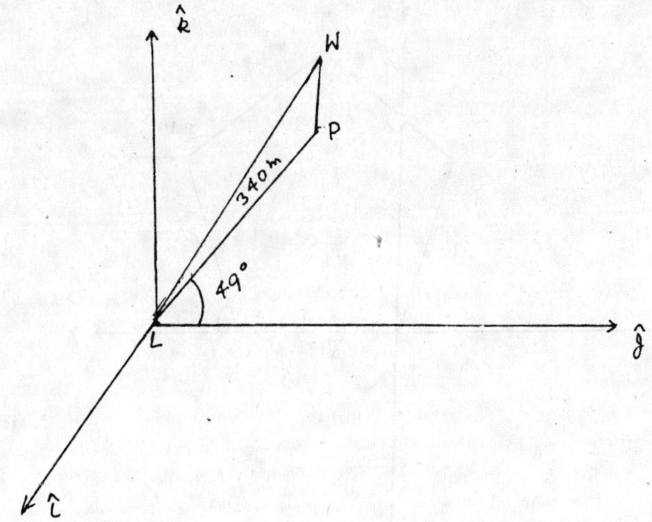

(a) Component of vector LP along x–axis

$$LP_x = -(340 \text{ m}) \sin 49° = -260 \text{ m}$$

and component of vector LP along y–axis

$$LP_y = (340 \text{ m}) \cos 49° = 220 \text{ m}$$

(b) $PL_x = (340 \text{ m}) \sin 49° = 260 \text{ m}$

$PL_y = -(340 \text{ m}) \cos 49° = -220 \text{ m}$

(c) The coordinates of W are x = –260 m, y = 220 m, and z = 35 m.

13. $\mathbf{r} = x\mathbf{i} + y\mathbf{j}$

(a) $r = \sqrt{x^2 + y^2}$

(b) $\hat{\mathbf{r}} = \frac{\mathbf{r}}{r} = \frac{x\mathbf{i} + y\mathbf{j}}{\sqrt{x^2 + y^2}}$

15. $\mathbf{A} = 3\mathbf{i} + 4\mathbf{j}$ and $\mathbf{B} = -2\mathbf{i} - 6\mathbf{j} + 5\mathbf{k}$

(a) $\mathbf{A} + \mathbf{B} = 3\mathbf{i} + 4\mathbf{j} - 2\mathbf{i} - 6\mathbf{j} + 5\mathbf{k} = \mathbf{i} - 2\mathbf{j} + 5\mathbf{k}$

(b) $\mathbf{A} - \mathbf{B} = 3\mathbf{i} + 4\mathbf{j} + 2\mathbf{i} + 6\mathbf{j} - 5\mathbf{k} = 5\mathbf{i} + 10\mathbf{j} - 5\mathbf{k}$

(c) $\mathbf{B} - \mathbf{A} = -2\mathbf{i} - 6\mathbf{j} + 5\mathbf{k} - 3\mathbf{i} - 4\mathbf{j} = -5\mathbf{i} - 10\mathbf{j} + 5\mathbf{k}$

(d) $\mathbf{B} + \mathbf{A} = -2\mathbf{i} - 6\mathbf{j} + 5\mathbf{k} + 3\mathbf{i} + 4\mathbf{j} = \mathbf{i} - 2\mathbf{j} + 5\mathbf{k}$

17. $\mathbf{r}_Q = x_Q\mathbf{i} + y_Q\mathbf{j}$

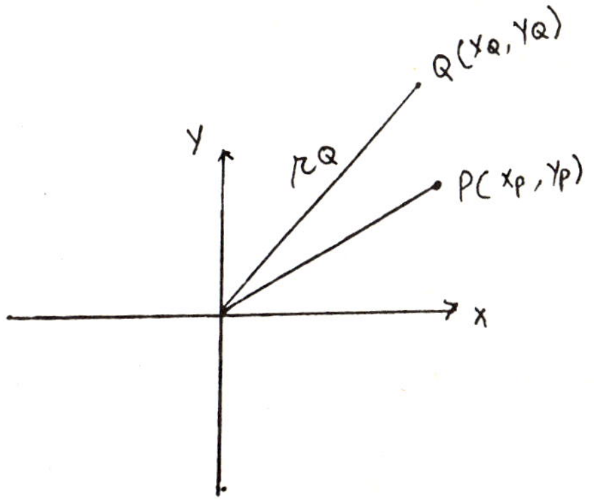

(a) $\mathbf{r}_P = x_P\mathbf{i} + y_P\mathbf{j}$

(b) $$\overrightarrow{PQ} = \mathbf{r}_Q - \mathbf{r}_P = x_Q\mathbf{i} + y_Q\mathbf{j} - x_P\mathbf{i} - y_P\mathbf{j} = (x_Q - x_P)\mathbf{i} + (y_Q - y_P)\mathbf{j}$$

(c) Position vectors $\mathbf{r}_Q$ and $\mathbf{r}_P$ were drawn from the same origin of coordinates.

19. A = 1.37 km
B = 0.85 km
C = 2.12 km

(Vectors A, B, and C not drawn to scale)

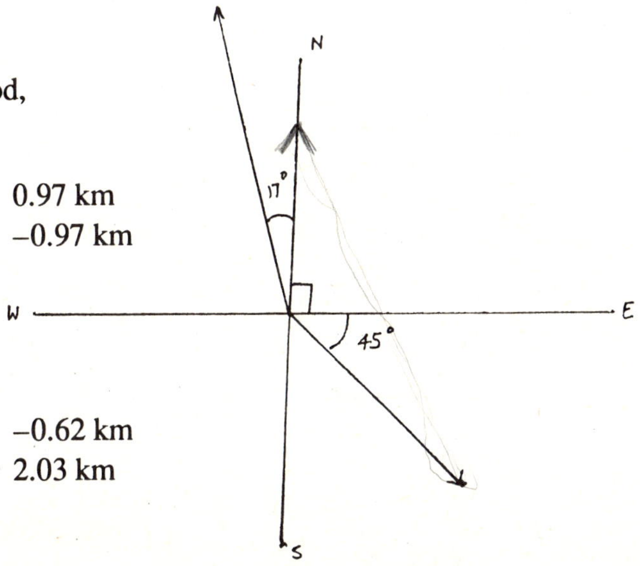

Using the components method, we get:

$$A_x = A\cos 45° = 0.97 \text{ km}$$
$$A_y = -A\sin 45° = -0.97 \text{ km}$$

$$B_x = 0$$
$$B_y = B = 0.85 \text{ km}$$

$$C_x = -C\sin 17° = -0.62 \text{ km}$$
$$C_y = C\cos 17° = 2.03 \text{ km}$$

Let **D** be the resultant vector

$$\mathbf{D} = D_x\mathbf{i} + D_y\mathbf{j}$$

$$\therefore\ D_x = A_x + B_x + C_x = 0.35 \text{ km}$$
$$D_y = A_y + B_y + C_y = 1.91 \text{ km}$$

$$\therefore\ D = \sqrt{D_x^2 + D_y^2} = 1.94 \text{ km}$$

$$\theta = \tan^{-1}\left(\frac{D_y}{D_x}\right) = 79.6°$$

21. Let vectors $V_1 \ldots V_8$ represent the sides of the octagon shown.

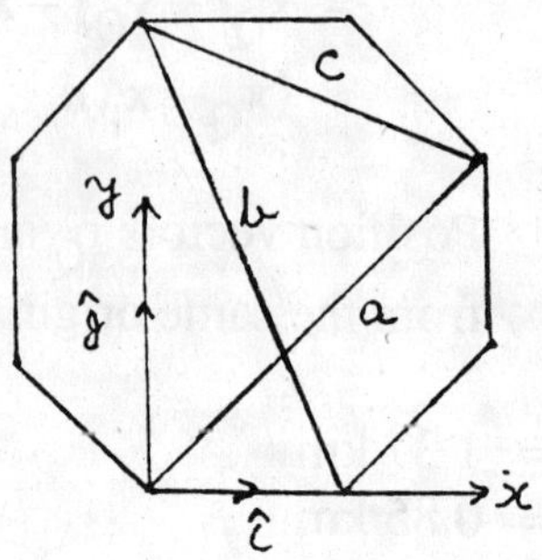

(a)
$$
\begin{aligned}
V_{1x} &= 25 \text{ mm} \\
V_{1y} &= 0 \\
V_{2x} &= 25 \cos 45° = 18 \text{ mm} \\
V_{2y} &= 25 \sin 45° = 18 \text{ mm} \\
V_{3x} &= 0 \\
V_{3y} &= 25 \text{ mm} \\
V_{4x} &= -25 \cos 45° = -18 \text{ mm} \\
V_{4y} &= 25 \sin 45° = 18 \text{ mm} \\
V_{5x} &= -25 \text{ mm} \\
V_{5y} &= 0 \\
V_{6x} &= -25 \cos 45° = -18 \text{ mm} \\
V_{6y} &= -25 \sin 45° = -18 \text{ mm} \\
V_{7x} &= 0 \\
V_{7y} &= -25 \text{ mm} \\
V_{8x} &= 25 \cos 45° = 18 \text{ mm} \\
V_{8y} &= -25 \sin 45° = -18 \text{ mm}
\end{aligned}
$$

(b)

$$
\left.
\begin{aligned}
a_x &= V_1 + V_{2x} = 43 \text{ mm} \\
a_y &= V_{2y} + V_3 = 43 \text{ mm} \\
a &= \sqrt{a_x^2 + a_y^2} = 60.81 \text{ mm}
\end{aligned}
\right\}
\quad \tan\theta_a = \frac{a_y}{a_x} \quad \text{or} \quad \theta_a = 45°
$$

$$
\left.
\begin{aligned}
b_x &= V_{2x} + V_{3x} + V_{4x} + V_{5x} = -25 \text{ mm} \\
b_y &= V_{2y} + V_{3y} + V_{4y} = 61 \text{ mm} \\
b &= \sqrt{b_x^2 + b_y^2} = 65.92 \text{ mm}
\end{aligned}
\right\}
\quad \tan\theta_b = \frac{b_y}{b_x} \quad \text{or} \quad \theta_b = 112.29°
$$

$$
\left.
\begin{aligned}
c_x &= V_{4x} + V_{5x} = -43 \text{ mm} \\
c_y &= V_{4y} = 18 \text{ mm} \\
c &= \sqrt{c_x^2 + c_y^2} = 46.61 \text{ mm}
\end{aligned}
\right\}
\quad \tan\theta_c = \frac{c_y}{c_x} \quad \text{or} \quad \theta_c = 157.29°
$$

23. For vectors **A** and **B** as shown in the figure, the resultant vector C is given as shown in example 2.5:

$$C = \sqrt{A^2 + B^2 + 2AB\cos\theta}$$

or $$C^2 = A^2 + B^2 + 2AB\cos\theta$$

or $$\cos\theta = \frac{C^2 - A^2 - B^2}{2AB}$$

or $$\theta = \cos^{-1}\left(\frac{C^2 - A^2 - B^2}{2AB}\right)$$

$$= 166^\circ \text{ with } A = 23 \text{ mm; } B = 18 \text{ mm; and } C = 7 \text{ mm}$$

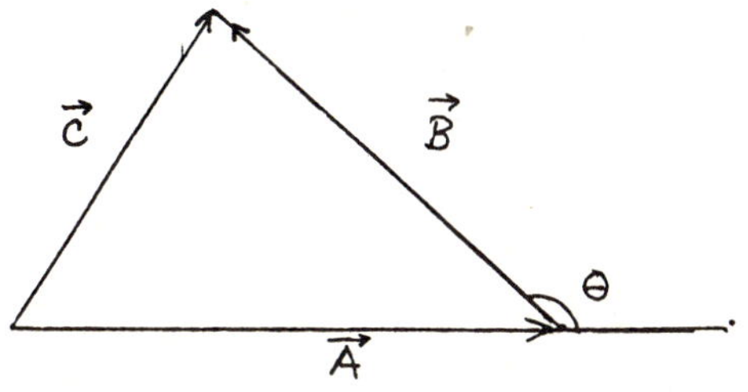

25. (a) $\mathbf{C} = \mathbf{A} + \mathbf{B}$

We know from the law of cosines that

$$C^2 = A^2 + B^2 + 2AB\cos\theta \quad \text{[see example 2.5]}$$

$$C^2 - A^2 - B^2 = 2AB\cos\theta$$

or $$\cos\theta = \frac{C^2 - A^2 - B^2}{2AB}$$

Now we know that $\cos\theta$ lies between –1 and 1.

$$\therefore \quad -1 \le \frac{C^2 - A^2 - B^2}{2AB} \le 1$$

From the above equation

$$A^2 + B^2 - 2AB \le C^2 \quad \text{and} \quad C^2 \le A^2 + B^2 + 2AB$$

or $$(A - B)^2 \le C^2 \quad \text{and} \quad C^2 \le (A + B)^2$$

Using vector notation

$$(\mathbf{A} - \mathbf{B}) \cdot (\mathbf{A} - \mathbf{B}) \le \mathbf{C} \cdot \mathbf{C} \quad \text{and} \quad \mathbf{C} \cdot \mathbf{C} \le (\mathbf{A} + \mathbf{B}) \cdot (\mathbf{A} + \mathbf{B})$$

or $$|\mathbf{A} - \mathbf{B}| \le |\mathbf{C}| \le |\mathbf{A} + \mathbf{B}|$$

(b) $|\mathbf{A} - \mathbf{B}| = C$

From the law of cosines,

$$C^2 = A^2 + B^2 + 2AB \cos \theta$$

$$\text{or} \quad \cos \theta = \frac{C^2 - A^2 - B^2}{2AB}$$

$$\text{as} \quad -1 \le \cos \theta \le 1$$

$$-1 \le \frac{C^2 - A^2 - B^2}{2AB} \le 1$$

$$\text{or} \quad -2AB \le C^2 - A^2 - B^2 = C^2 - (A^2 + B^2)$$

Therefore, from the given condition

$$(A - B)^2 = C^2$$

$$\text{or} \quad A^2 + B^2 - 2AB = C^2$$

$$\text{or} \quad C^2 - (A^2 + B^2) = -2AB$$

$$\text{or} \quad \frac{C^2 - (A^2 + B^2)}{2AB} = -1 = \cos \theta$$

$$\text{or} \quad \theta = 180°$$

(c) As shown in part (b),

$$-1 \le \frac{C^2 - (A^2 + B^2)}{2AB} \le 1$$

$$\text{where} \quad \frac{C^2 - (A^2 + B^2)}{2AB} = \cos \theta$$

$$\text{If} \quad A + B = C$$

$$A^2 + B^2 + 2AB = C^2$$

$$\text{or} \quad \frac{C^2 - (A^2 + B^2)}{2AB} = 1 = \cos \theta$$

$$\therefore \quad \theta = 0°$$

Problems:

1. (a) From Figure 2.28, the x, y, and z components of the unit vector $\hat{\mathbf{n}}$ are respectively given by

$$\cos\alpha, \quad \cos\beta, \quad \text{and} \quad \cos\gamma$$

This shows that the direction cosines $\cos\alpha$, $\cos\beta$, and $\cos\gamma$ are the components of the unit vector $\hat{\mathbf{n}}$.

(b) From part (a),

$$\hat{\mathbf{n}} = \mathbf{i}\cos\alpha + \mathbf{j}\cos\beta + \mathbf{k}\cos\gamma$$

$$1 = \hat{\mathbf{n}} \cdot \hat{\mathbf{n}}$$

$$\therefore \quad 1 = \cos^2\alpha + \cos^2\beta + \cos^2\gamma$$

(c) $\mathbf{F} = F\hat{\mathbf{n}}$

where $\hat{\mathbf{n}} = \mathbf{i}\cos\alpha + \mathbf{j}\cos\beta + \mathbf{k}\cos\gamma$

$$\therefore \quad F_x = F\cos\alpha$$

$$F_y = F\cos\beta$$

and $$F_z = F\cos\gamma$$

3. (a) We know from the law of cosines that

$$c = \sqrt{a^2 + b^2 - 2ab\cos\theta}$$

$$c^2 = a^2 + b^2 - 2ab\cos\theta$$

or $$\cos\theta = \frac{-c^2 + a^2 + b^2}{2ab}$$

with $b = \frac{1}{2}a$ and $c = \sqrt{a^2 + b^2} = \frac{\sqrt{5}\,a}{2}$

$$\cos\theta = \frac{-\frac{5a^2}{4} + a^2 + \frac{1}{4}a^2}{2ab} = 0$$

or $$\theta = 90°$$

(b) $c = \sqrt{a^2 - b^2} = \frac{\sqrt{3}\,a}{2}$

From cosine law,

$$\cos\theta = \frac{-c^2 + a^2 + b^2}{2ab}$$

or $$\cos\theta = \frac{-\frac{3a^2}{4} + a^2 + \frac{1}{4}a^2}{\not{2}a \cdot \frac{a}{\not{2}}} = \frac{1}{2}$$

or $$\theta = 60°$$

(c) With $c = \frac{3a}{2}$, cosine law gives

$$\cos\theta = \frac{-\frac{9a^2}{4} + a^2 + \frac{a^2}{4}}{\not{2}a \cdot \frac{a}{\not{2}}} = -1$$

or $$\theta = 180°$$

(d) With $c = \frac{1}{2}a$,

$$\cos\theta = \frac{-\frac{a^2}{4} + a^2 + \frac{a^2}{4}}{a^2} = 1$$

or $$\theta = 0°$$

5. (a) $\hat{\mathbf{A}} = \frac{\mathbf{A}}{A}$

or $\mathbf{A} = A\hat{\mathbf{A}}$

In this equation, A is the magnitude of vector **A**, which is a positive quantity.

Therefore, $\hat{\mathbf{A}}$ and **A** are parallel to each other.

(b) $\hat{\mathbf{r}} = \frac{\mathbf{r}}{r}$

where $\mathbf{r} = \mathbf{i}x + \mathbf{j}y + \mathbf{k}z$

$$\hat{\mathbf{r}} = \frac{\mathbf{i}x + \mathbf{j}y + \mathbf{k}z}{(x^2 + y^2 + z^2)^{1/2}}$$

$$\therefore \quad \hat{\mathbf{r}}_x = \frac{\mathbf{i}x}{(x^2 + y^2 + z^2)^{1/2}}$$

$$\hat{\mathbf{r}}_y = \frac{\mathbf{j}y}{(x^2 + y^2 + z^2)^{1/2}}$$

and $$\hat{\mathbf{r}}_z = \frac{\mathbf{k}z}{(x^2 + y^2 + z^2)^{1/2}}$$

7. (a) $\mathbf{r} = \mathbf{i}x + \mathbf{j}y$

with $x = 3\text{ m}, \quad y = -4\text{ m}$

$$\mathbf{r} = 3\mathbf{i} + (-4\mathbf{j})\text{ m}$$

(b) $$\hat{\mathbf{r}} = \frac{\mathbf{r}}{r} = \frac{\mathbf{i}x + \mathbf{j}y}{(x^2 + y^2)^{1/2}} = \frac{3\mathbf{i} - 4\mathbf{j}}{5}$$

(c) $\frac{4}{5}\mathbf{i} + \frac{3}{5}\mathbf{j}$

(d) $\mathbf{k}$

9. (a) $\mathbf{G} = (1.0\text{ m})\mathbf{i} + (2.0\text{ m})\mathbf{j} + (3.0\text{ m})\mathbf{k}$

Magnitude of vector **G** is given by

$$G = \sqrt{(1.0\text{ m})^2 + (2.0\text{ m})^2 + (3.0\text{ m})^2}$$

or $$G = 3.741\text{ m}$$

(b) $\mathbf{G} = G\hat{\mathbf{G}}$

Let the unit vector $\hat{\mathbf{G}}$ make angles α, β, and γ with the x, y, and z coordinate axes.

$$\therefore \quad \hat{\mathbf{G}} = G\cos\alpha\,\mathbf{i} + G\cos\beta\,\mathbf{j} + G\cos\gamma\,\mathbf{k}$$

or $$\mathbf{G} = G(\cos\alpha\,\mathbf{i} + \cos\beta\,\mathbf{j} + \cos\gamma\,\mathbf{k})$$

or the direction cosines (cos α, cos β, cos γ) are the components of **G**.

CHAPTER 3 MOTION IN ONE DIMENSION

Questions:

1. If one is interested in the height reached by the coin, it is legitimate to treat it as a particle. Some error will be incurred in not considering the rotational motion of the coin. If one is interested in whether it falls heads up or tails up, the rotation of the coin is the most important aspect of its motion, and it is not possible to treat it as a particle.

3. In (a), a functional relation is implied; in (b), a product is implied.

5. Statement (a) is meaningless because the direction of the velocity is not specified.
Statement (b) is meaningless because the velocity is given in incorrect units.
Statement (d) is incorrect because speeds are always positive.
Statement (f) is incomplete because the direction along which the component is taken is not specified.

7. Speed is the magnitude of the velocity vector, and must be positive.

9. The car's odometer measures the total distance travelled, regardless of direction. It does not measure the displacement.

11. If the speed is decreasing, the object is decelerating. If the velocity component is decreasing, the object is decelerating only if the velocity component is positive.

13. The acceleration of the rock is constant throughout its flight and has a component of $-9.8\ \text{m/s}^2$ along the **j** direction.

15. Referring to equation 3–12, the final speed depends only on the initial speed and the height of the building. Thus the final speed is the same regardless of whether the rock is thrown upward or downward.

17. v is the slope of the graph and a is positive if the graph is concave upwards; thus $v(t_1) < 0$, $v(t_2) = 0$, $v(t_3) > 0$, and a is positive at all three times.

19. $v(t_1) = v(t_2)$ and $|v(t_1)| = |v(t_2)|$ in Fig. 3–23(a)
$v(t_1) > v(t_2)$ and $|v(t_1)| < |v(t_2)|$ in Fig. 3–23(b)
$v(t_1) < v(t_2)$ and $|v(t_1)| < |v(t_2)|$ in Fig. 3–23(c)
$v(t_1) < v(t_2)$ and $|v(t_1)| > |v(t_2)|$ in Fig. 3–23(d)

21. Since $a = dv/dt$, $v(t)$ would have to be a quadratic expression.

23.

Symbol	Represents	Type	SI unit
r	displacement	vector	m
x_o	initial coordinate	component	m
v	velocity component	component	m/s
|**v**|	speed	magnitude of a vector	m/s
a	acceleration component	component	m/s^2

Exercises:

1. From the given data,

(a) $x_i = -16$ m

(b) $x_f = 37$ m

(c) $\mathbf{r}_i = -16$ m**i**

(d) $\mathbf{r}_f = 37$ m**i**

(e) $\Delta\mathbf{r} = \mathbf{r}_f - \mathbf{r}_i = (37\text{ m} - (-16\text{ m}))\mathbf{i} = 53\text{ m}\mathbf{i}$

3. (a) $x(t) = 52 \sin((.44\text{ s}^{-1})t)$

t (s)	x (mm)	t (s)	x (mm)	t (s)	x (mm)
0	0				
1	22	6	25	11	−52
2	40	7	3.2	12	−44
3	50	8	−19	13	−28
4	51	9	−38	14	−6.4
5	42	10	−49	15	16

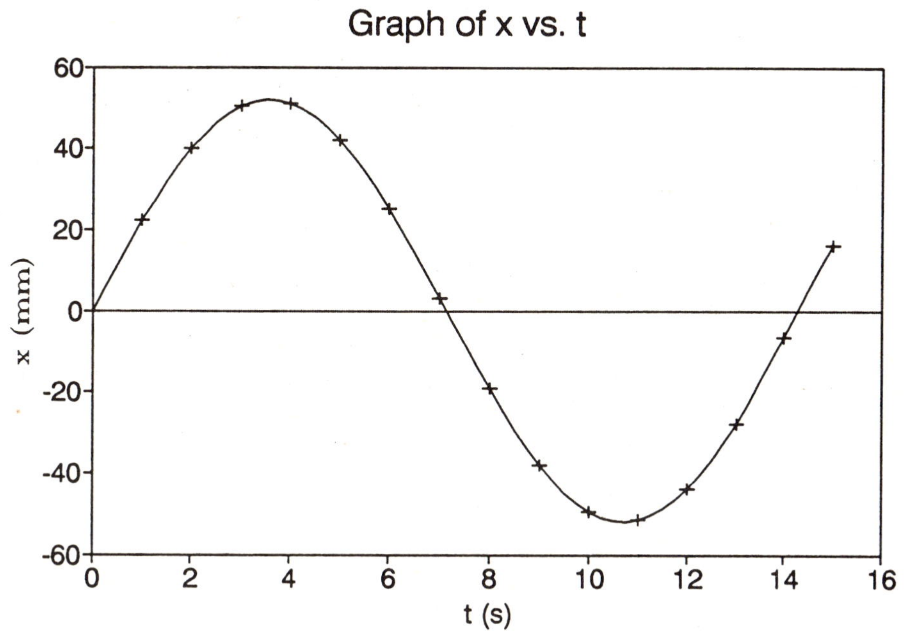

(b) Between 0 and 10 s, the distance travelled is 153 mm and (c) the displacement is -49 mm**i**.

5. (a) To convert from ft/s to m/s, multiply by:

$$\frac{1 \text{ m}}{3.28 \text{ ft}} = 0.305 \text{ m/ft}$$

(b) $25\,\frac{\text{m}}{\text{s}} \times 3.28\,\frac{\text{ft}}{\text{m}} = 82.0 \text{ ft/s}$

7. (a) For light from the sun to reach the earth, it would take

$$t = d/v = \frac{1 \times 10^8 \text{ mi} \times 1600\,\frac{\text{m}}{\text{mi}}}{3 \times 10^8 \text{ m/s}}$$

$$= 500 \text{ s} \quad \text{or} \quad 9 \text{ minutes}$$

(b) Accordingly, the distance from the earth to the sun is 9 light minutes.

9. The distance to the transformer would be equal to the distance travelled by sound in 0.50 s:

$$d = vt = \left(340\,\frac{\text{m}}{\text{s}}\right)(0.50 \text{ s})$$

$$= 170 \text{ m}$$

11. Given $x(t) = (2.2\ \text{m/s}^3)t^3 - 18\ \text{m}$, then

(a) $$x(3.0) = 41\ \text{m}$$
$$x(1.0) = -16\ \text{m}$$
$$\bar{v} = \frac{(41-(-16))\ \text{m}}{(3.0-1.0)\ \text{s}} = 29\ \text{m/s}$$

(b) $$\frac{dx}{dt} = (6.6\ \text{m/s}^3)t^2$$
$$\therefore \quad v(2.0) = 26\ \text{m/s}$$

13. From the graph, $x(1.0\ \text{s}) = 3.0\ \text{m}$, $x(5.0\ \text{s}) = 6.2\ \text{m}$, $x(4.0\ \text{s}) = 7.0\ \text{m}$, $x(3.0\ \text{s}) = 6.4\ \text{m}$, $x(2.0\ \text{s}) = 5.0\ \text{m}$.

$$\therefore \quad \bar{v}_{1,5} = \frac{x(5\ \text{s}) - x(1\ \text{s})}{4.0\ \text{s}} = \frac{3.2\ \text{m}}{4.0\ \text{s}} = 0.80\ \text{m/s}$$
$$\bar{v}_{1,4} = \frac{x(4\ \text{s}) - x(1\ \text{s})}{3.0\ \text{s}} = \frac{4.0\ \text{m}}{3.0\ \text{s}} = 1.3\ \text{m/s}$$
$$\bar{v}_{1,3} = \frac{3.4\ \text{m}}{2.0\ \text{s}} = 1.7\ \text{m/s}$$
$$\bar{v}_{1,2} = \frac{2.0\ \text{m}}{1.0\ \text{s}} = 2.0\ \text{m/s}$$

Drawing a tangent at $t = 1.0$ s, $v(1.0\ \text{s}) = 2.4$ m/s

15. In analyzing the photograph, the positions of the ball are 0.010 s apart, but this does not apply to the image of the ball on the tee. The distance between the first and second positions after it has left the tee in the photograph is 22 mm, and the height of the golfer is approximately 50 mm. Estimating the actual height of the golfer as 1.8 m gives the distance travelled by the ball as 0.79 m and thus the speed of the ball is

$$v = \frac{d}{t} = \frac{0.79\ \text{m}}{0.010\ \text{s}} \sim 80\ \text{m/s}$$

17. The ball's positions along the meter stick at 0.10 sec intervals are –.02, .11, .24, .37, .50, .63, and .76 m. These are at equal intervals of .13 m, so the speed of the ball is

$$v = \frac{d}{t} = \frac{.13\ \text{m}}{.10\ \text{s}} = 1.3\ \text{m/s}$$

(a) If the ball is travelling to the right, then

$$x(t) = -0.02\ \text{m} + (1.3\ \text{m/s})t$$

(b) If the ball is travelling to the left, then

$$x(t) = 0.76\ \text{m} - (1.3\ \text{m/s})t$$

19. (a) $\mathbf{a} = \frac{v_f - v_i}{t_f - t_i} = \frac{23 \text{ m/s}\,\mathbf{i} - 18 \text{ m/s}\,\mathbf{i}}{5.8 \text{ s}} = 0.9\,\mathbf{i} \text{ m/s}^2$

(b) $\mathbf{a} = -0.9\,\mathbf{i} \text{ m/s}^2$

21. From the graph, $v_x(0.0) = 0.5$ m/s, $v(3.0) = 0.0$ m/s, and $v(9.0) = 1.2$ m/s.

∴ between t = 0.0 s and t = 3.0 s, $\bar{a} = \frac{-0.5 \text{ m/s}}{3.0 \text{ s}} = -.2 \text{ m/s}^2$

between t = 3.0 s and t = 9.0 s, $\bar{a} = \frac{1.2 \text{ m/s} - 0.0 \text{ m/s}}{9.0 \text{ s} - 3.0 \text{ s}} = 0.5 \text{ m/s}^2$

From the slope,

$a(5.0 \text{ s}) = 0.25 \text{ m/s}^2$

23. Given $x(t) = (-1.6 \text{ m/s}^3)t^3 + (2.1 \text{ m/s}^2)t^2 - 42 \text{ m}$

(a) $v(t) = \frac{dx}{dt} = (-4.8 \text{ m/s}^3)t^2 + (4.2 \text{ m/s}^2)t$

$a(t) = \frac{dv}{dt} = (-9.6 \text{ m/s}^3)t + 4.2 \text{ m/s}^2$

(b) $a(4.1 \text{ s}) = (-9.6 \text{ m/s}^3)(4.1 \text{ s}) + 4.2 \text{ m/s}^2 = -35.2 \text{ m/s}^2$

(c) $a(0) = 4.2 \text{ m/s}^2$

25. $4^h\,40^m = 4\text{h}(3600 \text{ s/h}) + 40 \text{ min}(60 \text{ s/min}) = 16800 \text{ s}$

$v = \frac{ds}{dt} = \frac{60 \times 10^3 \text{ m}}{16800 \text{ s}} = 3.57 \text{ m/s}$

27. Given $\mathbf{v} = 2.2 \text{ m/s}\,\mathbf{i}$

$\mathbf{r}_o = 15 \text{ m}\,\mathbf{i}$

(a) $\mathbf{r} = \mathbf{r}_o + \mathbf{v}t$

$= (15 \text{ m} + 2.2 \text{ m/s } t)\mathbf{i}$

(b) $\mathbf{r}(28 \text{ s}) = [15 \text{ m} + 2.2 \text{ m/s } (28 \text{ s})]\mathbf{i} = (76 \text{ m})\mathbf{i}$

(c) $\mathbf{r}(t) = 51 \text{ m}\,\mathbf{i}$

$15 \text{ m} + 2.2 \text{ m/s } t = 51 \text{ m}, \quad t = \frac{51 \text{ m} - 15 \text{ m}}{2.2 \text{ m/s}} = 16 \text{ s}$

29. Given $a = 3.6\ m/s^2$. Assume $v_o = 0$, $x_o = 0$

(a) $v = (3.6\ m/s^2)t$

(b) $v(24) = (3.6\ m/s^2)(24\ s) = 86\ m/s$

(c) $x = \int_0^t v\ dt = (1.8\ m/s^2)t^2$

(d) $x(24) = 1.8\ m/s^2(24\ s)^2 = 1.0 \times 10^3\ m$

31. Given $a = 3.4\ m/s^2$, $v_o = 0$,

$$v^2 = v_o^2 + 2ax$$

(a) When $x = 20$ m,

$$v^2 = 0^2 + 2(3.4\ m/s^2)(20\ m) = 136\ m^2/s^2$$

$$v = \sqrt{136\ m^2/s^2} = 12\ m/s$$

(b) When $x = 40$ m,

$$v = \sqrt{2(3.4\ m/s^2)(40\ m)} = 16\ m/s$$

33. Given $v_o = 18\ m/s$, $v = 25\ m/s$, $x = 85\ m$,

$$v^2 = v_o^2 + 2ax$$

$$\therefore \quad a = \frac{v^2 - v_o^2}{2x} = \frac{(25\ m/s)^2 - (18\ m/s)^2}{2(85\ m)} = 1.8\ m/s^2$$

35. Given $x_o = 0$, $v_o = 0$, $a = 3.7\ m/s^2$,

$x = x_o + v_o t + \frac{1}{2}at^2$ reduces to $x = \frac{1}{2}at^2$

(a) $x = \frac{1}{2}at^2 \Rightarrow t = \sqrt{\frac{2x}{a}} = \sqrt{\frac{2(5.0\ m)}{3.7\ m/s^2}} = 1.6\ s$

(b) $t = \sqrt{\frac{2x}{a}} = \sqrt{\frac{2(10\ m)}{3.7\ m/s^2}} = 2.3\ s$

37. Given $x_0 = 18$ m, $v_0 = 16$ m/s, $a = -2.3$ m/s^2, $x = 65$ m,

$$x = x_0 + v_0 t + \frac{1}{2} a t^2$$

$$65 \text{ m} = 18 \text{ m} + 16 \text{ m/s } t - \frac{2.3 \text{ m/s}^2}{2} t^2$$

or $$2.3 \text{ m/s}^2 t^2 - 32 \text{ m/s } t + 94 \text{ m} = 0$$

Using the quadratic formula,

$$t = \frac{32 \text{ m/s} \pm \sqrt{(32 \text{ m/s})^2 - 4\,(94 \text{ m})(2.3 \text{ m/s}^2)}}{2\,(2.3 \text{ m/s}^2)} = 9.7 \text{ s or } 4.2 \text{ s}$$

The reason there are two possible answers is that, under the conditions given, the car would reach $x = 65$ m in 4.2 s and continue to decelerate, reach 0 velocity, and, if the acceleration is still constant, begin to travel backwards reaching $x = 65$ m a second time when $t = 9.7$ s. Since the deceleration results from applying the brakes, this second part of the trip does not occur and the only valid answer is $t = 4.2$ s.

39. Estimate the length of the barrel of the rifle as 1 m. Assuming constant acceleration,

$$v^2 = v_0^2 + 2ax$$

Here $v = 700$ m/s, $v_0 = 0$, $x = 1$ m

$$\therefore \quad a = \frac{(700 \text{ m/s})^2}{2\,(1 \text{ m})} = 2 \times 10^5 \text{ m/s}^2$$

41. Assuming constant acceleration, $x = x_0 + v_0 t + \frac{1}{2} a t^2$

$$\therefore \quad x_0 = x\,(0.0 \text{ s}) = 3.0 \text{ m}$$

$$x_2 = 3.0 \text{ m} + 2 \text{ s } v_0 + 2 \text{ s}^2 a = 15.2 \text{ m} \quad \text{or} \quad v_0 + (1 \text{ s})\, a = 6.1 \text{ m/s}$$

$$x_4 = 3.0 \text{ m} + 4 \text{ s } v_0 + 8 \text{ s}^2 a = 40.2 \text{ m} \quad \text{or} \quad v_0 + (2 \text{ s})\, a = 9.3 \text{ m/s}$$

solving $a = 3.2$ m/s^2, $v_0 = 2.9$ m/s, and $x = (3.0 \text{ m} + 2.9 \text{ m/s } t + 1.6 \text{ m/s}^2\, t^2)$

43. Given $v_0 = 0$, $v = 25$ m/s, $t = 4.0$ s,

$$v = v_0 + at$$

$$\therefore \quad a = \frac{v - v_0}{t} = \frac{25 \text{ m/s}}{4 \text{ s}} = 6.3 \text{ m/s}^2$$

In terms of g, $a = (6.3 \text{ m/s}^2) \dfrac{g}{9.8 \text{ m/s}^2} = 0.64\, g$

45. Given $v_o = 0$, $y = 3.5$ m, $t = 0.83$ s, where the y–axis is directed downwards with the origin at the point of release,

$$y = v_o t + \frac{1}{2} a t^2$$

$$\therefore \quad a = \frac{2y}{t^2} = \frac{2\,(3.5\text{ m})}{(.83\text{ s})^2} = 10\text{ m/s}^2$$

47. Given $v_o = 12$ m/s, $y_o = 1.8$ m,

(a) $v = v_o - g t$

$$\therefore \quad v = 0 \text{ when } t = v_o/g = 12\text{ m/s}/9.8\text{ m/s}^2$$

$$= 1.2\text{ s} = t_m$$

$$y = y_o + v_o t - \frac{1}{2} g t^2$$

$$\therefore \quad y(t_m) = y_m = 1.8\text{ m} + 12\text{ m/s}(1.2\text{ s}) - \frac{1}{2}\,9.8\text{ m/s}^2(1.2\text{ s})^2$$

$$= 9.1\text{ m}$$

(b) If $v = 5$ m/s, then $t = \frac{v_o - v}{g} = \frac{12\text{ m/s} - 5\text{ m/s}}{9.8\text{ m/s}^2} = \frac{7\text{ m/s}}{9.8\text{ m/s}^2} = 0.7\text{ s}$

If $v = -5$ m/s, then $t = \frac{12\text{ m/s} - (-5\text{ m/s})}{9.8\text{ m/s}^2} = \frac{17\text{ m/s}}{9.8\text{ m/s}^2} = 1.7\text{ s}$

Note that these times are equally spaced on either side of t_m.

(c) $v^2 = v_o^2 - 2g(y - y_o)$

$$\therefore \quad y - y_o = \frac{v_o^2 - v^2}{2g} = \frac{(12\text{ m/s})^2 - (5\text{ m/s})^2}{2\,(9.8\text{ m/s}^2)} = 6.1\text{ m}$$

$$y = 1.8\text{ m} + 6.1\text{ m} = 7.9\text{ m}$$

Both events occur at the same y since $(5\text{ m})^2 = (-5\text{ m})^2$.

49. Given $y = 5.1$ m, $y_o = 0$, $v = 0.70$ m/s,

$$v^2 = v_o^2 - 2g(y - y_o)$$

$$v_o^2 = v^2 + 2g(y - y_o)$$

$$= (0.70\text{ m/s})^2 + 2(9.8\text{ m/s}^2)(5.1\text{ m}) = 100\text{ m}^2/\text{s}^2$$

$$\therefore \quad v_o = 10\text{ m/s}$$

51. Given $v_o = 0$, $v = 30$ km/h, $x = 3$ m

Change v to SI units:

$$v = 30\,\frac{\text{km}}{\text{hr}} \times \frac{1000\text{ m}}{\text{km}} \times \frac{\text{hr}}{3600\text{ s}} = 8.3\text{ m/s}$$

$$v^2 = v_o^2 + 2ax$$

$$a = \frac{v^2 - v_o^2}{2x} = \frac{(8.3\text{ m/s})^2}{2\,(3\text{ m})} = 11\text{ m/s}^2$$

53. Taking the origin at the point of release,

$$t_m = \frac{v_o}{g}, \qquad h_m = \frac{v_o^2}{2g},$$

so $$g = \frac{v_o^2}{2h}$$

Since v_o is the same in both situations,

$$\frac{g_X}{g} = \frac{h}{h_X} = \frac{8.4\text{ m}}{12.6\text{ m}} = 0.67$$

or $$g_X = (0.67)(9.8\text{ m/s}^2) = 6.5\text{ m/s}^2$$

Assumptions:

(1) Atmospheric resistance is negligible.

(2) The radius of planet X is several orders of magnitude larger than h_X.

55. From Fig. 3–32, the slope at t = 3.5 s is v = – 10 m/s. The acceleration is negative since the graph is concave downwards. Since the velocity and the acceleration are in the same direction, the speed is increasing, so the object is accelerating.

57. The acceleration is zero when the graph of x vs t has a point of inflection. In Fig. 3–32, this occurs when t is approximately 2.0 s.

Problems:

1. During the constant acceleration phase

$$v = at$$

$$x = \frac{1}{2}at^2$$

and $v^2 = 2ax$,

so when $x = 15$ m,

$$t_{15} = \sqrt{\frac{2(15\text{ m})}{a}} \quad \text{and} \quad v = \sqrt{2(15\text{ m})a}$$

For the remaining 85 m, the velocity remains constant, so the time for this portion is

$$t_{85} = \frac{85\text{ m}}{v} = \frac{85\text{ m}}{\sqrt{(30\text{ m})a}}$$

The total time elapsed is 10.0 s,

$$\text{so} \quad 10.0\text{ s} = \frac{85\text{ m}}{\sqrt{(30\text{ m})a}} + \frac{\sqrt{(30\text{ m})}}{\sqrt{a}} = \frac{(21\text{ m}^{1/2})}{\sqrt{a}}$$

$$a = \frac{(21\text{ m}^{1/2})^2}{(10\text{ s})^2} = 4.4\text{ m/s}^2 \quad \text{(d)}$$

Inserting this value into the expressions obtained previously,

$$t_{15} = \sqrt{\frac{30\text{ m}}{4.4\text{ m/s}^2}} = 2.6\text{ s}$$

$$v = \sqrt{30\text{ m}(4.4\text{ m/s}^2)} = 11.5\text{ m/s}$$

$$t_{85} = \frac{85\text{ m}}{\sqrt{30\text{ m}(4.4\text{ m/s}^2)}} = 7.4\text{ s}$$

3. Given $v_{Ao} = 18$ m/s, $a_A = 0$, $v_{Bo} = 0$, $a_B = 4.6$ m/s^2,

$$x_A = v_{Ao}t = (18\text{ m/s})t$$

$$x_B = v_{Bo}t + \frac{1}{2}a_Bt^2 = \frac{1}{2}(4.6\text{ m/s}^2)t^2$$

(a) $x_A = x_B$ when $(18\text{ m/s})t = (2.3\text{ m/s}^2)t^2$

i.e., $t = 0$ or $t = \dfrac{18\text{ m/s}}{2.3\text{ m/s}^2} = 7.8$ s

(b) $x_B = (2.3\text{ m/s}^2)(7.8\text{ s})^2 = 140$ m

(c) $v_B = (4.6\text{ m/s}^2)(7.8\text{ s}) = 36$ m/s

5. The distance required to reach flying speed is found from:

$$v^2 = 2ax$$

where $v = 85$ m/s and $a = 4.0$ m/s^2

$$\therefore \quad x = v^2/2a = (85 \text{ m/s})^2/2(4.0 \text{ m/s}^2) = 900 \text{ m}$$

The distance required to stop if $v_o = 85$ m/s and $a = -5.0$ m/s^2 is given by

$$x = -v_o^2/2a = (85 \text{ m/s})^2/2(5.0 \text{ m/s}^2) = 720 \text{ m}$$

Thus the total length of the runway should be at least 1620 m.

7. (a) Using $x_f = x_o + v_o t_f + \frac{1}{2} a t_f^2$

and $x_i = x_o + v_o t_i + \frac{1}{2} a t_i^2$,

we get $x_f - x_i = v_o(t_f - t_i) + \frac{1}{2} a(t_f^2 - t_i^2)$

Now $\bar{v} = \dfrac{x_f - x_i}{t_f - t_i} = v_o + \dfrac{1}{2} a \dfrac{(t_f^2 - t_i^2)}{t_f - t_i}$

$= v_o + \frac{1}{2} a(t_f + t_i)$

Also $v_i = v_o + at_i$ and $v_f = v_o + at_f$,

thus $\frac{1}{2}(v_i + v_f) = v_o + \frac{1}{2} a(t_i + t_f) = \bar{v}$

(b) Using $a = 2.0$ m/s, $v_o = 1.0$ m/s, $t_i = 1.0$ s, and $t_f = 3.0$ s,

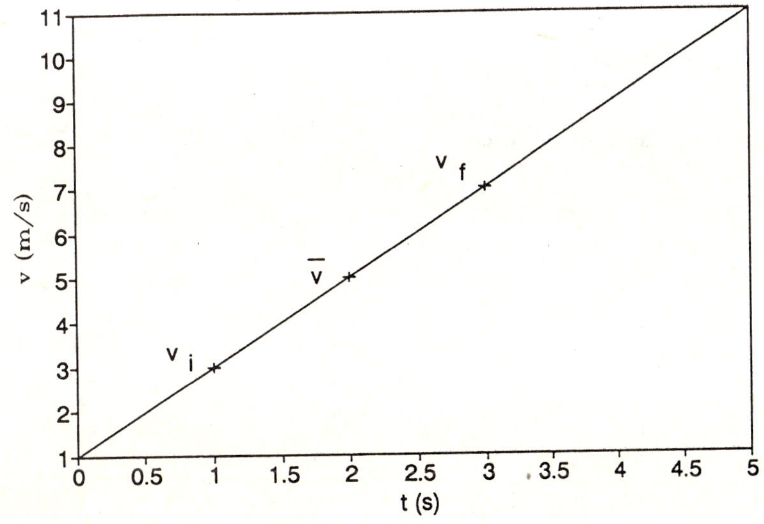

Note that $\bar{v} = v\left(\frac{1}{2}(t_i + t_f)\right)$.

9. To deal with this problem, it is necessary to establish a method for calculating successive positions of the particle. A simple scheme uses the following equations:

$$x(t+h/2) = x(t) + v(t)h/2$$

$$x(t+h) = x(t) + v(t)h + \frac{1}{2}a(t+h/2)h^2$$

The first equation calculates the position of the particle half a time interval later in terms of the initial position and the initial velocity (assumed constant). This allows the calculation of the acceleration at time $t + h/2$, and this acceleration (assumed constant) is used in the second equation to calculate the new position of the particle. To complete the process so it can be carried indefinitely, the velocity at $t + h$ can be calculated from

$$v(t+h) = v(t) + a(t+h/2)h$$

and the process can be started over again.

This is set up in a spreadsheet as follows:

A heading and column labels occupy the first four rows. The calculations begin in row 5.

Column A contains the time. Cells B5 and C5 contain the initial values of x and v. Cells D5 and E5 calculate $x(t + h/2)$ and $a(t + h/2)$. Columns D and E are copies of the formulas in D5 and E5. Cells B6 and C6 calculate $x(t + h)$ and $v(t + h)$. The rest of columns B and C are copies of the formulas in B6 and C6.

Below is a sample spreadsheet for t between 0 and 3 s, using a time interval of 0.1 s. Column F is a calculation of

$$\sin(t/2)$$

which is the exact solution to the problem.

The formulas are printed out in the next table. Note that there is an absolute reference to cell E1, which contains the time interval.

	A	B	C	D	E	F
1	Integration of a=−.25*x			h=	0.1	
2						
3	t	x	v	x (t+.5*h)	a (t+.5*h)	
4						
5	0	0	0.5	0.025	−0.00625	0
6	0.1	0.049968	0.499375	0.074937	−0.01873	0.049979
7	0.2	0.099812	0.497501	0.124687	−0.03117	0.099833
8	0.3	0.149406	0.494384	0.174126	−0.04353	0.149438
9	0.4	0.198627	0.490031	0.223129	−0.05578	0.198669
10	0.5	0.247351	0.484452	0.271574	−0.06789	0.247403
11	0.6	0.295457	0.477663	0.319340	−0.07983	0.295520
12	0.7	0.342824	0.469680	0.366308	−0.09157	0.342897
13	0.8	0.389335	0.460522	0.412361	−0.10309	0.389418
14	0.9	0.434871	0.450213	0.457382	−0.11434	0.434965
15	1	0.479321	0.438778	0.501260	−0.12531	0.479425
16	1.1	0.522572	0.426247	0.543885	−0.13597	0.522687
17	1.2	0.564517	0.412650	0.585150	−0.14628	0.564642
18	1.3	0.605051	0.398021	0.624952	−0.15623	0.605186
19	1.4	0.644072	0.382397	0.663191	−0.16579	0.644217
20	1.5	0.681482	0.365817	0.699773	−0.17494	0.681638
21	1.6	0.717189	0.348323	0.734606	−0.18365	0.717356
22	1.7	0.751104	0.329958	0.767601	−0.19190	0.751280
23	1.8	0.783140	0.310768	0.798678	−0.19966	0.783326
24	1.9	0.813218	0.290801	0.827758	−0.20693	0.813415
25	2	0.841264	0.270107	0.854769	−0.21369	0.841470
26	2.1	0.867206	0.248738	0.879643	−0.21991	0.867423
27	2.2	0.890980	0.226746	0.902318	−0.22557	0.891207
28	2.3	0.912527	0.204189	0.922737	−0.23068	0.912763
29	2.4	0.931793	0.181120	0.940849	−0.23521	0.932039
30	2.5	0.948729	0.157599	0.956609	−0.23915	0.948984
31	2.6	0.963293	0.133684	0.969977	−0.24249	0.963558
32	2.7	0.975449	0.109434	0.980920	−0.24523	0.975723
33	2.8	0.985166	0.084911	0.989412	−0.24735	0.985449
34	2.9	0.992420	0.060176	0.995429	−0.24885	0.992712
35	3	0.997194	0.035290	0.998958	−0.24973	0.997494

```
A1:  'Integration of a=-.25*x
D1:  'h=
E1:  0.1
A3:  ^t
B3:  ^x
C3:  ^v
D3:  'x(t+.5*h)
E3:  'a(t+.5*h)
F3:  'sin(t/2)
A5:  0
B5:  0
C5:  0.5
D5:  +B5+C5*$E$1/2
E5:  -0.25*D5
F5:  @SIN(A5/2)
A6:  0.1
B6:  +B5+C5*$E$1+E5*$E$1^2/2
C6:  +C5+E5*$E$1
D6:  +B6+C6*$E$1/2
E6:  -0.25*D6
F6:  @SIN(A6/2)
A7:  0.2
B7:  +B6+C6*$E$1+E6*$E$1^2/2
C7:  +C6+E6*$E$1
D7:  +B7+C7*$E$1/2
E7:  -0.25*D7
F7:  @SIN(A7/2)
A8:  0.3
B8:  +B7+C7*$E$1+E7*$E$1^2/2
C8:  +C7+E7*$E$1
D8:  +B8+C8*$E$1/2
E8:  -0.25*D8
F8:  @SIN(A8/2)
A9:  0.4
B9:  +B8+C8*$E$1+E8*$E$1^2/2
C9:  +C8+E8*$E$1
D9:  +B9+C9*$E$1/2
E9:  -0.25*D9
F9:  @SIN(A9/2)
```

11. Given $y_o = 0.49$ m, $y_m = 1.13$ m

To determine v_o:

$$y = y_o + v_o t - \frac{1}{2} g t^2$$
$$v = v_o - g t$$
$$v = 0 \quad \text{when} \quad t = v_o/g$$
$$y_m = y_o + v_o^2/2g$$

$$\therefore \quad v_o = \sqrt{2g(y_m - y_o)} = \sqrt{2(9.8\ \text{m/s}^2)(1.13\ \text{m} - .49\ \text{m})}$$
$$= 3.54\ \text{m/s}$$

To determine time to reach $y = 0$:

$$y = y_o + v_o t - \frac{1}{2} g t^2$$

$$\therefore \quad (4.9\ \text{m/s}^2)t^2 - (3.54\ \text{m/s})t - 0.49\ \text{m} = 0_m$$

$$t = \frac{3.54\ \text{m/s} \pm \sqrt{(3.54\ \text{m/s})^2 + 4(4.9\ \text{m/s}^2)(.49\ \text{m})}}{9.8\ \text{m/s}^2}$$

$$= 0.84\ \text{s} \qquad \text{(positive root applies)}$$

In terms of rotations, this corresponds to 8.4 rotations; thus the coin lands heads up (0 – .5 rotations = heads; .5 – 1.0 = tails, etc.).

13. Given $x = A \sin \omega t$

(a) $v = \frac{dx}{dt} = A\omega \cos \omega t$

(b) $a = \frac{dv}{dt} = -A\omega^2 \sin \omega t$

(c) $a/x = -\omega^2 \quad \therefore \quad a = -\omega^2 x$

Given $A = 10.0$ mm, $\omega = .628$ rad/s

t (s)	v (mm/s)	a (mm/s^2)
0	6.28	0
1	5.08	−2.32
2	1.94	−3.75
3	−1.94	−3.75
4	−5.08	−2.32
5	−6.28	−0.01
6	−5.09	2.31
7	−1.95	3.75
8	1.93	3.75
9	5.07	2.33
10	6.28	0.01

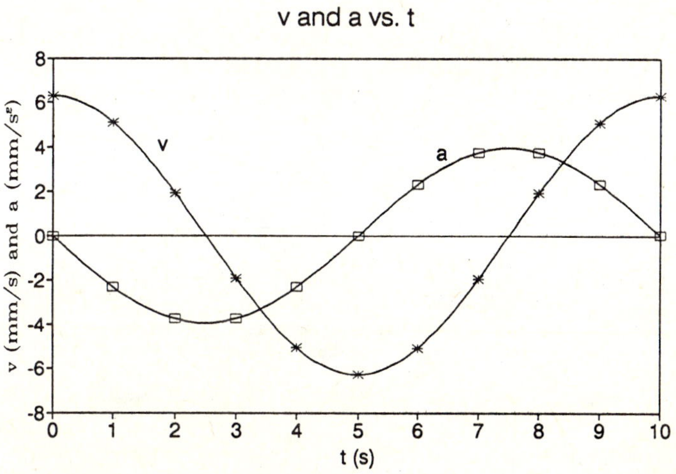

15. Given $x = bt^2 - ct^3$, $x(4\ s) = 32\ m$, $v(4\ s) = 0\ m/s$,

$$v = \frac{dx}{dt} = 2bt - 3ct^2$$

(a) $\therefore$

$$32.0\ m = b(4.0\ s)^2 - c(4.0\ s)^3$$

$$0\ m/s = 2b(4.0\ s) - 3c(4.0\ s)^2$$

$$\therefore \quad b = \frac{3c(4.0\ s)^2}{2(4.0\ s)} = (6.0\ s)c$$

$$32.0\ m = (6.0\ s)c(4.0\ s)^2 - c(4.0\ s)^3$$

$$= (32.0\ s^3)c$$

$$\therefore \quad c = \frac{32.0\ m}{32.0\ s^3} = 1.00\ m/s^3, \quad b = 6.00\ m/s^2$$

(b)

$$x = (6.00\ m/s^2)t^2 - (1.00\ m/s^3)t^3$$

$$v = \frac{dx}{dt} = (12.0\ m/s^2)t - (3.00\ m/s^3)t^2$$

$$a = \frac{d^2x}{dt^2} = (12.0\ m/s^2) - (6.00\ m/s^3)t$$

$$\therefore \quad a = 0 \quad \text{when} \quad t = 2.00\ s$$

t (s)	x (m)	v (m/s)	a (m/s²)
0	0	0	12
1	5	9	6
2	16	12	0
3	27	9	−6
4	32	0	−12

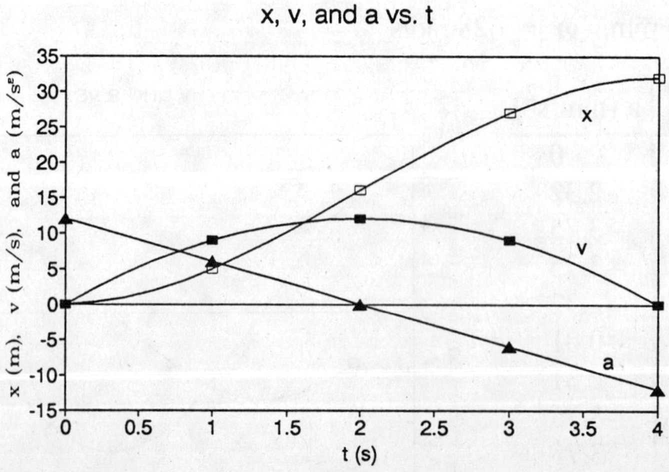

17. Using $y = v_o t - \frac{1}{2} g t^2$, the times that the marble is at the same elevation, y, are given by

$$y - v_o t + \frac{1}{2} g t^2 = 0$$

$$t = \frac{v_o \pm \sqrt{v_o^2 - 2gy}}{g}$$

Thus $\Delta t_1 = \frac{2}{g}\sqrt{v_o^2 - 2gy_1}$

$$\Delta t_2 = \frac{2}{g}\sqrt{v_o^2 - 2gy_2}$$

$$(\Delta t_2)^2 - (\Delta t_1)^2 = \frac{4}{g^2}(2gy_1 - 2gy_2)$$

$$= \frac{8}{g}\Delta y$$

where $\Delta y = y_1 - y_2$

Thus $g = 8\Delta y / [(\Delta t_2)^2 - (\Delta t_1)^2]$

19. Given $x = \lambda(t/\tau - 1 + e^{-t/\tau})$

(a) $v = \frac{dx}{dt} = \lambda\left(\frac{1}{\tau} - \frac{1}{\tau} e^{-t/\tau}\right) = \frac{\lambda}{\tau}(1 - e^{-t/\tau})$

(b) $a = \frac{dv}{dt} = \frac{\lambda}{\tau^2} e^{-t/\tau}$

(c) Given $\lambda = 133$ m, $\tau = 5.77$ s

t (s)	x (m)	v (m/s)	a (m/s²)
0.00	0.00	0.00	3.99
1.00	1.89	3.67	3.36
2.00	7.14	6.75	2.82
3.00	15.23	9.35	2.38
4.00	25.69	11.53	2.00
5.00	38.16	13.36	1.68
6.00	52.32	14.90	1.41
7.00	67.89	16.20	1.19
8.00	84.65	17.29	1.00
9.00	102.41	18.21	0.84
10.00	121.01	18.98	0.71

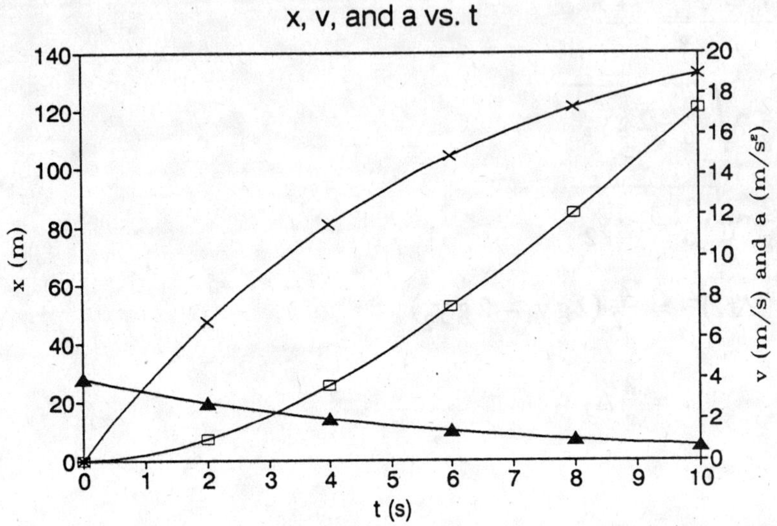

(d) λ/τ is the final velocity, $v(\infty)$

(e) λ/τ^2 is the initial acceleration, $a(0)$

21. (a) Speed of ball immediately before it hits the floor:

$$v^2 - v_o^2 = -2gy$$

Here $y = -2.0$ m, $v_o = 0$ m/s, so

$$v = \sqrt{2(9.8 \text{ m/s}^2)(2.0 \text{ m})} = 6.3 \text{ m/s}$$

(b) Speed immediately after bounce:

Here $v = 0$, $y = 1.0$ m, so

$$v_o = \sqrt{2(9.8 \text{ m/s}^2)(1.0 \text{ m})} = 4.4 \text{ m/s}$$

(c) If maximum compression is 2.0 mm, estimate deceleration by assuming it to be constant, with $v_o = 6.3$ m/s, $v = 0$, and $y = -.0020$ m.

$$a = \frac{v^2 - v_o^2}{2y} = \frac{-(6.3 \text{ m/s})^2}{2(-.0020 \text{ m})} = 9800 \text{ m/s}^2$$

(d) Similarly, for acceleration in the upward phase of the bounce, with $v_0 = 0$, $v = 4.4$ m/s, and $y = .0020$ m,

$$a = \frac{(4.4\ \text{m/s})^2}{2\,(.0020\ \text{m})} = 4800\ \text{m/s}^2$$

In both cases, the acceleration is directed upward.

(e) The formulas used in (c) and (d) assume constant acceleration.

23. $y_{th} = b \ln\left(\frac{v_0 t}{b} + 1\right)$

$b = .142$ m

$v_0 = 233$ m/s

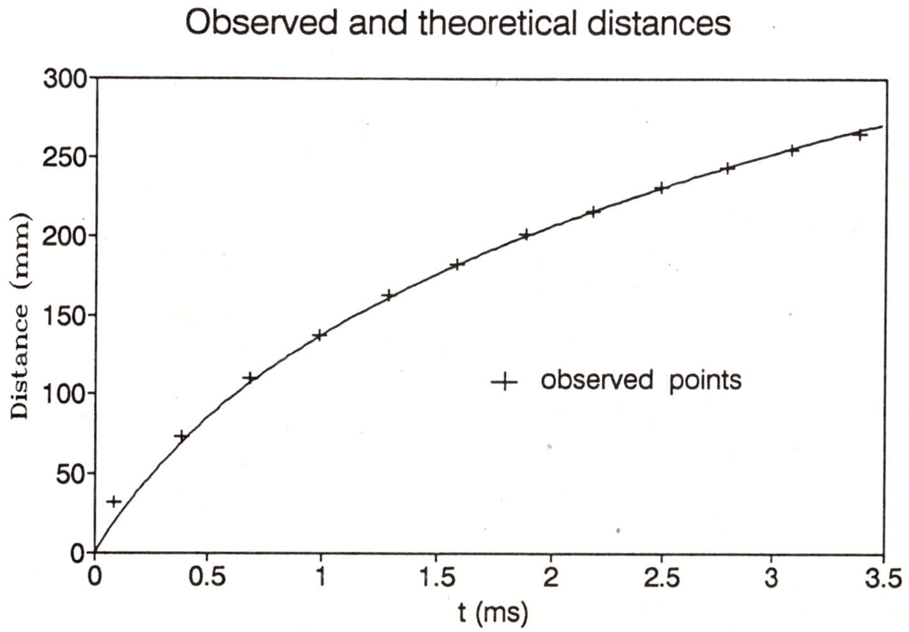

The two curves are similar and get closer as t gets larger.

CHAPTER 4 MOTION IN TWO DIMENSIONS

Questions:

1. (a) The position vector locates an object relative to the origin of a reference frame, and there is no particular relationship between this and the velocity vector.

(b) The displacement, $\overrightarrow{\Delta r} = \mathbf{r}_f - \mathbf{r}_i$, and is directed from an object's initial position to its final position, and is in the same direction as the average velocity.

(c) As seen from Fig. 4d, the velocity vector **v** is parallel to a line tangent to the path and points in the direction of motion.

(d) By definition

$$\mathbf{a} = \frac{\overrightarrow{\Delta v}}{dt},$$

so that **a** has the same direction as $\overrightarrow{\Delta v}$, where a is the acceleration of the object and $\overrightarrow{\Delta v}$ is its change in velocity.

(e) For a straight line path, the lines tangent to the path and acceleration are in the same direction.

3. As shown in Fig. 4–6d, the acceleration is partly directed inward, or toward the concave side of the path.

5. The average acceleration is defined as the rate of change of velocity

$$\mathbf{a} = \frac{\overrightarrow{\Delta v}}{\Delta t}.$$

The acceleration is the limiting value of the average acceleration as the time interval approaches zero.

If an object moving in a straight line comes to rest, its acceleration also becomes zero. On the other hand, an object thrown up is constantly acted upon by the acceleration due to gravity, although it comes to rest for a second.

7. As discussed in question 5, an object thrown up comes to rest for a second, although it is acted upon by the acceleration due to gravity that is constantly pulling it down.

Since acceleration is defined as the rate of change of velocity, for a nonzero acceleration the velocity cannot be zero for the entire time interval.

9. Lighter objects will definitely be influenced by air resistance. It will be valid to neglect air resistance for golf ball, 5 m, and baseball, 5 m.

11. $\mathbf{v}_0 = (3 \text{ m/s})\mathbf{i} + (2 \text{ m/s})\mathbf{j}$

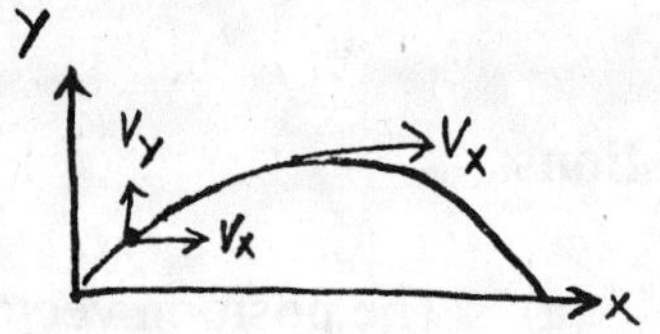

(a) At the top of the trajectory, the vertical component of the velocity becomes zero, and the velocity is in the horizontal direction.

$$\therefore \quad \mathbf{v}_0 = v_x\mathbf{i} = (3 \text{ m/s})\mathbf{i}$$

(b) The horizontal component of the velocity remains constant throughout the motion of the projectile. As stated in part (a), the vertical component of the velocity becomes zero at the top of the trajectory, but it is acted upon by the acceleration due to gravity which is pulling it down.

$$\therefore \quad \mathbf{a} = -g\mathbf{j} = (-9.8 \text{ m/s}^2)\mathbf{j}$$

13. The vertical component of the velocity can be parallel or antiparallel to the acceleration but, because of the constant value of the horizontal component of the velocity, the acceleration v_0 can never be parallel to the velocity. The acceleration which is the acceleration due to gravity is perpendicular to the velocity at the top of the trajectory as at that point $v_{oy} = 0$ and $v_{ox} = v_0 \cos 40°$ is perpendicular to **g**.

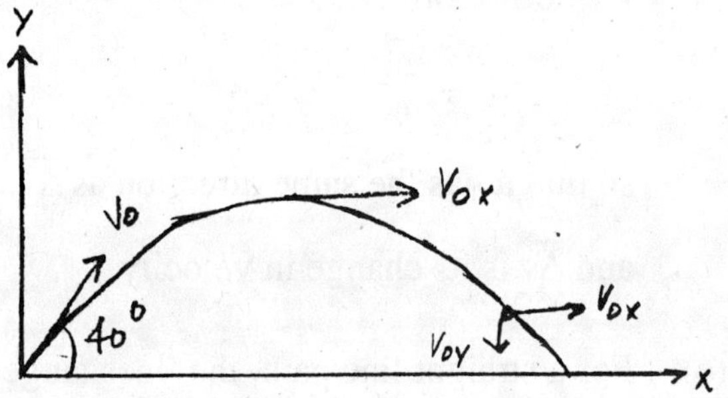

15. (i) An object at the top of the trajectory of a projectile motion has a constant horizontal component of the velocity and is acted upon by the acceleration due to gravity.

(ii) An object going around in a circle with a constant speed is being accelerated at every point of its motion as it is constantly changing direction.

17. (a) As shown in Fig. 4–14, the position vector is perpendicular to the velocity vector.

(b) From the same Fig. 4–14, it can be seen that the position vector is antiparallel to the acceleration vector and

(c) velocity vector is perpendicular to the acceleration vector.

19. $\mathbf{v} = r\vec{\omega}$ and $\mathbf{a} = r\vec{\alpha}$, where ω and α are the angular velocity and acceleration, respectively. For a constant value of ω and α, if the radius is halved, the velocity and acceleration will also be halved subsequently.

21. When the bob comes to the mean position $\theta = 0°$, and in this case the direction of the acceleration is indicated by arrow 3. Arrow 2 best indicates the other positions whether the bob is swinging to the left or right.

23. The object is travelling with a constant speed, but since its direction is changing while following the spiral path its velocity is not constant. The magnitude of acceleration changes as the radius changes, and it increases as the radius decreases.

25. (a) Two observers who are moving at constant velocity relative to one another and each measuring the velocity of an object will find that their velocity measurements yield different results. Figure 4–18 shows two reference frames for the two observers. Each observer is at rest in his frame. The two frames are moving at constant velocity relative to one another along their common x–direction. Therefore, when two frames of reference are in motion relative to one another, the velocity of an object measured relative to these frames is different, and the difference is the relative velocity of the frames.

(b) If the two observers move with a constant relative velocity, each observer measures same acceleration for a particle. Therefore, the object's acceleration as measured by the two observers will be the same.

27. No

Exercises:

1. (a) radius = 45.0 m

At $t = 0$ s, $x = 45$ m.

At $t = 16.8$ s,

the jogger is at M.

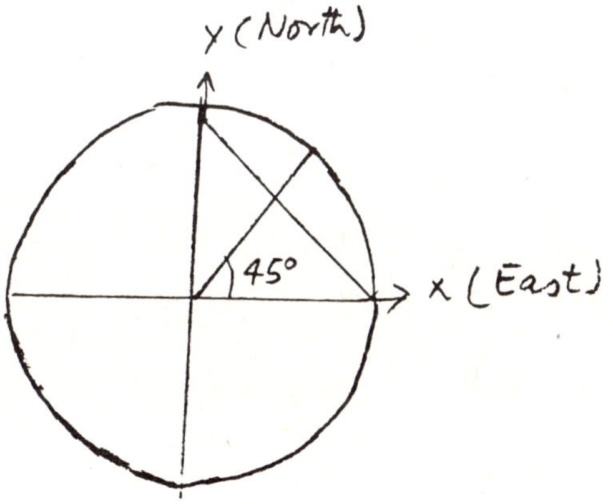

Therefore, the position vector

$$\vec{OM} = (OM \cos 45°)\mathbf{i} + (OM \sin 45°)\mathbf{j}$$

or $$\vec{OM} = (45\text{ m} \cos 45°)\mathbf{i} + (45\text{ m} \sin 45°)\mathbf{j}$$

or $$\vec{OM} = (31.82\text{ m})\mathbf{i} + (31.82\text{ m})\mathbf{j}$$

(b) At t = 33.6 s, y = 45 m, and the jogger is at point y. The displacement

$$xy = \sqrt{(Ox)^2 + (Oy)^2}$$

$$xy = \sqrt{(45\text{ m})^2 + (45\text{ m})^2}$$

or $$xy = 63.64\text{ m}$$

(c) 63.64 m

3. (a) $$v_x = \frac{dx}{dt} = \frac{-45\text{ m}}{33.6\text{ s}} = -1.34\text{ m/s}$$

$$v_y = \frac{dy}{dt} = \frac{45\text{ m}}{33.6\text{ s}} = 1.34\text{ m/s}$$

$$\therefore \quad \mathbf{v} = -(1.34\text{ m/s})\mathbf{i} + (1.34\text{ m/s})\mathbf{j}$$

(b) $$v_x = \frac{dx}{dt} = \frac{-24.4\text{ m}}{16.8\text{ s}} = -1.45\text{ m/s}$$

$$v_y = \frac{dy}{dt} = \frac{24.4\text{ m}}{16.8\text{ s}} = 1.45\text{ m/s}$$

$$\therefore \quad \mathbf{v} = -(1.45\text{ m/s})\mathbf{i} + (1.45\text{ m/s})\mathbf{j}$$

5. (a) At C, v_C = 23 m/s, distance between A and C = 300 m

Let v_A be the velocity of the car at A ($v_A = 0$)

Use $$v_C^2 - v_A^2 = 2\,a\,y \quad (y = 300\text{ m})$$

$$v_C^2 = 2\,a\,(300\text{ m})$$

$$(23\text{ m/s})^2 = 2\,a\,(300\text{ m})$$

$$a = 0.88\text{ m/s}^2$$

Using points B and C,

$$v_C^2 - v_B^2 = 2\,a\,y \quad (y = 150\text{ m})$$

$$(23\text{ m/s})^2 - v_B^2 = 2\,(0.88\text{ m/s}^2)\,(150\text{ m})$$

$$v_B = 16.28\text{ m/s} + \text{y–axis or } \mathbf{j}$$

(b) Acceleration at B = 0.88 m/s + y–axis or **j**

(c) $v_D = 23\text{ m/s}$

(d) $\mathbf{a}_{C-E} = 0$

7. $\mathbf{r} = [(3.5\text{ m/s})t + 4.2\text{ m}]\mathbf{i} + [(5.1\text{ m/s})t]\mathbf{j}$

(a) $\mathbf{v} = \frac{d\vec{r}}{dt}$

$\mathbf{v} = 3.5\text{ m/s}\,\mathbf{i} + 5.1\text{ m/s}\,\mathbf{j}$

(b) The magnitude of the position vector is given by

t (s)	**r**	r (m)
0	4.2 m **i**	4.2
0.5	5.95 m **i** + 2.55 m **j**	6.47
1.0	7.7 m **i** + 5.1 m **j**	9.24
1.5	9.45 m **i** + 7.65 m **j**	12.16
2.0	11.2 m **i** + 10.2 m **j**	15.15
2.5	12.95 m **i** + 12.75 m **j**	18.17
3.0	14.7 m **i** + 15.3 m **j**	21.2

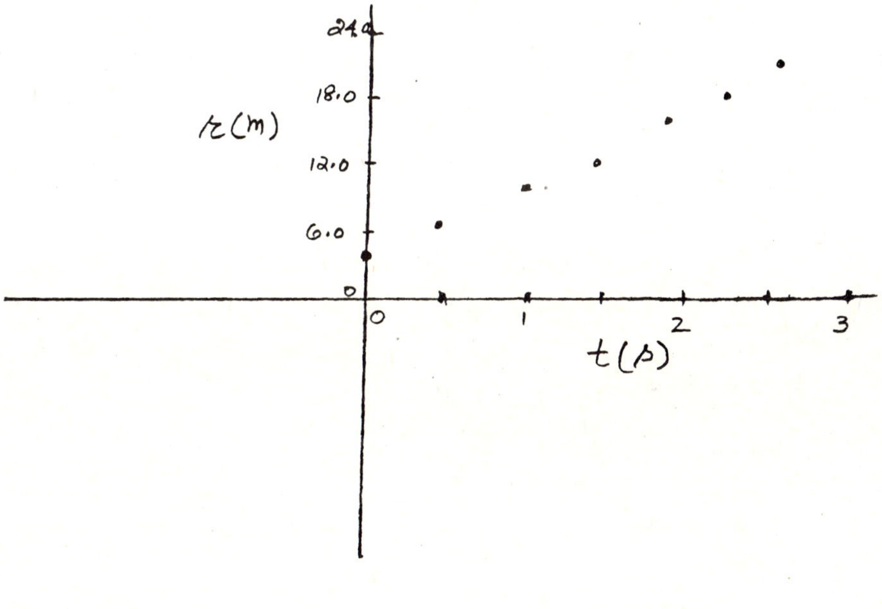

$\mathbf{v} = \frac{d\vec{r}}{dt}$

9. $v_x = v \cos 29°$

or $v_x = (16 \text{ m/s}) \cos 29°$

$v_x = 13.99 \text{ m/s}$

and $v_y = (16 \text{ m/s}) \sin 29°$

$v_y = 7.8 \text{ m/s}$

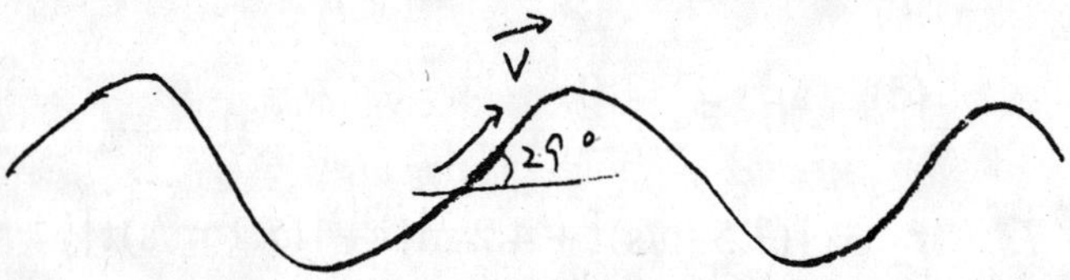

11. (a) $\mathbf{a} = 1.81 \text{ m/s}^2 (\mathbf{i} \cos 15° - \mathbf{j} \sin 15°)$

$\therefore \quad a_x = 1.75 \text{ m/s}^2$

$a_y = -0.468 \text{ m/s}^2$

$v_x = (1.75 \text{ m/s}^2)\,t$

$v_y = (-0.47 \text{ m/s}^2)\,t$

$x = \frac{1}{2}(1.75 \text{ m/s}^2)t^2 = (0.874 \text{ m/s}^2)t^2$

$y = \frac{1}{2}(-0.468 \text{ m/s}^2)t^2 = (-0.234 \text{ m/s}^2)t^2$

(b) $\mathbf{a} = 1.81 \text{ m/s}^2\,\mathbf{i}$

$\therefore \quad a_x = 1.81 \text{ m/s}^2$

$a_y = 0$

$v_x = (1.81 \text{ m/s}^2)\,t$

$v_y = 0$

$x = (0.905 \text{ m/s}^2)t^2$

$y = 0$

(c) This motion is one-dimensional since it is motion along a straight line.

13. $v_o = 36 \text{ m/s}, \; \theta = 62°$

The horizontal component of the velocity remains constant throughout the projectile's motion.

$v_x = v_o \cos 62°$

$v_x = 36 \text{ m/s} \cos 62°$

$= 16.9 \text{ m/s}$

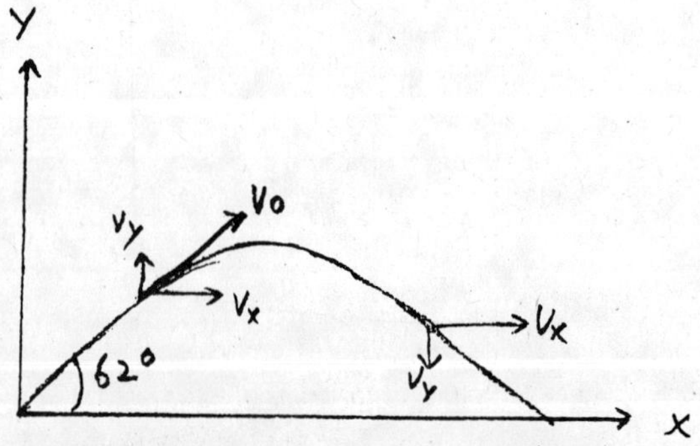

The vertical component is acted upon by gravity

$$v_y = v_o \sin 62° - gt$$

$$v_y = 36\text{ m/s} \sin 62° - (9.8\text{ m/s}^2)t$$

or $$v_y = 31.8\text{ m/s} - (9.8\text{ m/s}^2)t$$

Using distance = (Average velocity)(Time),

$$x = (v_o \cos 62°)t = (16.9\text{ m/s})t$$

$$y = (v_o \sin 62°)t - \frac{1}{2} g t^2$$

$$y = (31.8\text{ m/s})t - (4.9\text{ m/s}^2)t^2$$

15. Use equation 4–15:

$$y = (\tan \theta)x - \frac{g}{2(v_o \cos \theta)^2} x^2$$

$$-1.5\text{ m} = (\tan 16°)17\text{ m} - \frac{9.8\text{ m/s}^2}{2(v_o \cos 16°)^2}(17\text{ m})^2$$

Solving the above equation,

$$v_o = 16\text{ m/s}$$

17. $R = 48$ m, $v_o = 33$ m/s

As shown in example 4–4, the range is given by

$$R = \frac{v_o^2 \sin 2\theta}{g}$$

$$48\text{ m} = \frac{(33\text{ m/s})^2 \sin 2\theta}{9.8\text{ m/s}^2}$$

$$\sin 2\theta = \frac{(48\text{ m})(9.8\text{ m/s}^2)}{(33\text{ m/s})^2}$$

or $$\sin 2\theta = 0.43$$

$$\theta = 12.8°$$

19. $v_o = 52$ m/s, $\theta = 75°$

(a) At the maximum height,

$$v_y = v_o \sin \theta - gt = 0$$

or $$t = \frac{v_o \sin \theta}{g} = \frac{(52\text{ m/s}) \sin 75°}{9.8\text{ m/s}^2}$$

or $$t = 5.1\text{ s}$$

(b) The maximum height is given by (see self test 4–3)

$$h_m = \frac{(v_o \sin\theta)^2}{2g}$$

or $$h_m = \frac{(52\text{ m/s} \sin 75^\circ)^2}{2*9.8\text{ m/s}^2}$$

or $$h_m = 128.7\text{ m}$$

(c) The horizontal distance or range is given by

$$R = \frac{v_o^2 \sin 2\theta}{g}$$

$$R = \frac{(52\text{ m/s})^2 \sin 150^\circ}{9.8\text{ m/s}^2}$$

or $$R = 138\text{ m}$$

21. $g = 10.0\text{ m/s}^2$

At the maximum height, $v_y = 0$

$$\therefore \quad v_o \sin\theta_o - gt = 0 \qquad (1)$$

Using t = 2 s (from graph) in equation (1),

$$v_o \sin\theta_o - (10\text{ m/s}^2)(2\text{s}) = 0$$

or $$v_o \sin\theta_o = 20\text{ m/s} \qquad (2)$$

At t = 2 s, the rock attains the maximum height and it has a horizontal displacement x = 40 m. Therefore, at t = 4 s, it will have a displacement x = 80 m.

$$x = (v_o \cos\theta_o)t \qquad (3)$$

$$80\text{ m} = (v_o \cos\theta_o)4\text{s}$$

or $$v_o \cos\theta_o = 20\text{ m/s} \qquad (4)$$

Dividing equation (2) by equation (4),

$$\tan\theta_o = 1$$

$$\theta_o = \tan^{-1}(1) = 45^\circ$$

Using equation (2),

$$v_o \sin 45^\circ = 20\text{ m/s}$$

or $$v_o = 28.3\text{ m/s}$$

23. (a)

$$f = 33.3 \text{ rev/min} = 0.56 \text{ rev/s}, \quad r = 0.13 \text{ m}$$

$$\omega = 2\pi f = 3.52 \text{ rad/s}$$

$$a_C = \frac{v^2}{r} = r\omega^2 \quad (\text{use } v = r\omega)$$

or $$a_C = (0.13 \text{ m})(3.52 \text{ rad/s})^2$$

or $$a_C = 1.6 \text{ m/s}^2$$

(b)

$$f = 45 \text{ rev/min} = 0.75 \text{ rev/s}$$

$$\omega = 2\pi f = 4.71 \text{ rad/s}$$

$$a_C = r\omega^2 = (0.13 \text{ m})(4.71 \text{ rad/s})^2$$

or $$a_C = 2.9 \text{ m/s}^2$$

25. (a) $v_D = 23$ m/s northeast

(b) $v_H = 23$ m/s southwest

(c) The centripetal acceleration at point D is

$$a_D = \frac{v_D^2}{r} = \frac{(23 \text{ m/s})^2}{200 \text{ m}}$$

or $$a_D = 2.65 \text{ m/s}^2$$

In terms of the unit vectors,

$$\mathbf{a}_D = 2.65 \text{ m/s}^2\,\mathbf{i} - 2.65 \text{ m/s}^2\,\mathbf{j}$$

Similarly,

$$a_H = \frac{(v_H)^2}{r} = \frac{(23 \text{ m/s})^2}{300 \text{ m}}$$

or $$a_H = 1.76 \text{ m/s}^2$$

In terms of the unit vectors,

$$\mathbf{a}_H = (1.76 \text{ m/s}^2)\cos 45^\circ\,\mathbf{i} - (1.76 \text{ m/s}^2)\sin 45^\circ\,\mathbf{j}$$

or $$\mathbf{a}_H = 1.25 \text{ m/s}^2\,\mathbf{i} - 1.25 \text{ m/s}^2\,\mathbf{j}$$

27. Radius of the circle $r = 0.96$ m

Time period $T = 1.1$ s

(a) $$\text{Speed} = \frac{\text{Distance}}{\text{Time}}$$

$$v = \frac{2\pi r}{T} = \frac{2\pi(0.96\text{ m})}{1.1\text{ s}}$$

or $$v = 5.5\text{ m/s}$$

(b) Centripetal acceleration

$$a_C = \frac{v^2}{r} = \frac{(5.48\text{ m/s})^2}{0.96\text{ m}}$$

or $$a_C = 31.3\text{ m/s}^2$$

29. Radius of the earth $R = 6.37 \times 10^6$ m

(a) Velocity $$v = \frac{2\pi R}{T}$$

or $$v = \frac{2\pi(6.37\times 10^6\text{ m})}{86{,}400\text{ s}} \quad (1\text{ day} = 86{,}400\text{ s})$$

or $$v = 463.26\text{ m/s}$$

Centripetal acceleration

$$a_C = \frac{v^2}{R} = \frac{(463.26\text{ m/s})^2}{6.37\times 10^6\text{ m}}$$

or $$a_C = 0.034\text{ m/s}^2$$

or $$a_C = 3.44\times 10^{-3}\text{ g}$$

(b) The radius of the earth's orbit about the sun $= 1.5 \times 10^{11}$ m.

Using $$v = \frac{2\pi R}{T} \quad (T = 3.154\times 10^7\text{ s})$$

or $$v = \frac{2\pi(1.5\times 10^{11}\text{ m})}{3.154\times 10^7\text{ s}}$$

or $$v = 3\times 10^4\text{ m/s}$$

$$a_C = \frac{v^2}{R} = \frac{(3\times 10^4\text{ m/s})^2}{1.5\times 10^{11}\text{ m}}$$

$$a_C = 6\times 10^{-3}\text{ m/s}^2$$

or $$a_C = 0.61\times 10^{-3}\text{ g}$$

(c) In this case,

$$R = 2.8 \times 10^{20}\ \text{m}$$

$$v = 2.5 \times 10^{5}\ \text{m/s}$$

$$a_C = \frac{v^2}{R} = \frac{(2.5 \times 10^5\ \text{m/s})^2}{2.8 \times 10^{20}\ \text{m}}$$

or $$a_C = 2.2 \times 10^{-10}\ \text{m/s}^2$$

or $$a_C = 0.224 \times 10^{-10}\ \text{g}$$

31. We know that the centripetal acceleration is given by

$$a_C = \frac{v^2}{r} \qquad (1)$$

(a) For one complete cycle

$$v = \frac{2\pi r}{T} \qquad (T = \text{Time period})$$

$$\therefore \quad a_C = \frac{(2\pi r)\,v}{rT}$$

or $$a_C = \frac{2\pi v}{T} \qquad (2)$$

(b) Using T = 5.1 s and v = 11 m/s in equation (2),

$$a_C = \frac{2\pi(11\ \text{m/s})}{5.1\ \text{s}}$$

or $$a_C = 13.6\ \text{m/s}^2$$

33. (a) Suzie's speed relative to Becky is 6 m/s.
(b) Suzie's velocity relative to Becky is 6 m/s south.
(c) Becky's speed relative to Suzie is 6 m/s.
(d) Becky's velocity relative to Suzie is 6 m/s north.

35. The resultant v of vectors v_1 and v_2 is determined using the law of cosines

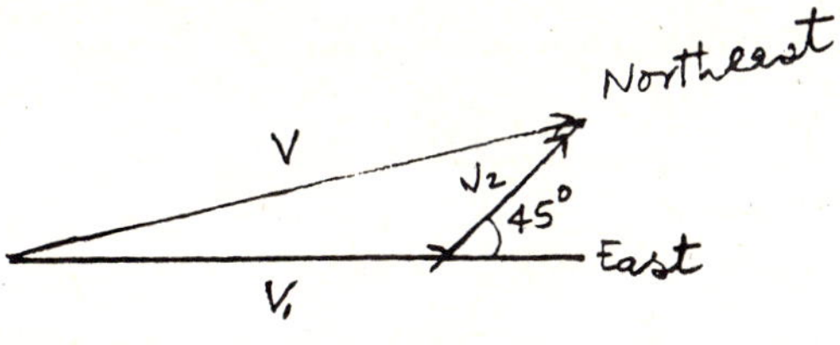

$$v = \sqrt{v_1^2 + v_2^2 + 2v_1v_2 \cos 45°}$$

$$v = \sqrt{(3.4\ \text{m/s})^2 + (1.2\ \text{m/s})^2 + 2(3.4\ \text{m/s})(1.2\ \text{m/s}) \cos 45°}$$

or $v = 4.3 \text{ m/s}$

$$\sin\theta = \frac{v_2 \sin 135^\circ}{v} = \frac{(1.2 \text{ m/s})\sin 135^\circ}{4.3 \text{ m/s}}$$

or $\theta = 11.3^\circ$

37. (a) Velocity of the boat relative to water is given by:

The **j** component of $v = \sqrt{(7.8 \text{ m/s})^2 - (2.3 \text{ m/s})^2} = 7.5 \text{ m/s } \mathbf{j}$

$$v_{BW} = -(2.3 \text{ m/s})\,\mathbf{i} + (7.5 \text{ m/s})\,\mathbf{j}$$

(b) $\text{Velocity} = \frac{\text{Distance}}{\text{Time}}$ $(v = \frac{s}{t},\ s = 1800 \text{ m},\ v = 7.5 \text{ m/s})$

$$\therefore \quad t = \frac{1800 \text{ m}}{7.5 \text{ m/s}} = 240 \text{ s} = 4.0 \text{ min}$$

(c) The **j** component of $v = \sqrt{(7.8 \text{ m/s})^2 - (4.6 \text{ m/s})^2} = 6.3 \text{ m/s}$

$$\therefore \quad v_{BW} = -(4.6 \text{ m/s})\,\mathbf{i} + (6.3 \text{ m/s})\,\mathbf{j}$$

and $t = \frac{1800 \text{ m}}{6.3 \text{ m/s}} = 4.8 \text{ min}$

41. Read the commentary on Galileo Galilei carefully and make an estimate of the radius of Venus's orbit.

43. $v_o = 9.3 \text{ m/s}$, range $R = 4.7 \text{ m}$

(a) From example 4–4, the range is given by

$$R = \frac{v_o^2 \sin 2\theta}{g}$$

$$4.7 \text{ m} = \frac{(9.3 \text{ m/s})^2 \sin 2\theta}{9.8 \text{ m/s}^2}$$

or $\sin 2\theta = 0.5325$

$$\theta = 16.1^\circ$$

(b) The horizontal displacement is given by

$$x = (v_o \cos\theta_o)t$$

$$4.7 \text{ m} = (9.3 \text{ m/s} \cos 16.1^\circ)t$$

or $t = 0.53 \text{ s}$

45. $y_{GB} = y_{GM} + 1.3\text{ m/s } t$

$v_{GB} = v_{GM} + 1.3\text{ m/s}$

$a_{GB} = a_{GM} = -g$

$y_{GM} = v^o_{GM} t - \frac{1}{2} g t^2; \qquad y_{GB} = (v^o_{GM} + 1.3\text{ m/s})t - \frac{1}{2} g t^2$

$v_{GM} = v^o_{GM} - gt; \qquad v_{GB} = (v^o_{GM} + 1.3\text{ m/s}) - gt$

(a) Ball reaches maximum height with respect to Millie when $v_{GM} = 0$; i.e.,

$t_M = \frac{v^o_{GM}}{g}$. At that point:

$$v_{GM} = 0, \qquad v_{GB} = 1.3\text{ m/s}$$

$$a_{GM} = -9.8\text{ m/s}^2, \quad a_{GB} = -9.8\text{ m/s}^2$$

(b) Ball reaches maximum height with respect to Bill when $v_{GB} = 0$; i.e.,

$t_B = (v^o_{GM} + 1.3\text{ m/s})/g$. At that point:

$$v_{GM} = -1.3\text{ m/s}, \qquad v_{GB} = 0$$

$$a_{GM} = -9.8\text{ m/s}^2, \quad a_{GB} = -9.8\text{ m/s}^2$$

(c) Event (b) occurs at $\Delta t = t_B - t_M$ after event (a); i.e.,

$$\Delta t = \frac{1.3\text{ m/s}}{9.8\text{ m/s}^2} = 0.13\text{ sec}$$

47. The equations for the trajectory of the ball are:

$$x = v_{ox} t$$

$$y = v_{oy} t - \frac{1}{2} g t^2$$

Eliminating t, $t = \frac{x}{v_{ox}}$

$$y = x \frac{v_{oy}}{v_{ox}} - \frac{1}{2} \frac{g x^2}{v_{ox}^2}$$

$$= x \tan \theta_o - \frac{1}{2} \frac{g x^2}{v_o^2} \sec^2 \theta_o$$

Thus, $$v_o^2 = \frac{g\,x^2 \sec^2\theta_o}{2\,(x\tan\theta_o - y)}$$

Now $\theta_o = 60°$, and $y = 1.5$ m when $x = 4.3$ m,

so $$v_o = \left[\frac{(9.8\text{ m/s}^2)\,(4.3\text{ m})^2 \sec^2 60°}{2\,((4.3\text{ m})\tan 60° - (1.5\text{ m}))}\right]^{1/2}$$

$$= 7.81\text{ m/s}$$

49. $v_o = 43$ m/s, $\theta_o = 41°$

$x = 20$ m, $h = ?$

Use $x = (v_o \cos\theta)t$

or $$t = \frac{x}{v_o\cos\theta} = \frac{20\text{ m}}{43\text{ m/s}\cos 41°}$$

or $t = 0.62$ sec

Use $$h = (v_o \sin\theta_o)t - \frac{1}{2}gt^2$$

$$h = [(43\text{ m/s})\sin 41°](0.62\text{ s}) - \frac{1}{2}(9.8\text{ m/s}^2)(0.62\text{ sec})^2$$

or $h = 17.49\text{ m} - 1.88\text{ m}$

or $h = 15.61$ m

51. Use $$v = \frac{2\pi R}{T}$$

$$v = \frac{2\pi(20{,}000\text{ m})}{1\text{ s}}$$

or $v = 125668$ m/s

$$\therefore \quad a_C = \frac{v^2}{R}$$

$$a_C = \frac{(125668\text{ m/s})^2}{20{,}000\text{ m}}$$

$$a_C = 7.9 \times 10^5\text{ m/s}^2$$

Problems:

1. (a)

$$x = R\cos\omega t$$

$$v_x = \frac{dx}{dt} = -R\,\omega\sin\omega t$$

$$y = R\sin\omega t$$

$$v_y = \frac{dy}{dt} = R\,\omega\cos\omega t$$

(b)

$$a_x = \frac{dv_x}{dt} = \frac{d}{dt}(-R\,\omega\sin\omega t)$$

or $$a_x = -R\,\omega^2\cos\omega t$$

$$a_y = \frac{dv_y}{dt} = \frac{d}{dt}(R\,\omega\cos\omega t)$$

$$a_y = -R\,\omega^2\sin\omega t$$

(c) From (a) and (b)

$$a_x = -\omega^2 x$$

$$a_y = -\omega^2 y$$

$\therefore$ $$\mathbf{a} = \mathbf{i}\,a_x + \mathbf{j}\,a_y$$

$$= -\omega^2(\mathbf{i}\,x + \mathbf{j}\,y)$$

or $$\mathbf{a} = -\omega^2\mathbf{r}$$

(d) $$R = 45.0\text{ m},\quad \omega = 46.75\text{ m rad/s}$$

$\therefore$ $$x = 45.0\text{ m}\cos 0.4675\,t$$

$$y = 45.0\text{ m}\sin 0.4675\,t$$

or $$x^2 + y^2 = 2025\text{ m}^2 \qquad \text{(circle)}$$

at $$t = 0,\quad x = 45\text{ m},\quad y = 0$$

$$t = 134.4\text{ sec},\quad x = (45\text{ m})\cos 0.4675(134.4\text{ s}) = 45\text{ m}$$

at $$t = 134.4\text{ sec},\quad y = 0 \qquad \text{(circle)}$$

(e) $v = \sqrt{v_x^2 + v_y^2}$

From part (a),

$$v = \sqrt{(-R\,\omega \sin \omega t)^2 + (R\,\omega \cos \omega t)^2}$$

or $$v = \pm R\,\omega$$

at $$t = 16.8 \text{ sec}, \quad \omega t = \pi/4 \text{ radians}$$

$$\tan \theta = \frac{v_y}{v_x} = -\frac{R\,\omega \cos \omega t}{R\,\omega \sin \omega t} = -\cot \omega t$$

with $$\omega t = \pi/4, \quad \theta = 135°$$

For acceleration

$$\tan \theta = \frac{a_y}{a_x} = \frac{-R\,\omega^2 \sin \omega t}{-R\,\omega^2 \cos \omega t} = \tan \omega t$$

at $$\omega t = \pi/4, \quad \theta = \pi/4$$

3. (a) The maximum height $h_m = \dfrac{(v_o \sin \theta_o)^2}{2g}$ (1)

Range $R = \dfrac{v_o^2 \sin 2\theta_o}{g} = \dfrac{2 v_o^2 \sin \theta_o \cos \theta_o}{g}$ (2)

Dividing equation (1) by (2),

$$\frac{h_m}{R} = \frac{v_o^2 \sin^2 \theta_o \cdot g}{2g \cdot 2 v_o^2 \sin \theta_o \cos \theta_o} = \frac{\tan \theta_o}{4}$$

or $$\tan \theta_o = \frac{4\,h_m}{R}$$

$$\theta_o = \tan^{-1} \frac{4\,h_m}{R}$$

(b) $h_m = 6.2 \text{ m}, \quad R = 32 \text{ m}$

From part (a),

$$\theta_o = \tan^{-1} \frac{4\,h_m}{R}$$

or $$\theta_o = \tan^{-1} \frac{4\,(6.2)\,\text{m}}{32\,\text{m}}$$

or $$\theta_o = 37.8°$$

(c) $h_m = R$

$\theta_o = \tan^{-1} 4$

$\theta_o = 76°$

(d) For maximum range ($R = R_m$), $\theta_o = 45°$,

$$45° = \tan^{-1}\frac{4\,h_m}{R_m}$$

or $\frac{4\,h_m}{R_m} = 1$, or $h_m = \frac{R_m}{4}$

5. $v_o = 34$ m/s, $\theta_o = 25°$

Use equation 4–15 for the trajectory of a projectile

$$y = (\tan\theta_o)x - \frac{g}{2\,(v_o\cos\theta_o)^2}\,x^2 \qquad (1)$$

$ab = 4.2$ m

$x = bc\cos 25° = x_o\cos 25°$

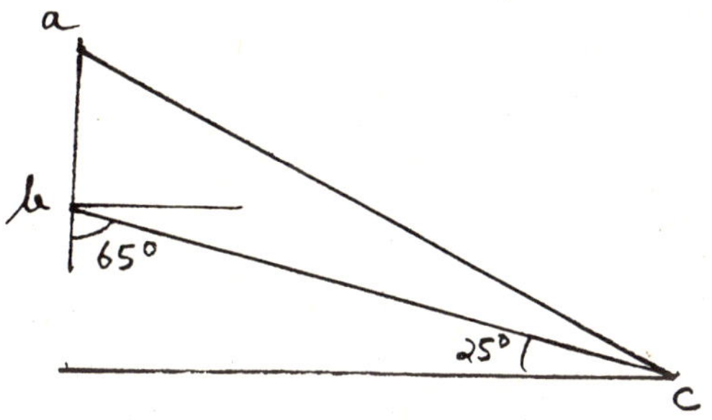

Equation (1) gives

$x_o^2 - 121\,x_o - 1206 = 0$

$x_o = 130 \text{ m} = bc$

Also $(ac)^2 = (bc)^2 + (ab)^2 + 2(bc)(ab)\cos 65°$

or $(ac)^2 = (130\text{ m})^2 + (4.2\text{ m})^2 + 2(130\text{ m})(4.2\text{ m})(0.4226)$

$(ac)^2 = 17379.12\text{ m}^2$

$ac = 131.8$ m

7. The x–component of the velocity remains constant.

$v_x = v_o\cos\theta_o = v_{x_o}$

The y–component of v_o is affected by gravity

$v_y = v_{y_o} - gt$

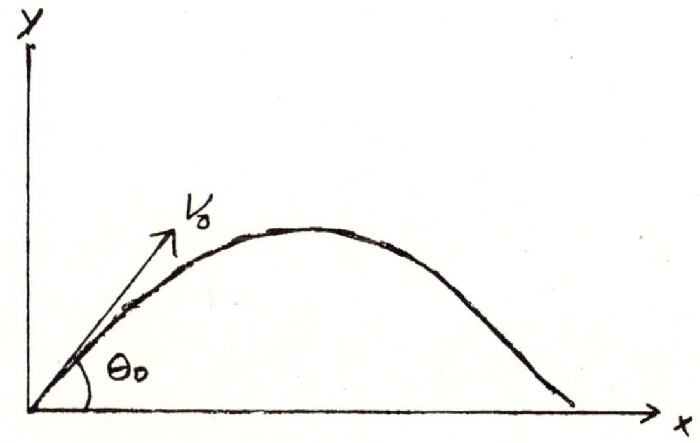

and $v_{y_o} = v_o \sin\theta_o$

$$\frac{v_y}{v_x} = \tan\theta_o = \frac{v_{y_o} - g\,t}{v_{x_o}}$$

or $\theta_o = \tan^{-1}\left(\frac{v_{y_o} - g\,t}{v_{x_o}}\right)$

9. (a) From Fig. 4–14a,

$$\Delta v = 2\,v \sin\frac{\Delta\theta}{2}$$

$$a = \frac{\Delta v}{\Delta t} = \frac{2\,v\sin\frac{\Delta\theta}{2}}{\Delta t} \qquad (1)$$

Also $\Delta r = R\,\Delta\theta$

or $\frac{\Delta r}{\Delta t} = v = R\,\frac{\Delta\theta}{\Delta t}$

or $\frac{\Delta\theta}{\Delta t} = \frac{v}{R}$

Therefore, (1) gives

$$a = \frac{2\,v\sin\frac{\Delta\theta}{2}}{\Delta t}$$

$$= \frac{2\,v\sin\frac{\Delta\theta}{2}}{\Delta t} \bullet \frac{\Delta\theta}{\Delta\theta}$$

or $|a| = \frac{v^2}{R}\left(\frac{2\sin\frac{\Delta\theta}{2}}{\Delta\theta}\right) \qquad (2)$

(b) For $\Delta\theta = \pi/2$, equation (2) gives

$$|a| = 0.92\,\frac{v^2}{R}$$

(c) $|a| = 0.974\,\frac{v^2}{R}$

(d) $|a| = 0.9958\,\frac{v^2}{R}$

(e) $|a| = 0.9999\,\frac{v^2}{R}$

(f) $$\lim_{\Delta\theta \to 0} \frac{\sin\left(\frac{\Delta\theta}{2}\right)}{\Delta\theta} = \frac{1}{2}$$

(g) From (a) using (f)

$$|a| = \frac{v^2}{R}$$

11. $v_o = 33$ m/s and $\theta_o = 32°$

Use equation 4–14

$$y = (v_o \sin\theta_o)t - \frac{1}{2}gt^2$$

For this problem y = 0

$$0 = (33 \text{ m/s} \sin 32°)t - \frac{1}{2}(9.8 \text{ m/s}^2)t^2$$

or $$(4.9 \text{ m/s}^2)t^2 - (17.49 \text{ m/s})t = 0$$

or $t = 0$ or 3.57 sec

Since the ball is struck after 0.50 sec later, the outfielder runs for 3.07 sec.

The range is given by equation 4–18

$$R = \frac{v_o^2 \sin 2\theta_o}{g} = \frac{(33 \text{ m/s})^2 \sin 64°}{9.8 \text{ m/s}^2} = 99.9 \text{ m}$$

Since the outfielder is 118 m from the plate, the distance that he runs is 118 m minus the range.

Therefore, distance run = 118 m – 99.9 m = 18.1 m

Using $$s = v_o t + \frac{1}{2}at^2$$

$$s = \frac{1}{2}at^2$$

$$18.1 \text{ m} = \frac{1}{2}a(3.07 \text{ s})^2$$

or $$a = 3.9 \text{ m/s}^2$$

13. (a) Let v_1 be the velocity of the girl relative to the water

$$v_1 = 1.43 \text{ m/s}$$

Let v_2 be the velocity of the water relative to the earth

$$v_2 = 0.85 \text{ m/s}$$

Let v_3 be the velocity of the girl relative to the earth

$$\therefore \quad v_3 = v_1 + v_2$$

$$= 1.43 \text{ m/s} + 0.85 \text{ m/s}$$

$$\text{or} \quad v_3 = 2.28 \text{ m/s} \qquad \text{(swimming downstream)}$$

$$v_3 = 1.43 \text{ m/s} - 0.85 \text{ m/s}$$

$$= 0.58 \text{ m/s} \qquad \text{(swimming upstream)}$$

Time required for the girl to swim 50 m and return

$$t = \frac{50 \text{ m}}{2.28 \text{ m/s}} + \frac{50 \text{ m}}{0.58 \text{ m/s}}$$

$$\text{or} \quad t = 108.1 \text{ sec}$$

(b) $$v_3 = \sqrt{v_1^2 - v_2^2} = 1.15 \text{ m/s}$$

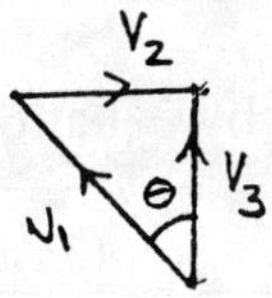

$$t = \frac{100 \text{ m}}{1.15 \text{ m/s}} = 87 \text{ sec}$$

(c) Trip in (a) is longer by

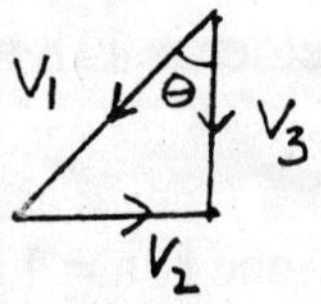

$$108.1 \text{ sec} - 87 \text{ sec} = 21.1 \text{ sec}$$

15. The acceleration of the object falling vertically is given by

$$a_y = -g + bv^2$$

$$b = 0.002 \text{ m}^{-1}$$

Also, as the object's speed approaches the terminal speed, its acceleration approaches zero. Therefore, $a_y = 0$

$$\text{or} \quad 0 = -g + bv^2$$

$$v = \sqrt{\frac{g}{b}} = \sqrt{\frac{9.8 \text{ m/s}^2}{0.002 \text{ m}^{-1}}}$$

$$\text{or} \quad v = 70 \text{ m/s}$$

17. (a) Side ab of triangle abc represents the speed of the car.

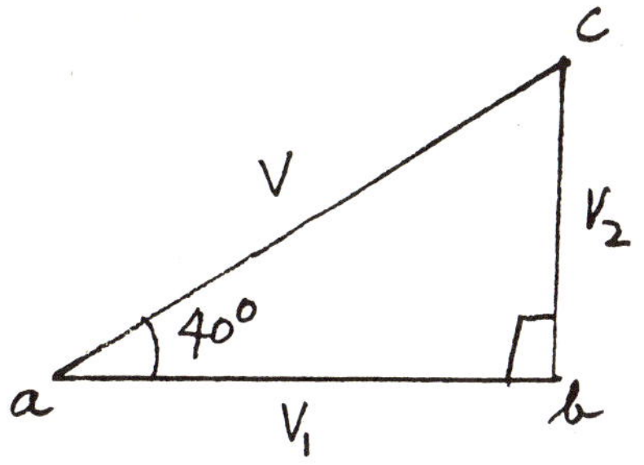

$$v_1 = 10 \text{ m/s}$$

and $\angle cab = 40°$

v and v_2 represent the speed of the rain relative to the car and the earth.

$$\frac{v_2}{v_1} = \tan 40°$$

$$v_2 = v_1 \tan 40° = (10 \text{ m/s}) \tan 40°$$

$$v_2 = 8.4 \text{ m/s}$$

(b) $\frac{v_2}{v} = \sin 40°$

$$v = \frac{v_2}{\sin 40°} = \frac{8.4 \text{ m/s}}{\sin 40°}$$

$$v = 13.07 \text{ m/s}$$

19. (a)

$$x = vt - R \sin \frac{vt}{R} \quad (1)$$

$$y = R - R \cos \frac{vt}{R} \quad (2)$$

also $T = \frac{2\pi R}{v} = t$

Therefore, equation (1) gives

$$x = \cancel{v}\left(\frac{2\pi R}{\cancel{v}}\right) - R \sin \frac{\cancel{v}}{\cancel{R}}\left(\frac{2\pi \cancel{R}}{\cancel{v}}\right)$$

or $x = 2\pi R - R \sin 2\pi$

Equation (2) gives

$$y = R - R \cos \frac{\cancel{v}}{\cancel{R}}\left(\frac{2\pi \cancel{R}}{\cancel{v}}\right)$$

or $y = R - R \cos 2\pi$

Evaluate x and y using equations (1) and (2) at different values of t and then plot the desired graph.

(b) From equation (1),

$$v_x = \frac{dx}{dt}$$

or $$v_x = v - v\cos\frac{vt}{R} \qquad (3)$$

Similarly, from equation (2),

$$v_y = \frac{dy}{dt}$$

or $$v_y = v\sin\frac{vt}{R}$$

(c) $$a_x = \frac{dv_x}{dt} = \frac{v^2}{R}\sin\frac{vt}{R}$$

$$a_y = \frac{dv_y}{dt} = \frac{v^2}{R}\cos\frac{vt}{R}$$

(d) $$a = \sqrt{a_x^2 + a_y^2}$$

or $$a = \left[\left(\frac{v^2}{R}\sin\frac{vt}{R}\right)^2 + \left(\frac{v^2}{R}\cos\frac{vt}{R}\right)^2\right]^{1/2}$$

or $$a = \frac{v^2}{R}$$

21. Make different measurements from the photograph of the bouncing golf ball on the first page of Chapter 4 and answer parts a–e of this problem.

23. (a) From Fig. 4–31,

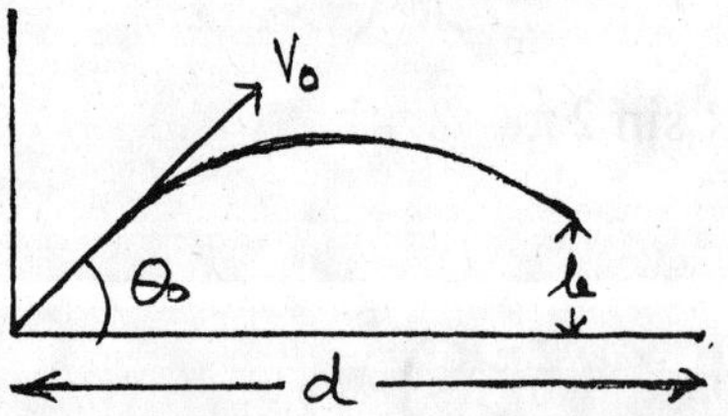

$$d = v_x t = (v_0\cos\theta_0)t \qquad (1)$$

$$b = v_y t - \frac{1}{2}gt^2 = (v_0\sin\theta_0)t - \frac{1}{2}gt^2 \qquad (2)$$

From (1),

$$t = \frac{d}{v_o \cos\theta_o}$$

Therefore, (2) gives

$$b = (v_o \sin\theta_o)\left(\frac{d}{v_o \cos\theta_o}\right) - \frac{1}{2} g \left(\frac{d}{v_o \cos\theta_o}\right)^2$$

or $$b = d \tan\theta_o - \frac{1}{2}\frac{g d^2}{v_o^2 \cos^2\theta_o}$$

or $$2b\, v_o^2 \cos^2\theta_o = 2d\, v_o^2 \tan\theta_o \cos^2\theta_o - g d^2$$

or $$2\, v_o^2 \cos^2\theta_o (d \tan\theta_o - b) = g d^2$$

$$v_o^2 = \frac{g d^2}{2 \cos^2\theta_o (d \tan\theta_o - b)}$$

(b) For $d = 35$ m, $b = 4.0$ m, and $\theta_o = 41°$,

$$v_o = 20 \text{ m/s}$$

25. (a) When the moon is between the earth and the sun as shown by position 1 in the figure, the centripetal acceleration of the moon relative to the sun is given by

$$a_c = \frac{v^2}{R - r}$$

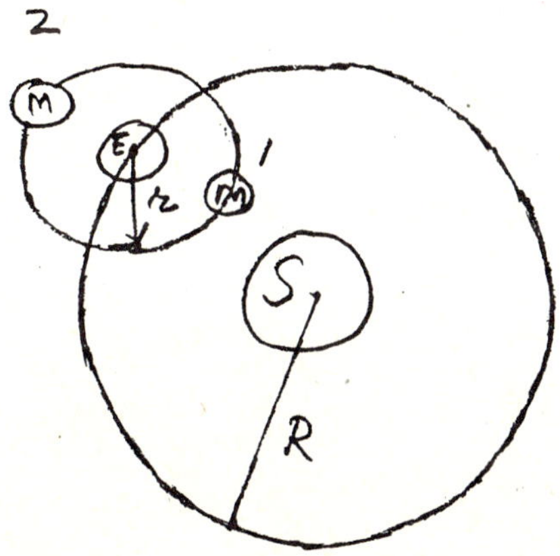

(b) In this case, the position of the moon is shown by 2 in the figure.

$$a_c = \frac{v^2}{R + r}$$

CHAPTER 5 NEWTON'S LAWS OF MOTION

Questions:

1. Newton was describing the vector character of force and, in particular, the parallelogram addition of forces.

The distance travelled by a particle of mass m under the influence of a constant force F in time t is $d = \frac{1}{2}\frac{F}{m}t^2$. What Newton was saying is that if a force $\mathbf{F}_1$ results in a displacement $\mathbf{d}_1$ and a force $\mathbf{F}_2$ in a displacement $\mathbf{d}_2$, then the force $\mathbf{F} = \mathbf{F}_1 + \mathbf{F}_2$ results in a displacement $\mathbf{d}_1 + \mathbf{d}_2$ in the same time t. Mathematically,

if $\mathbf{d}_1 = \frac{1}{2}\frac{\mathbf{F}_1}{m}t^2$ and $\mathbf{d}_2 = \frac{1}{2}\frac{\mathbf{F}_2}{m}t^2$

then $\mathbf{d}_1 + \mathbf{d}_2 = \frac{1}{2}\frac{t^2}{m}\left(\mathbf{F}_1 + \mathbf{F}_2\right)$

3. It would be a violation of Newton's first law of motion for an object to follow a curved path if the net force on it were zero.

5. There is no radial force on the skate. The only force acting on the skate is friction, which prevents it from moving laterally. The merry-go-round, on the other hand, is accelerating. Thus, relative to the merry-go-round, the skate appears to have a radial acceleration, whereas it is really the merry-go-round that is accelerating.

7. Both objects have the same downward acceleration due to gravity, but they have the same retarding *force* due to air resistance. The retarding acceleration is therefore smaller on object a (F/m_a). Hence object a reaches the ground first.

9. The car is slowed by the frictional force between the road and the tires. If the road is icy, the frictional force is much smaller and it becomes more difficult to slow the car.

11. Assuming my mass to be 70 kg, then my weight is

$$70\text{ kg} \times 9.80\frac{\text{m}}{\text{s}^2} \times \frac{1\text{ lb}}{4.45\text{ N}} = 154\text{ lb}$$

The weight of one slug is 1 sl × 32 ft/s^2 = 32 lb. Thus, my mass is

$$\frac{154\text{ lb}}{32\text{ lb/slug}} = 4.8\text{ slugs}$$

13. String b is more likely to break because it will reach the maximum tension with the smaller stretch.

15. The merry-go-round is not an inertial frame. The string hangs at an angle to the vertical and the net force on the rock is not zero. While the vertical component of the tension balances the weight of the rock, the horizontal component causes the rock to have the same acceleration as the merry-go-round.

17. In a gradual pull, the tension $T_a = T_b + W$ where W is the weight of the block. Thus, if both strings are equally strong, string a breaks because the tension reaches the breaking value first.

In a sudden jerk, $T_a = T_b + W - ma$ where m is the mass of the block and a is its acceleration. Assuming that $a > g$, $T_a < T_b$ so b will reach breaking stress first.

19. The tensions are all 25 N and the readings on the spring scales are also 25 N.

21. By Newton's third law, the force exerted by the cart on the horse is equal and opposite to the force exerted by the horse on the cart. Thus the force is horizontal, of magnitude F and directed towards the south.

23. If my mass is 70 kg, the force I exert on the earth is an upward force of 690 N.

25. I would expect my mass to remain constant. However, because the acceleration due to gravity on another planet is likely to be different from that on earth, I would expect my weight to be different.

27. The ecologists' statement indicates that changes never occur in isolation. For example, eradicating insects over a large area affects plants because of pollination, kills birds because their food supply is removed, etc. Newton's third law is also a statement that certain changes cannot occur in isolation. It is a more specific statement since it says that one object cannot accelerate without any other change in the universe. Thus, it is the more fundamental statement.

Exercises:

1. The mass of a 72 kg person in slugs is

$$72\text{ kg} \times \frac{1\text{ sl}}{14.59\text{ kg}} = 4.9\text{ sl}$$

3. $1 \text{ ton} = 1 \text{ ton} \times 2000 \frac{\text{lb}}{\text{ton}} \times \frac{4.45 \text{ N}}{\text{lb}} = 8900 \text{ N}$

Thus,

(a) $1.6 \text{ ton} = 1.6 \text{ ton} \times 8900 \frac{\text{N}}{\text{ton}} = 14000 \text{ N}$

(b) $5.6 \text{ MN} = 5.6 \times 10^6 \text{ N} \times \frac{1 \text{ ton}}{8900 \text{ N}} = 630 \text{ tons}$

5. (a) By Pythagoras' theorem,

$$F_1 = \sqrt{(2.4 \text{ N})^2 + (6.1 \text{ N})^2} = 6.6 \text{ N}$$

(b) By trigonometry, the angle between $\mathbf{F}_1$ and the x–axis is

$$\theta_1 = \tan^{-1}\left(\frac{-2.4}{6.1}\right) = 110°$$

(c and d) In resolving the ambiguity in the inverse tangent, one must take into account that the tangent of θ_1 is negative because the x–component of $\mathbf{F}_1$ is negative. Therefore, $\mathbf{F}_1$ is in the second quadrant. The same result can be obtained more directly using the rectangular to polar conversion in a calculator which already takes this into account. Using this method,

$$F_2 = 13 \text{ N}, \qquad \theta_2 = -49°$$

The net force is $\mathbf{F} = (6.1\,\mathbf{i} - 3.6\,\mathbf{j})\text{N}$ and

$$F = 7.1 \text{ N} \qquad \theta = -31°$$

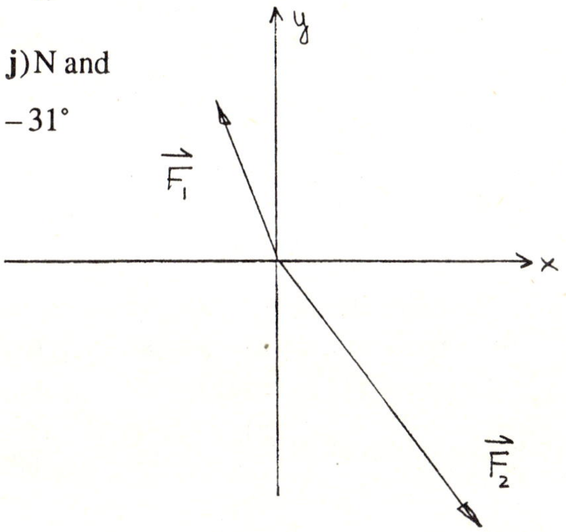

7. Taking the positive x–axis towards the east,

$$\mathbf{F}_A = (130\,\mathbf{j})\text{N}, \qquad F_B = 180 \text{ N}, \qquad \theta_B = 58°$$

Using polar to rectangular conversion,

$$\mathbf{F}_B = (95\,\mathbf{i} + 153\,\mathbf{j})\text{N}$$

Thus,

$$F = F_A + F_B = (95\,\mathbf{i} + 283\,\mathbf{j})\text{N}$$

and using rectangular to polar conversion,

$$F = 300\text{ N}, \qquad \theta = 19° \text{ east of north}$$

9. The two forces are the force exerted by the earth on the skydiver and the air resistance. Since the skydiver is not accelerating, the net force is zero. Thus, the first force is 720 N downward and the second is 720 N upward.

11. To find the acceleration:

$$v = at, \qquad a = v/t = \frac{22\text{ m/s}}{10\text{ s}} = 2.2\text{ m/s}^2$$

Applying Newton's second law,

$$F = ma = 830\text{ kg} \times 2.2\text{ m/s}^2$$
$$= 1830\text{ N}$$

13. Assuming constant acceleration,

$$a = \frac{v^2 - v_o^2}{2x} = \frac{0^2 - (1 \times 10^7\text{ m/s})^2}{2\,(1 \times 10^{-14}\text{ m})} = -.5 \times 10^{28}\text{ m/s}^2$$

The force is therefore

$$F = ma = (2 \times 10^{-27}\text{ kg})(.5 \times 10^{28}\text{ m/s}^2)$$
$$= 10\text{ N}$$

15. Estimate the velocity of the ball at 90 mph = 140 km/hr = 40 m/s. If the pitcher's arm is 70 cm long and swings 180°, it travels a total distance of about 2 m. Thus, the acceleration of the ball is

$$a = \frac{v^2}{2x} = \frac{(40\text{ m/s})^2}{2\,(2\text{ m})} = 400\text{ m/s}^2$$

On a 0.3 kg ball, the force would be

$$F = ma = (0.3\text{ kg})(400\text{ m/s}^2) \approx 100\text{ N}$$

17. $$a = \frac{v_2 - v_1}{t_2 - t_1} = \frac{81\text{ m/s} - (-48\text{ m/s})}{0.01\text{ s}} = 12900\text{ m/s}^2$$

$$F = ma = (.15\text{ kg})(12900\text{ m/s}^2)$$
$$= 1930\text{ N}$$

19. By Newton's third law, the forces occur in equal and opposite pairs. Thus, the force exerted by the Milky Way on the Andromeda galaxy is 7×10^{28} N and the force exerted by the earth on the sun is 3.5×10^{22} N.

21. By Newton's third law, assuming no net external force on the system

$$(m_1 + m)a_1 = m_2a_2$$

$$\therefore \quad m_1 + m = \frac{m_2a_2}{a_1} = \frac{(1.0\text{ kg})(1.14\text{ m/s}^2)}{0.51\text{ m/s}^2} = 2.2\text{ kg}$$

$$\therefore \quad m = 1.2\text{ kg}$$

23. The acceleration of the two masses is

$$a = \frac{F}{m} = \frac{12\text{ N}}{6.0\text{ kg}} = 2.0\text{ m/s}^2$$

If cart 1 pushes on cart 2, it must exert a force $m_2a = 4.0$ N on cart 2, with an equal and opposite force exerted by cart 2 on cart 1. If cart 2 is being pushed against cart 1, the forces become $m_1a = 8.0$ N.

25. Since the force exerted by the planet on a 71 kg space traveller is 950 N, the acceleration of the traveller in free fall would be

$$a = \frac{F}{m} = \frac{950\text{ N}}{71\text{ kg}} = 13.4\text{ m/s}^2$$

27. $W = m_eg$

$$= (9.11 \times 10^{-31}\text{ kg})(9.8\text{ m/s}^2)$$

$$= 89 \times 10^{-31}\text{ N}$$

29. The weight measured by the spring scale is equal to the upward force F exerted by the scale on the person. The other force on the person is the real weight of the person, in this case $77\text{ kg}(9.8\text{ m/s}^2) = 750$ N. The net force causes the person to have the same acceleration as the elevator. Thus

(a) $F - 750\text{ N} = 77\text{ kg}(2.8\text{ m/s}^2)$; $\quad F = 970$ N

(b) $750\text{ N} - F = 77\text{ kg}(3.1\text{ m/s}^2)$; $\quad F = 510$ N

(c) $750\text{ N} - F = 0$; $\quad F = 750$ N

31. The maximum acceleration results when the tension has its maximum value.

$$T - W = ma$$

$$\therefore \quad a = \frac{310\ \text{N} - 24\ \text{kg}\,(9.8\ \text{m/s}^2)}{24\ \text{kg}} = 3.1\ \text{m/s}^2$$

The time required is thus given by

$$x = \frac{1}{2}at^2,$$

$$t = \sqrt{\frac{2x}{a}} = \sqrt{\frac{2\,(4.6\ \text{m})}{3.1\ \text{m/s}^2}} = 1.7\ \text{s}$$

33. (a)

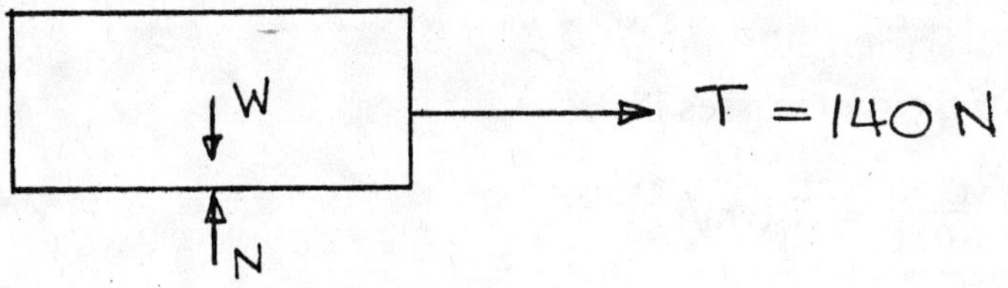

(b) Since friction is negligible, N is vertical. Since the sled remains in contact with the horizontal surface, the net vertical force is zero. Thus

$$N = W = 32\ \text{kg}\,(9.8\ \text{m/s}^2) = 310\ \text{N}$$

and $\quad \mathbf{N} = (310\ \mathbf{j})\text{N}$

(c) $a_x = \dfrac{F_x}{m} = \dfrac{140\ \text{N}}{32\ \text{kg}} = 4.4\ \text{m/s}^2$

$a_y = 0$

$\therefore \quad \mathbf{a} = (4.4\ \mathbf{i})\ \text{m/s}^2$

(d) $x = \dfrac{1}{2}at^2 = \dfrac{1}{2}(4.4\ \text{m/s}^2)(1.3\ \text{s})^2$

$= 3.7\ \text{m}$

35. Along the slope,

$$T\cos 20° - W\sin 15° = 0$$

so $\quad T = mg\sin 15°/\cos 20°$

$= (2.4)(9.8)(.275) = 6.5\ \text{N}$

Perpendicular to the slope,

$$T\sin 20° + R = W\cos 15°$$

so $\quad R = 22.7\ \text{N} - 2.2\ \text{N} = 20.5\ \text{N}$

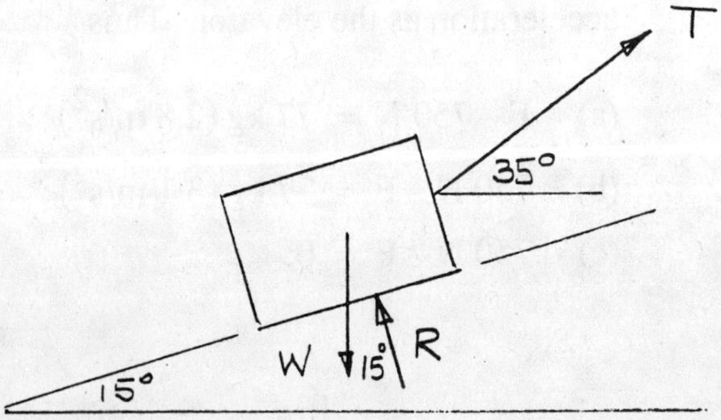

37. $\theta = 27.5°$

$W = (.200\text{ kg})(9.8\text{ m/s}^2) = 1.96\text{ N}$

Vertically, $W = T\cos\theta$

$\therefore \quad T = 1.96/\cos 27.5° = 2.21\text{ N}$

Horizontally, $F = T\sin\theta$

$\therefore \quad F = 2.21\sin 27.5° = 1.02\text{ N}$

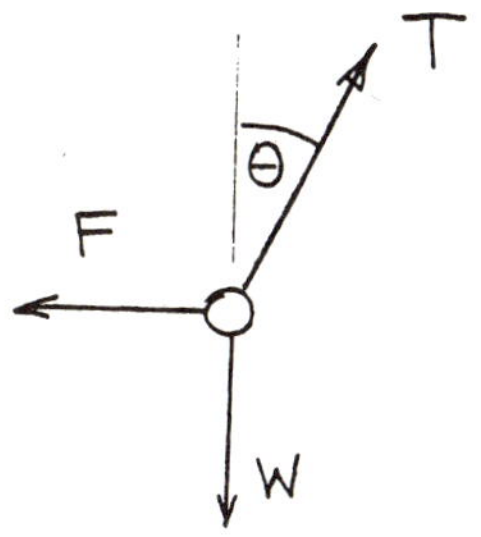

39. For equilibrium,

$$2T\sin\theta = W = mg$$

$$\therefore \quad T = \frac{mg}{2\sin\theta}$$

$$\text{Thus, } T = \frac{(.026\text{ kg})(9.8\text{ m/s}^2)}{2\sin\theta} = \frac{.127\text{ N}}{\sin\theta}$$

$$T(5°) = \frac{.127\text{ N}}{.087} = 1.45\text{ N}$$

$$T(.5°) = \frac{.127\text{ N}}{.0087} = 14.5\text{ N}$$

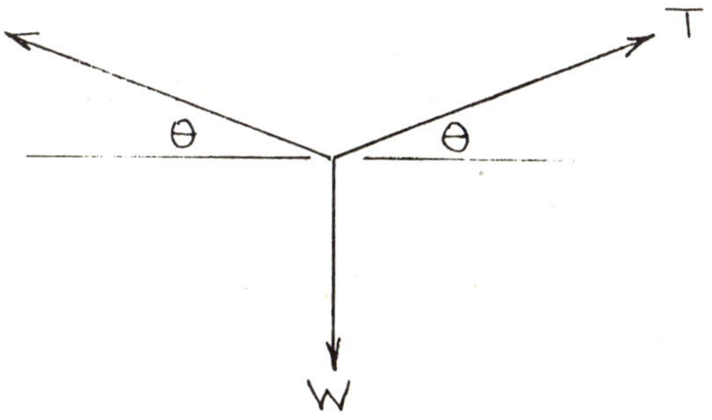

41. Taking the x–axis towards the north and the y–axis vertically upward, the horizontal force is

(a) $F_x = (940)(1.9) = 1790\text{ N}$

and the vertical force is

(b) $F_y = (940)(9.8) = 9200\text{ N}$

(c) The total force is obtained using rectangular to polar conversion:

$F = 9400\text{ N}, \quad \theta = 79°$

43. From the free body diagram,

$$T - W\sin\theta = ma$$

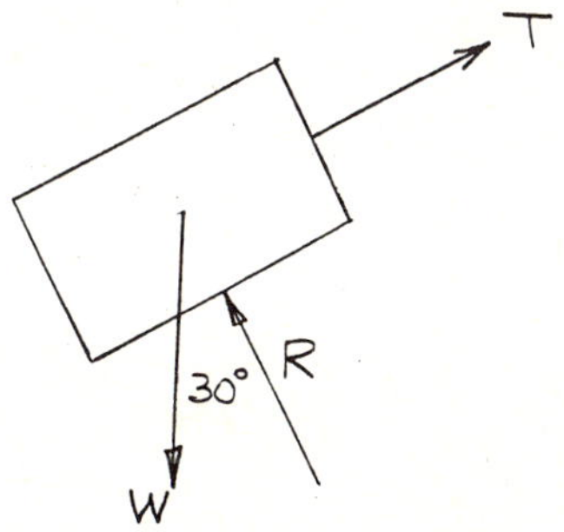

$$\therefore \quad T = (3.7\text{ kg})(9.8\text{ m/s}^2)\sin 30° + (3.7\text{ kg})(1.7\text{ m/s}^2)$$

$$= 24\text{ N}$$

45. x–direction:

$$F_2 \cos 45° = F_3 \cos 30°$$

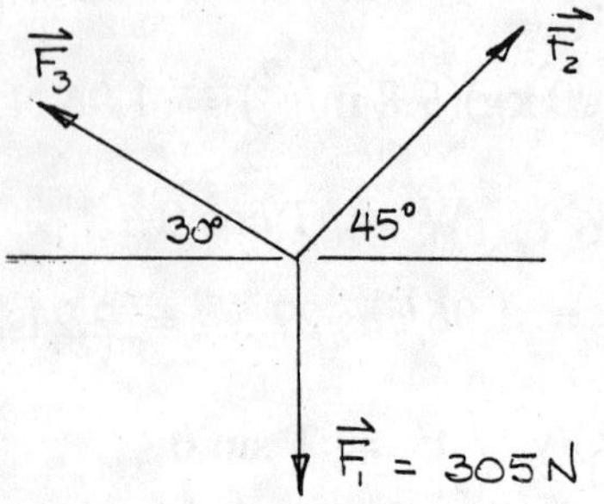

y–direction:

$$F_1 = F_2 \sin 45° + F_3 \sin 30°$$

$$\therefore \quad F_1 = F_2\left(\sin 45° + \frac{\cos 45° \sin 30°}{\cos 30°}\right) = 1.11\ F_2$$

$$\therefore \quad F_2 = \frac{305\ \text{N}}{1.11} = 273\ \text{N} \quad \text{and} \quad F_3 = F_2 \frac{\cos 45°}{\cos 30°} = 223\ \text{N}$$

47. (a) $F_x = ma_x = (55\ \text{kg})(3.4\ \text{m/s}^2) = 190\ \text{N}$

(b) $F_y = mg = (55\ \text{kg})(9.8\ \text{m/s}^2) = 540\ \text{N}$

(c) $F = \sqrt{F_x^2 + F_y^2} = \sqrt{(190\ \text{N})^2 + (540\ \text{N})^2} = 570\ \text{N}$

49. $a = 4.1\ \text{m/s}^2$

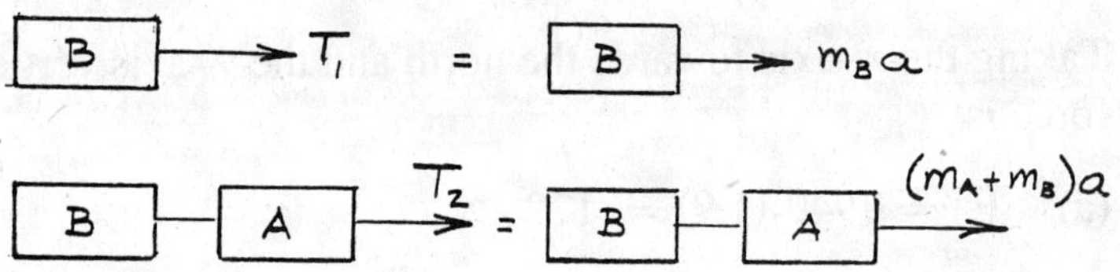

$T_1 = m_B a$

$T_2 = (m_A + m_B)a$

(a) $m_A = 2.0\ \text{kg} \quad m_B = 1.0\ \text{kg}$

$$\therefore \quad T_2 = (3.0\ \text{kg})(4.1\ \text{m/s}^2) = 12.3\ \text{N}$$

$$T_1 = (1.0\ \text{kg})(4.1\ \text{m/s}^2) = 4.1\ \text{N}$$

(b) $m_A = 1.0\ \text{kg} \quad m_B = 2.0\ \text{kg}$

$$\therefore \quad T_2 = 12.3\ \text{N}$$

$$T_1 = (2.0\ \text{kg})(4.1\ \text{m/s}^2) = 8.2\ \text{N}$$

51. The maximum deceleration results when the tension has its maximum value

$$T - W = ma$$

$$1800\ \text{N} - 1750\ \text{N} = \frac{1750\ \text{N}}{9.8\ \text{m/s}^2}\, a$$

$$\therefore \quad a = \frac{50\ \text{N}\,(9.8\ \text{m/s}^2)}{1750\ \text{N}} = 0.28\ \text{m/s}^2 \quad \text{(upward)}$$

$$v^2 - v_o^2 = 2(-a)\,b$$

$$b = v_o^2/2a = (.50\ \text{m/s})^2/2\,(.28\ \text{m/s}^2)$$

$$= 0.45\ \text{m}$$

Problems:

1. Taking the positive x–axis towards the west and the y–axis vertically downwards,

$$\mathbf{F} = (4.36\,\mathbf{i} + 1.98\,\mathbf{j}) \times 10^{20}\ \text{N}$$

$$\mathbf{a} = \frac{\mathbf{F}}{m} = \frac{(4.36\,\mathbf{i} + 1.98\,\mathbf{j}) \times 10^{20}\ \text{N}}{7.35 \times 10^{22}\ \text{kg}}$$

$$= (.00593\,\mathbf{i} + .00269\,\mathbf{j})\ \text{m/s}^2$$

$$a = .00652\ \text{m/s}^2 \qquad \theta = 24.4°$$

This acceleration is relative to an inertial frame of reference whose axes instantaneously were aligned with the west and downward directions of the observer on earth.

3.

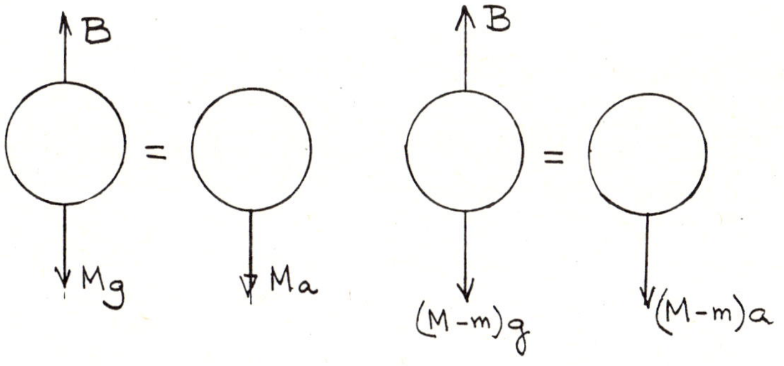

$$Mg - B = Ma \qquad\qquad B - (M - m)g = (M - m)a$$

Solving each equation for B gives

$$B = M(g - a) = (M - m)(g + a)$$

$$\therefore \quad m(g+a) = M[(g+a)-(g-a)]$$

$$= 2Ma$$

or $$m = \frac{2Ma}{g+a}$$

where it is assumed that air resistance is negligible and that the buoyant force is not changed when the ballast is dropped.

Given $M = 400$ kg, $a = 0.2$ m/s^2,

then $$m = \frac{2(400\text{ kg})(0.2\text{ m/s}^2)}{9.8\text{ m/s}^2 + .2\text{ m/s}^2} = 16\text{ kg}$$

5.

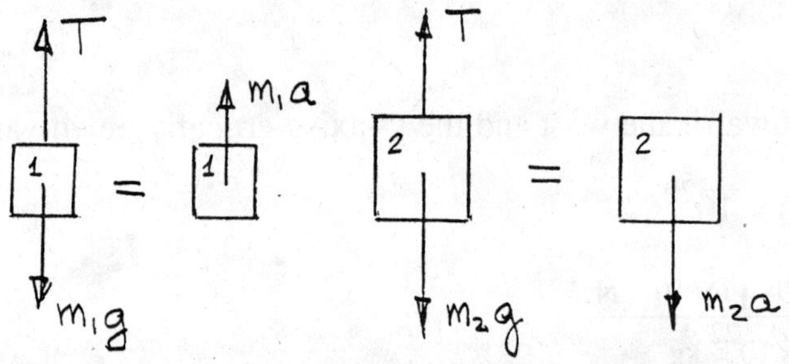

$$T - m_1 g = m_1 a \qquad\qquad m_2 g - T = m_2 a$$

Eliminating T gives

$$(m_2 - m_1)g = (m_2 + m_1)a$$

$$\therefore \quad g = \frac{(m_2+m_1)a}{m_2 - m_1}$$

Given $m_2 = 4.85$ kg, $m_1 = 4.65$ kg, a is found from

$$x = \frac{1}{2}at^2$$

with $x = 0.50$ m, $t = 2.5$ s.

$$a = \frac{2x}{t^2} = \frac{2(0.50\text{ m})}{(2.5\text{ s})^2} = 0.16\text{ m/s}^2$$

$$g = \frac{4.85\text{ kg} + 4.65\text{ kg}}{4.85\text{ kg} - 4.65\text{ kg}}(.16\text{ m/s}^2) = 7.6\text{ m/s}^2$$

7.

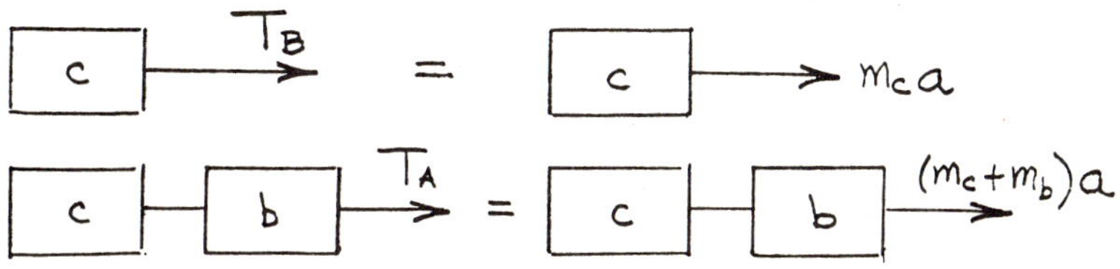

$$T_A = (m_b + m_c)\,a = (3.0\text{ kg})(2.0\text{ m/s}^2) = 6.0\text{ N}$$

$$T_B = m_c a = (2.0\text{ kg})(2.0\text{ m/s}^2) = 4.0\text{ N}$$

9. (a)

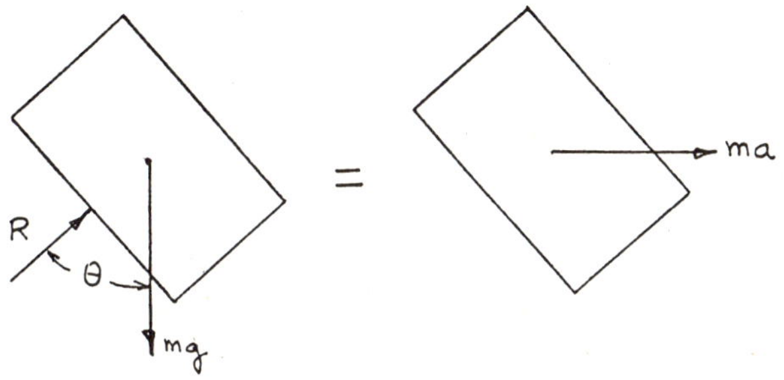

In the horizontal direction, $F_x = ma_x$

$$R\sin\theta = ma$$

while in the vertical, $\sum F_y = 0$

$$R\cos\theta = mg$$

Eliminating R,

$$a = g\tan\theta$$

(b) If $a > g\tan\theta$, the cart rides up the wedge.

11. The accelerations of the sleds are

$$a_A = \frac{T}{M_A} \qquad \text{and} \qquad a_B = \frac{T}{M_B}$$

(a) If T = constant,

$$a_A = \frac{5.0\text{ N}}{50\text{ kg}} = 0.10\text{ m/s}^2$$

$$a_B = \frac{5.0\text{ N}}{25\text{ kg}} = 0.20\text{ m/s}^2$$

(b) Since both sleds start from rest,

$$v_A = \frac{1}{M_A}\int_0^t T\,dt \qquad \text{and} \qquad v_B = \frac{1}{M_B}\int_0^t T\,dt$$

Thus, $\dfrac{v_A}{v_B} = \dfrac{M_B}{M_A} = 0.5$

(c) Since $v_A/v_B = 0.5$, we know $x_A/x_B = 0.5$, and

$$x_A + x_B = 6\text{ m}$$

$$x_B(0.5 + 1) = 6\text{ m}$$

which gives $x_B = 4\text{ m}$, $x_A = 2\text{ m}$

(d) The results in both (b) and (c) would not change if the tension were not held constant.

13.

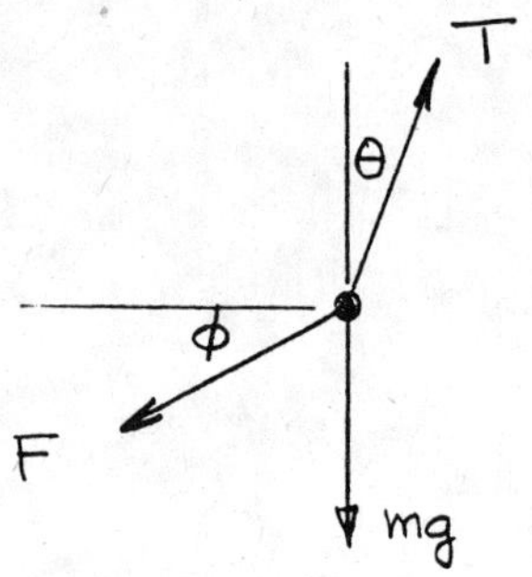

x–direction: $T \sin\theta = F \cos\phi$

y–direction: $T \cos\theta = F \sin\phi + mg$

Eliminate T:

$$F \cos\phi \cot\theta = mg + F \sin\phi$$

or $$F = \frac{mg}{\cos\phi \cot\theta - \sin\phi}$$

and $$T = \frac{F \cos\phi}{\sin\theta} = \frac{mg \cos\phi}{\sin\theta(\cos\phi \cot\theta - \sin\phi)}$$

$$= \frac{mg}{\cos\theta - \sin\theta \tan\phi}$$

15. (a and b) The velocity of the ball just before it strikes the surface is $v = \sqrt{2gh}$. Assuming constant acceleration, the distance d required to stop is

$$d = v^2/2a$$

So, $$a = v^2/2d = gh/d$$

The force exerted by the surface is

$$F = W + ma = mg(1 + h/d)$$

upward.

(c) For a tennis ball with m = 0.06 kg, h = 2 m, and d = 1 mm

$$F = (0.06\text{ kg})(9.8\text{ m/s}^2)\left(1 + \frac{2\text{ m}}{.001\text{ m}}\right)$$

$$= 1000\text{ N}$$

(d) The force on a billiard ball would be larger because the mass is larger and especially because d would be much smaller.

17. (a) Let the force between the gliders be F(t). Then $a_1 = F/m_1$, $a_2 = F/m_2$, from which

$$v_1 = \frac{1}{m_1}\int F\,dt, \qquad v_2 = \frac{1}{m_2}\int F\,dt$$

and $m_1v_1 = m_2v_2$. Likewise,

$$\ell_1 = \int v_1\,dt, \qquad \ell_2 = \int v_2\,dt$$

$$\therefore \quad m_1\ell_1 = m_2\ell_2$$

(b) If $m_1 = 1.0$ kg, $\ell_1 = 0.40$ m, $\ell_2 = 0.20$ m

then $$m_2 = \frac{m_1\ell_1}{\ell_2} = \frac{(1.0\text{ kg})(0.40\text{ m})}{0.20\text{ m}} = 2.0\text{ kg}$$

19.

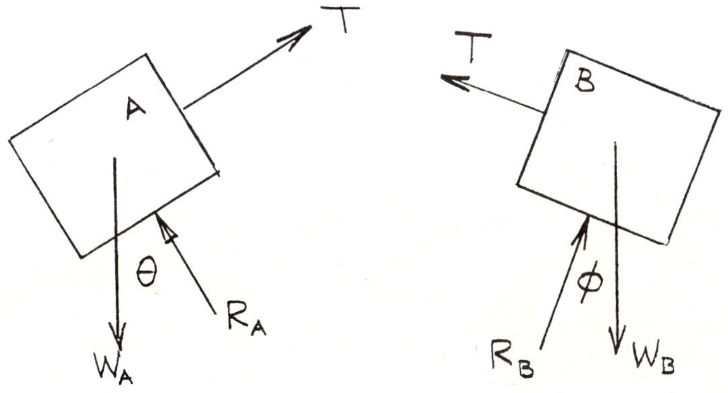

(a) Assuming A moves uphill, the equations of motion for A and B along their respective slopes are:

$$T - W_A \sin\theta = m_A a$$

$$W_B \sin\phi - T = m_B a$$

Adding the two equations gives

$$W_B \sin\phi - W_A \sin\theta = (m_A + m_B)a$$

or $$a = g(m_B \sin\phi - m_A \sin\theta)/(m_A + m_B)$$

(b) When $\theta = \phi = 90°$

$$a = g(m_B - m_A)/(m_A + m_B)$$

(Note: If $m_A \sin\theta > m_B \sin\phi$, then A moves downhill.)

21.

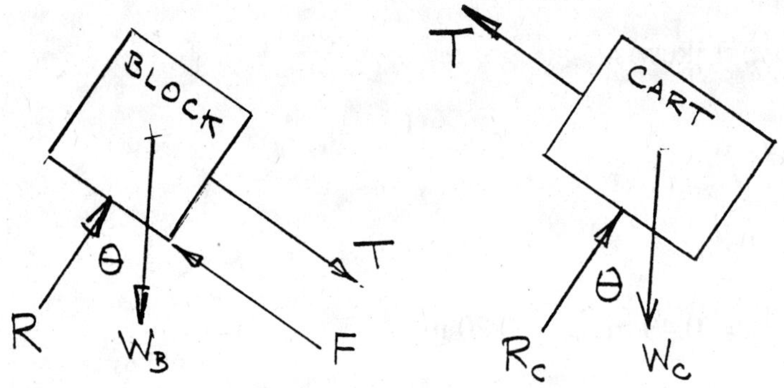

(a) For the cart,

$$T = W_C \sin\theta = m_C g \sin\theta$$

(b) For the block:

$$F = T + W_B \sin\theta$$

$$= (W_B + W_C)\sin\theta = (m_B + m_C)g \sin\theta$$

(c) $R = W_B \cos\theta = m_B g \cos\theta$

CHAPTER 6 APPLICATIONS OF NEWTON'S LAWS OF MOTION

Questions:

1. The contact force is the product of the normal force and the coefficient of friction.

3. You cannot slide the crate by exerting a horizontal force on it. You need to apply the force so that it has a vertical component, which reduces the force of friction between the crate and the floor and increases the friction between your shoes and the floor.

5. The value of the coefficient of kinetic friction will not be independent of the initial speed of the boat.

7. The coefficient of friction for melting ice is less than that for ice which is very cold and not melting.

9. $\frac{\mu_s}{\mu_k} >> 1$

11. (a) For $v = v_d$, the bubble is centered.

    (b) For $v > v_d$, the bubble is right of center.

    (c) For $v < v_d$, the bubble is left of center.

13. Given the same initial speed, the acceleration of the heavier car will be the same as the lighter one. Therefore, the heavier car cannot come to a stop in a shorter distance than the lighter one.

15. The car moves forward because of the frictional force between the tires and the road. The maximum static frictional force is proportional to the normal force exerted by the road on the tires. By increasing the normal force on the drive tires, the maximum static frictional force is increased and there is less chance of the tires sliding.

17. If the plane is turning right and the ball is left of center, the banking angle should be decreased to balance the centripetal force.

19. The object's weight in an earth–surface frame is proportional to its mass because of the dependence of centrifugal force on the mass of the object.

21. The coordinate frame at rest on Noino's equator is more nearly an inertial frame than a frame at rest on earth's equator. The reason for this is that the acceleration at Noino's equator is one half the acceleration at the earth's equator.

23. In the absence of static friction, the path of the car will be a straight line.

25. The train is either on a hill so that the floor is not horizontal or it is accelerating on level ground and the floor is horizontal.

Exercises:

1. (a) $\sum F_y = 0$

$T + F_N = W$

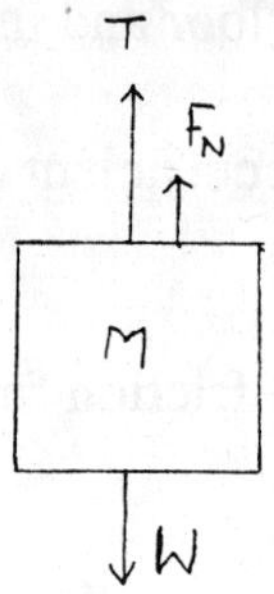

If $T = 52\ N$

$F_N = W - T$

$= (37\ kg)(9.8\ m/s^2) - 52\ N$

$= 362.6\ N - 52\ N$

or $F_N = 310.6\ N$

T = Tension force
W = Weight = mg
F_N = Normal force

(b) If $T = 170\ N$

$F_N = 192.6\ N$

(c) If $T = 360\ N$

$F_N = 2.6\ N$

3. $m = 41\ kg$, $\mu_s = 0.46$ and $\mu_k = 0.39$

(a) Using equation 6–3, we get

$$F_s \le \mu_s F_N$$

The minimum horizontal force $F = F_s$

and $F_N = mg = (41\ kg)(9.8\ m/s^2) = 401.8\ N$

$\therefore$ $F \le (0.46)(401.8\ N)$

$F \le 184.83\ N$

(b) In this case, $F = F_k$

and $F_k = \mu_k F_N = F$

or $F = (0.39)(401.8\ \text{N})$

$F = 156.7\ \text{N}$

5. Using equation 6–3, we get

$$F_s \leq \mu_s F_N$$

As the block just starts moving,

$$F_s = \mu_s F_N$$

also $F = F_s$ (F = horizontal force).

Given F = 28 N, $F_N = mg = (2.6\ \text{kg})(9.8\ \text{m/s}^2) = 25.48\ \text{N}$

(a) Therefore, $\mu_s = \dfrac{F}{F_N} = \dfrac{28\ \text{N}}{25.48\ \text{N}} = 1.1$

(b) In this case, $F = F_k$

and $F_k = \mu_k F_N$

Therefore, $\mu_k = \dfrac{F}{F_N} = \dfrac{19\ \text{N}}{25.48\ \text{N}} = 0.75$

7. Speed limit = 25 m/s
Stopping distance s = 57 m

Using $F_k = \mu_k F_N$, where $F_N = mg$

$$a = \frac{\mu_k F_N}{m} = 0.8\ g = (0.8)(9.8\ \text{m/s}^2)$$

or $a = 7.84\ \text{m/s}^2$

Using $v^2 = v_o^2 + 2as$, where $v = 0$

$$v_o^2 = -2as \quad [a = -7.84\ \text{m/s}^2 = \text{deceleration}]$$

or $v_o = \sqrt{2[7.84\ \text{m/s}^2][57\ \text{m}]}$

$v_o = 30\ \text{m/s}$

Yes, the driver was exceeding the limit by 5 m/s.

9. $\mu_k = 0.12, \; v_o = 5.2 \text{ m/s}$

$$a = \frac{F}{m} = \frac{F_k}{m} = \frac{\mu_k F_N}{m} = (0.12)(9.8 \text{ m/s}^2)$$

or $$a = 1.18 \text{ m/s}^2 \quad [\text{In this problem } a = -1.18 \text{ m/s}^2]$$

using $$v^2 = v_o^2 + 2as, \quad \text{where } v = 0$$

$$v_o^2 = 2as$$

$$s = \frac{v_o^2}{2a} = \frac{(5.2 \text{ m/s})^2}{2(1.18 \text{ m/s}^2)}$$

$$s = 11.46 \text{ m}$$

11. (a) The free body diagram for the crate is shown in Fig. 6–9b. Since the crate's velocity is constant,

$$\sum F_x = 0 \quad \text{or} \quad F_T \cos\theta - \mu_k F_N = 0 \qquad (1)$$

$$\sum F_y = 0 \quad \text{or} \quad F_T \sin\theta + F_N - mg = 0 \qquad (2)$$

From (1), $$F_T = \frac{\mu_k F_N}{\cos\theta}$$

Substituting this in equation (2) gives

$$\frac{\mu_k F_N}{\cos\theta} \cdot \sin\theta + F_N - mg = 0$$

or $$F_N = \frac{mg\cos\theta}{\cos\theta + \mu_k \sin\theta}$$

(b) $$F_c = \sqrt{F_N^2 + F_k^2}, \quad \text{where } F_k = \mu_k F_N$$

or $$F_c = F_N\sqrt{1 + \mu_k^2}$$

$$F_c = \frac{mg\cos\theta}{\cos\theta + \mu_k \sin\theta}\sqrt{1 + \mu_k^2}$$

(c) For $m = 45 \text{ kg}$, $\theta = 33°$ and $\mu_k = 0.63$

$$F_N = 310 \text{ N} \quad \text{and} \quad F_c = 370 \text{ N}$$

13. $\mu_s = 0.85$

Using equation 6–3,

$$F_s \le \mu_s F_N, \text{ where } F_N = \frac{mg}{2} \text{ for this problem.}$$

Also, $F_s = F = ma$

$$\therefore \quad ma \le \frac{\mu_s m g}{2}$$

or $$a \le \frac{\mu_s g}{2} = \frac{(0.85)(9.8 \text{ m/s}^2)}{2}$$

$$a = 4.17 \text{ m/s}^2 = \text{maximum acceleration of the car}$$

15. $m = 26$ kg

$\mu_s = 0.096; \ \mu_k = 0.072$

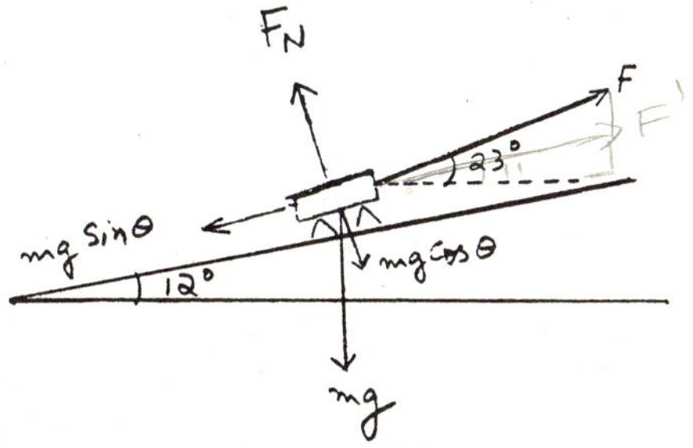

(a) The equation of motion is

$$F \cos\theta_2 = F_s + mg \sin\theta_1$$

or $$F \cos\theta_2 = \mu_s F_N + mg \sin\theta_1$$

$$F = \frac{\mu_s m g \cos\theta_1 + m g \sin\theta_1}{\cos\theta_2},$$

where $\theta_1 = 12°$ and $\theta_2 = 23° - 12° = 11°$

$$F = \frac{23.93 \text{ N} + 52.98 \text{ N}}{0.982} = 78 \text{ N}$$

(b) In this case,

$$F \cos\theta_2 - \mu_k F_N - mg \sin\theta_1 = 0$$

$$F = \frac{\mu_k m g \cos\theta_1 + m g \sin\theta_1}{\cos\theta_2}$$

or $$F = \frac{17.94 \text{ N} + 52.98 \text{ N}}{0.982}$$

$$F = 72 \text{ N}$$

17. $m = 6.4 \text{ kg}$ and $\mu_s = 0.76$

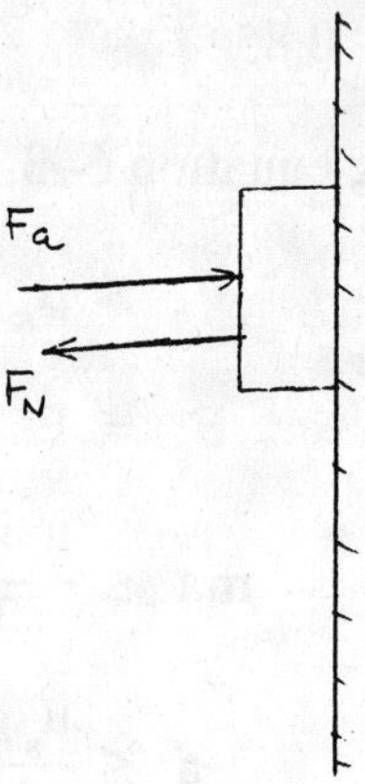

$\sum F_x = 0$ gives

$$F_a - F_N = 0 \quad (1)$$

$\sum F_y = 0$ gives

$$F_s - mg = 0 \quad (2)$$

Also, $F_s = \mu_s F_N$

Therefore, equation (2) becomes

$$\mu_s F_N = mg \quad (3)$$

From equation (1), $F_a = F_N$

Therefore, $\mu_s F_a = mg$

or $$F_a = \frac{mg}{\mu_s} = \frac{(6.4 \text{ kg})(9.8 \text{ m/s}^2)}{0.76}$$

$$F_a = 82.53 \text{ N}$$

19. $\theta = 5°$ and $\mu_r = 0.062$

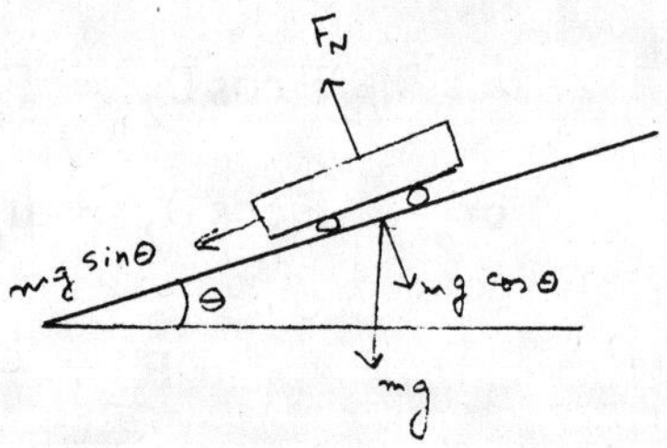

Using the law of rolling friction,

$$F_r = \mu_r F_N, \text{ where } F_N = mg\cos\theta$$

or $$F_r = \mu_r mg\cos\theta$$

Using Newton's 2nd law of motion ($F_{Net} = ma$), we get

$$mg\sin\theta - \mu_r mg\cos\theta = ma$$

or $$a = \frac{mg\sin\theta - \mu_r mg\cos\theta}{m}$$

or $$a = g\sin\theta - \mu_r g\cos\theta$$

$$a = (9.8 \text{ m/s}^2)\sin 5° - (0.062)(9.8 \text{ m/s}^2)\cos 5°$$

$$a = [0.854 - 0.605] \text{ m/s}^2$$

$$a = 0.25 \text{ m/s}^2$$

21. $m = 0.081\ \text{kg}$

$F_d = -(13\ \text{N} \cdot \text{s/m})\,v$

When v approaches v_t (terminal speed), F_d approaches mg, so that

$$b v_t = mg$$

or

$$v_t = \frac{mg}{b}$$

$$v_t = \frac{(0.081\ \text{kg})(9.8\ \text{m/s}^2)}{13\ \text{N} \cdot \text{s/m}}$$

$$v_t = 0.061\ \text{m/s}$$

23. $R = 310\ \text{m}, \quad v = 25\ \text{m/s}$

From example 6–6,

$$\theta_b = \tan^{-1} \frac{v_d^2}{Rg}$$

$$\theta_b = \tan^{-1} \frac{(25\ \text{m/s})^2}{(310\ \text{m})(9.8\ \text{m/s}^2)}$$

$$\theta_b = 11.63°$$

25. $v = 75\ \text{m/s}$

$\theta = 28°$

Using $\tan\theta = \dfrac{v^2}{Rg}$,

$$R = \frac{v^2}{g \tan\theta}$$

or

$$R = \frac{(75\ \text{m/s})^2}{(9.8\ \text{m/s}^2)\tan 28°}$$

$$R = 1079.5\ \text{m}$$

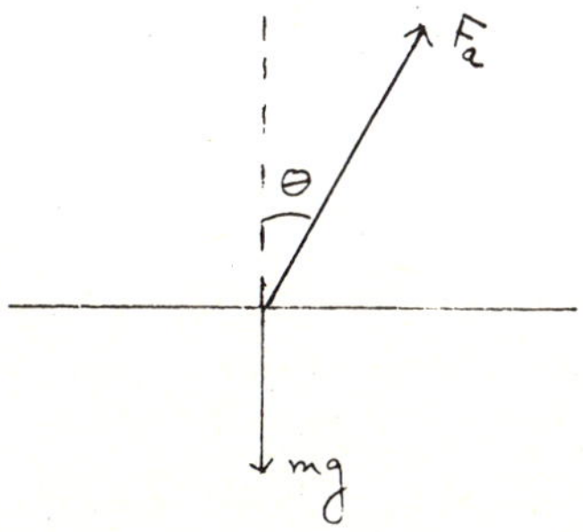

27. $m = 875\ \text{kg}, \quad R = 15\ \text{m}, \quad v = 7.5\ \text{m/s}$

(a) $a = \dfrac{v^2}{R}$ (centripetal acceleration)

$$a = \frac{(7.5\ \text{m/s})^2}{15\ \text{m}} = 3.75\ \text{m/s}^2$$

(b) $F_s = F_c = \dfrac{mv^2}{r} = (875\ \text{kg})(3.75\ \text{m/s}^2) = 3281.3\ \text{N}$

(c) $F_N = mg = (875\ \text{kg})(9.8\ \text{m/s}^2) = 8575\ \text{N}$

(d) $F_c = \sqrt{(F_s^2 + F_N^2)}$

or $F_c = \sqrt{(3281.3\ \text{N})^2 + (8575\ \text{N})^2}$

$F_c = 9181.37\ \text{N}$

(e) $\theta = \tan^{-1}\frac{F_s}{F_N} = \tan^{-1}\frac{3281.3}{8575} = 20.94°$

where θ is the angle between the contact force and the normal force.

29. $F_s = \mu_s F_N$

$$\mu_s = \frac{F_s}{F_N} = \frac{\frac{mv^2}{R}}{mg} = \frac{v^2}{Rg},$$

where $v = r\omega$

$$= \frac{(2\pi \times 0.092\ \text{m/rev})(33\tfrac{1}{2}\ \text{rev/min})}{(60\ \text{s/min})}$$

or $v = 0.32\ \text{m/s}$

$$\therefore \quad \mu_s = \frac{(0.32\ \text{m/s})^2}{(0.092\ \text{m})(9.8\ \text{m/s}^2)} = 0.11$$

31. (a) At the top of the vertical circle,

$$F_c + mg = \frac{mv^2}{R}$$

$$F_c = \frac{mv^2}{R} - mg$$

Since $F_c \geq 0$,

$$\frac{mv^2}{R} - mg \geq 0$$

or $v^2 \geq Rg$

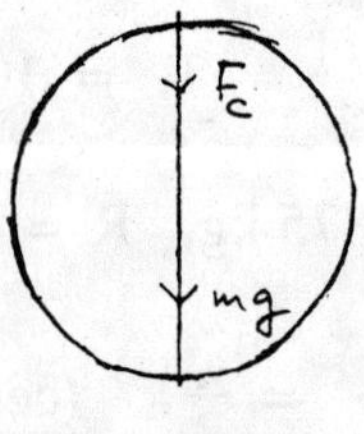

F_c is the contact force between the water and the bottom of the bucket.

(b) $v = \sqrt{Rg} = \sqrt{(1.0\ \text{m})(9.8\ \text{m/s}^2)}$

$v = 3.13\ \text{m/s}$

33. (a) At point A:

$$mg - F_{N_A} = \frac{mv^2}{r} \qquad (1)$$

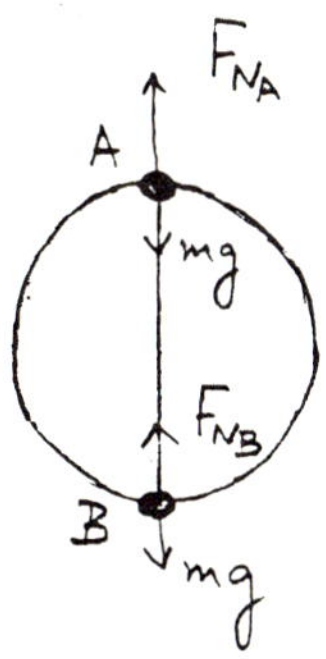

At point B:

$$F_{N_B} - mg = \frac{mv^2}{r} \qquad (2)$$

From equations (1) and (2),

$$mg - F_{N_A} = F_{N_B} - mg$$

or $$F_{N_B} = 2mg - F_{N_A}$$

$$F_{N_B} = 2(62\ \text{kg})(9.8\ \text{m/s}^2) - 210\ \text{N}$$

$$F_{N_B} = 1005.2\ \text{N}$$

(b) From equation (2),

$$(1005.2\ \text{N}) - (62\ \text{kg})(9.8\ \text{m/s}^2) = \frac{(62\ \text{kg})\,v^2}{7.1\ \text{m}}$$

or $$v = 6.75\ \text{m/s}$$

(c) $$T = \frac{2\pi r}{v} \qquad \text{[distance = (velocity)(Time)]}$$

$$T = \frac{2(3.1417)(7.1\ \text{m})}{6.75\ \text{m/s}}$$

$$T = 6.61\ \text{s}$$

35. From example 6–8,

$$F_e = F_w + (0.0339\ \text{m/s}^2)m$$

$$F_e = (12.75\ \text{N}) + (0.0339\ \text{m/s}^2)(1.3\ \text{kg})$$

$$F_e = 12.75\ \text{N} + 0.04\ \text{N}$$

$$F_e = 12.79\ \text{N}$$

37. (a) At the pole, $F = mg$

so $$g = \frac{12.8\ \text{N}}{1\ \text{kg}} = 12.8\ \text{m/s}^2$$

(b) At the equator,

$$F = mg - \frac{mv^2}{R}$$

so $$\frac{v^2}{R} = \frac{12.8\text{ N} - 10.1\text{ N}}{1\text{ kg}} = 2.7\text{ m/s}^2$$

$$v = \frac{2\pi R}{T}$$

$$\therefore \quad \frac{4\pi^2 R}{T^2} = 2.7\text{ m/s}^2$$

or $$R = \frac{2.7\text{ m/s}^2}{4\pi^2}(8.68 \times 10^3\text{ s})^2$$

$$R = 5.15 \times 10^6\text{ m}$$

39. $m = 75\text{ kg}$

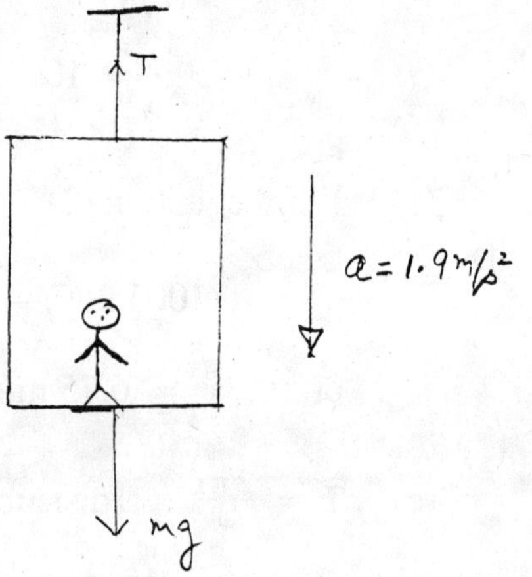

Using $F = ma$,

$$mg - T = ma$$

$$T = m(g - a)$$

$$T = (75\text{ kg})(9.8\text{ m/s}^2 - 1.9\text{ m/s}^2)$$

$$T = 592.5\text{ N}$$

$= $ (Person's weight magnitude)

Note if $a = 9.8\text{ m/s}^2$, the person will be weightless.

41. $m = 1.6\text{ kg}$

$\theta = 26.5°$, $\mu_s = 0.50$

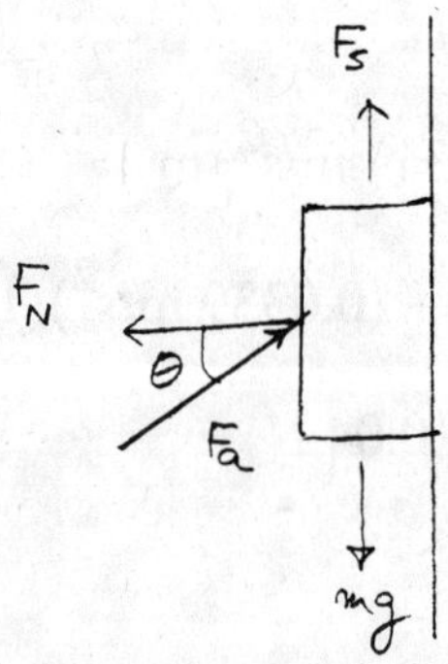

$$F_s = \mu_s F_N$$

Also, $$F_N = F_a \cos\theta$$

$$\therefore \quad F_s = \mu_s F_a \cos\theta$$

or $$F_a = \frac{F_s}{\mu_s \cos\theta}$$

$$F_a = \frac{mg}{\mu_s \cos\theta}$$

$$F_a = \frac{(1.6\text{ kg})(9.8\text{ m/s}^2)}{(0.50)\cos 26.5°}$$

$$F_a = 35.04\text{ N}$$

43. $v_1 = 3$ m/s, $d = 5$ m

From example 6–2, the coefficient of kinetic friction

$$\mu_k = \frac{v^2}{2gd} = \frac{(3\text{ m/s})^2}{2(9.8\text{ m/s}^2)(5\text{ m})}$$

or $$\mu_k = 0.09$$

Using the above equation again, we get

$$0.09 = \frac{(6\text{ m/s})^2}{2(9.8\text{ m/s}^2)d}$$

or $$d = 20.41\text{ m}$$

45. $L = 0.89$ m
$T = 1.4$ s

From the free body diagram,

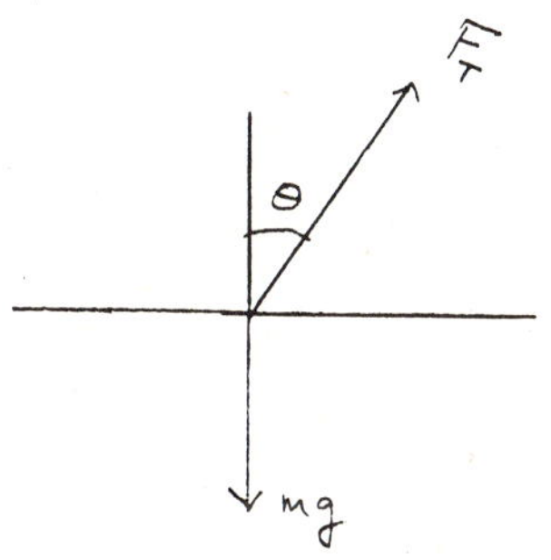

$$F_T \cos\theta = mg \qquad (1)$$

$$F_T \sin\theta = \frac{mv^2}{r} \qquad (2)$$

Dividing (2) by (1), we get

$$\tan\theta = \frac{v^2}{rg}$$

Substituting $v = \frac{2\pi r}{T}$ and $r = L\sin\theta$ gives

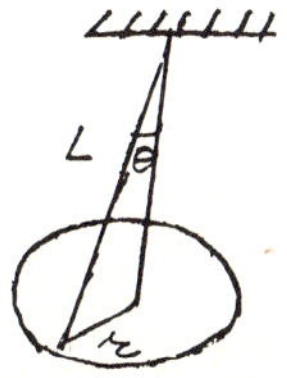

$$\cos\theta = \frac{gT^2}{4\pi^2 L}$$

or $$\cos\theta = \frac{(9.8\text{ m/s}^2)(1.4\text{ s})^2}{4\pi^2(0.89\text{ m})} = 0.547$$

$$\theta = 56.9°$$

47. $r = 178$ m, $\theta = 12°$

Using $\tan\theta = \frac{v^2}{rg}$,

$$v^2 = rg\tan\theta$$

$$v^2 = (178\text{ m})(9.8\text{ m/s}^2)\tan 12°$$

$$v = 19.26\text{ m/s}$$

49. $T = 5.8\ \text{s},\ r = 6.1\ \text{m},\ m = 1.4\ \text{kg}$

(a) At point A:

$$mg - F_{N_A} = \frac{mv^2}{r}$$

$$\therefore \quad F_{N_A} = mg - \frac{mv^2}{r} \qquad (1)$$

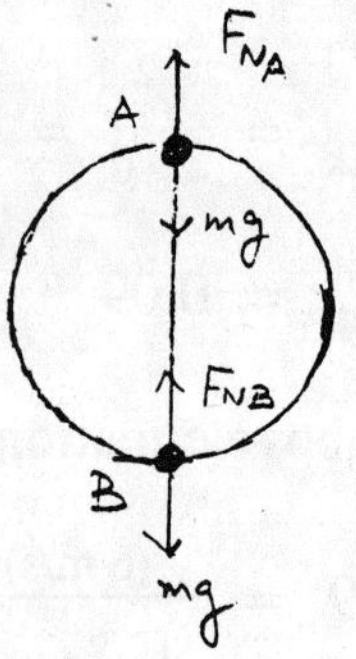

where $v = \frac{2\pi r}{T} = \frac{2\pi(6.1\ \text{m})}{5.8\ \text{s}}$

or $v = 6.61\ \text{m/s}$

Therefore, equation (1) gives

$$F_{N_A} = (1.4\ \text{kg})(9.8\ \text{m/s}^2) - \frac{(1.4\ \text{kg})(6.61\ \text{m/s})^2}{6.1\ \text{m}}$$

$$F_{N_A} = 3.69\ \text{N}$$

(b) At point B:

$$F_{N_B} - mg = \frac{mv^2}{r} \quad \text{or} \quad F_{N_B} = \frac{mv^2}{r} + mg$$

or $$F_{N_B} = \frac{(1.4\ \text{kg})(6.61\ \text{m/s})^2}{6.1\ \text{m}} + (1.4\ \text{kg})(9.8\ \text{m/s}^2)$$

$$F_{N_B} = 23.75\ \text{N}$$

Problems:

1. (a) $\mu_2 > \mu_1$

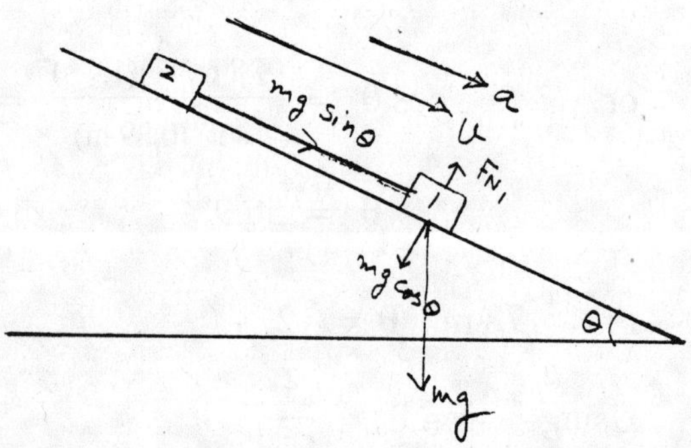

Using $F_s = \mu_s F_N$ gives

$$F_{s_1} = \mu_1 mg \cos\theta$$

$$F_{s_2} = \mu_2 mg \cos\theta$$

Using $F = ma$ gives

$$2mg \sin\theta - (\mu_1 + \mu_2)mg \cos\theta = 2ma \qquad (1)$$

For mass 1, $F = ma$ gives

$$mg \sin\theta - \mu_1 mg \cos\theta - F_T = ma \qquad (2)$$

Multiplying equation (2) by 2 and subtracting from equation (1) gives

$$\mu_1 \, mg\cos\theta - \mu_2 \, mg\cos\theta + 2F_T = 0$$

or $$F_T = \tfrac{1}{2}(\mu_2 - \mu_1)\, mg\cos\theta$$

(b) From equation (1),

$$a = g\left[\sin\theta - \tfrac{1}{2}(\mu_2 + \mu_1)\cos\theta\right]$$

(c) If v is constant, $a = 0$, $\theta = \theta_k$

From part (b),

$$0 = \left[\sin\theta_k - \tfrac{1}{2}(\mu_2 + \mu_1)\cos\theta_k\right]$$

$$\theta_k = \tan^{-1}\left[\tfrac{1}{2}(\mu_2 + \mu_1)\right]$$

3. $F_s \le \mu_s F_N$ (1)

$F_s = mg$ (2)

Using $F = ma$,

$$F_a - F_N = ma$$

and $$F_a = (m + M)a$$

$$\therefore \quad \frac{F_a - F_N}{m} = \frac{F_a}{m + M}$$

or $$F_N = \frac{MF_a}{m + M} \quad (3)$$

From equations (1), (2) and (3), we get

$$mg \le \mu_s \frac{MF_a}{m + M}$$

or $$F_a \ge (m + M)\left(\frac{mg}{\mu_s M}\right)$$

5. (a) $F_k = \mu_k F_N$ (1)

where $F_k = F_a \cos\theta$

$F_N + F_a \sin\theta = mg$

or $F_N = mg - F_a \sin\theta$

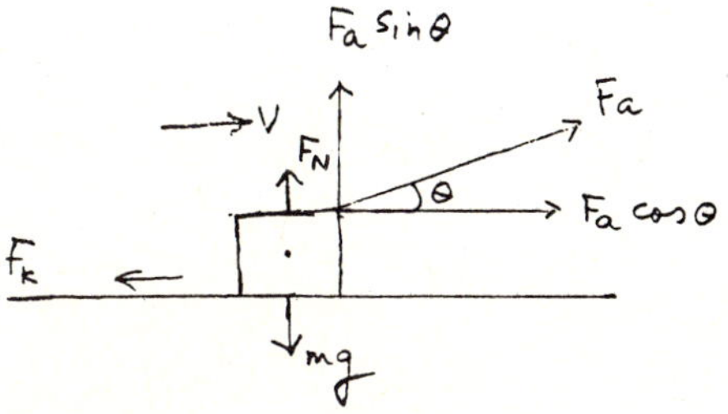

Using F_k and F_N in equation (1), we get

$$F_a \cos\theta = \mu_k[mg - F_a \sin\theta]$$

Therefore, $$F_a = \frac{\mu_k m g}{(\cos\theta + \mu_k \sin\theta)}$$

$$\frac{dF_a}{d\theta} = -\frac{\mu_k m g(\mu_k \cos\theta - \sin\theta)}{(\cos\theta + \mu_k \sin\theta)^2} = 0$$

or $$\tan\theta_m = \mu_k \quad \text{or} \quad \theta_m = \tan^{-1}\mu_k$$

(b) From part (a),

$$F_a = \frac{\mu_k m g}{(\cos\theta + \mu_k \sin\theta)}$$

Since $\tan\theta_m = \mu_k$,

$$\sin\theta_m = \frac{1}{(1+\mu_k^2)^{1/2}} \quad \text{and} \quad \cos\theta = \frac{\mu_k}{(1+\mu_k^2)^{1/2}}$$

Therefore, $$F_a = \frac{\mu_k m g}{\dfrac{\mu_k}{(1+\mu_k^2)^{1/2}} + \mu_k \dfrac{1}{(1+\mu_k^2)^{1/2}}}$$

or $$F_a = \frac{\mu_k m g}{(1+\mu_k^2)^{1/2}}$$

(c) $\mu_k = 0.70$, $m = 51$ kg

$$F_a = \frac{(0.7)(51\text{ kg})(9.8\text{ m/s}^2)}{[1+(0.7)^2]^{1/2}} = 286.61\text{ N}$$

(d) From part (a),

$$\theta_m = \tan^{-1}\mu_k$$

or $$\theta_m = \tan^{-1} 0.7 = 35°$$

(e) $F_k = \mu_k F_N$, where $F_N = mg$

or $$F_k = (0.7)(51\text{ kg})(9.8\text{ m/s}^2) = 349.86\text{ N}$$

as compared to $F_a = 286.61$ N

7. $\mu_s = 0.75, \quad v_o = 22\text{ m/s}$

$F_s = F = ma, \quad \text{where} \quad F_s = \mu_s F_N$

$\mu_s F_N = ma, \quad \text{where} \quad F_N = mg$

$$a = \frac{\mu_s m g}{m} = (0.75)(9.8\text{ m/s}^2)$$

$$a = 7.35\text{ m/s}^2$$

Using $\quad v^2 = v_o^2 + 2as, \quad \text{where} \quad v = 0$

$$0 = (22\text{ m/s})^2 - 2(7.35\text{ m/s}^2)s$$

or $\quad s = 32.93\text{ m}$

9. $m_A = m_B = 5.0\text{ kg} = m$

$\mu_k = 0.40$

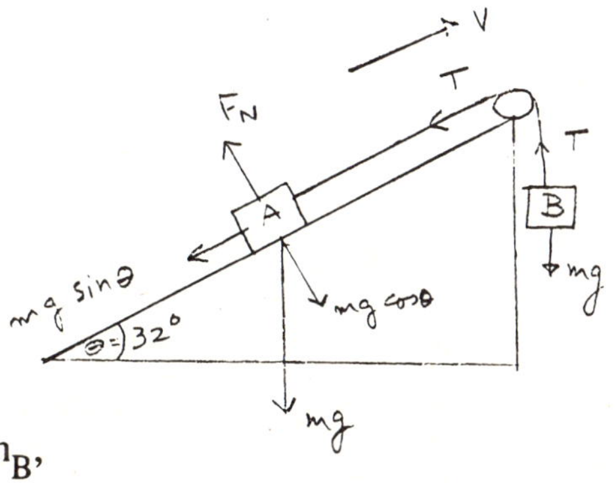

(a) Using F = ma for mass m_B,

$$mg - T = ma \qquad (1)$$

F = ma applied to mass m_A gives

$$T - mg\sin\theta - F_k = ma \qquad (2)$$

where $F_k = \mu_k F_N = \mu_k mg\cos\theta$

Therefore, equation (2) becomes

$$T - mg\sin\theta - \mu_k mg\cos\theta = ma \qquad (3)$$

Adding equations (1) and (3), we get

$$mg - mg\sin\theta - \mu_k mg\cos\theta = 2ma$$

$$a = \frac{g}{2}(1 - \sin\theta - \mu_k\cos\theta)$$

$$a = \frac{9.8\text{ m/s}^2}{2}(1 - \sin 32° - 0.4\cos 32°)$$

$$a = 0.64\text{ m/s}^2$$

(b) From equation (1),

$$T = mg - ma = m(g - a)$$

$$T = (5\text{ kg})[9.8\text{ m/s}^2 - 0.64\text{ m/s}^2]$$

$$T = 45.8\text{ N}$$

11.

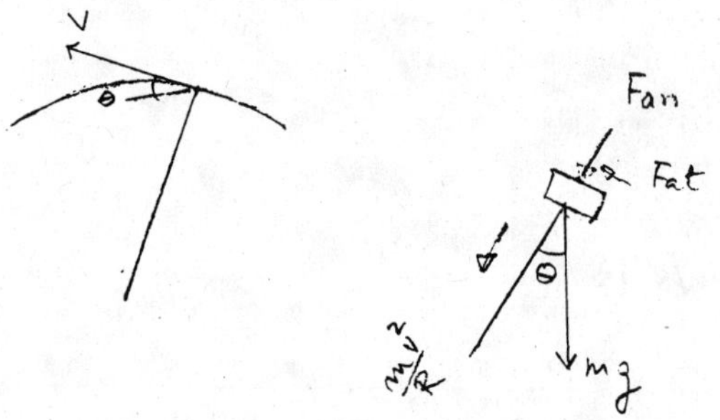

$$F_{an} + mg\cos\theta = \frac{mv^2}{R}$$

When $\theta = 0,\ F_{an} = 0$

Therefore, $\frac{mv^2}{R} = mg$

$$F_{an} = mg - mg\cos\theta = mg(1 - \cos\theta)$$

$$F_{at} = mg\sin\theta$$

$$F_a = \sqrt{F_{an}^2 + F_{at}^2} = mg\sqrt{(1-\cos\theta)^2 + \sin^2\theta}$$

$$= mg\sqrt{2 - 2\cos\theta} = mg\sqrt{4\sin^2\theta/2}$$

$$F_a = 2mg\sin\theta/2$$

13. $T = 8.2\text{ s},\ m = 0.69\text{ kg},\ r = 3.8\text{ m}$

(a) $v = \frac{2\pi r}{T} = \frac{2\pi(3.8\text{ m})}{8.2\text{ s}}$

$v = 2.91\text{ m/s}$

(b) $a = \frac{v^2}{r} = \frac{(2.91\text{ m/s})^2}{3.8\text{ m}}$

$a = 2.23\text{ m/s}^2$

(c) $F = \frac{mv^2}{r} = (0.69\text{ kg})(2.23\text{ m/s}^2)$

$F = 1.54\text{ N}$

15. (a)

$$F_s = \mu_s F_N \qquad (1)$$

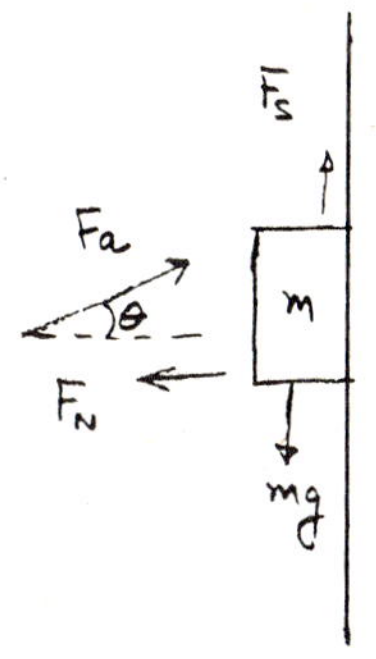

$$\sum F_x = 0 \text{ gives}$$

$$F_a \cos\theta - F_N = 0$$

or $$F_N = F_a \cos\theta \qquad (2)$$

$$\sum F_y = 0 \text{ gives}$$

$$F_a \sin\theta + F_s = mg$$

or $$F_s = mg - F_a \sin\theta \qquad (3)$$

Using equations (2) and (3) in equation (1), we get

$$mg - F_a \sin\theta = \mu_s F_a \cos\theta$$

$$F_a(\mu_s \cos\theta + \sin\theta) = mg$$

or $$F_{a,min} = \frac{mg}{(\mu_s \cos\theta + \sin\theta)}$$

(b) From part (a)

$$F_{a,min} = \frac{mg}{\sin\theta + \mu_s \cos\theta} \qquad (1)$$

$$\frac{dF_{a,min}}{d\theta} = -mg(\sin\theta + \mu_s \cos\theta)^{-2}(\cos\theta - \mu_s \sin\theta) = 0$$

or $$\frac{mg(\mu_s \sin\theta - \cos\theta)}{(\sin\theta + \mu_s \cos\theta)^2} = 0$$

or $$mg(\mu_s \sin\theta - \cos\theta) = 0$$

$$\mu_s \sin\theta = \cos\theta \qquad (2)$$

$$\theta = \tan^{-1}\left(\frac{1}{\mu_s}\right)$$

(c) Substituting equation (2) in equation (1) gives

$$F_{a,min} = \frac{mg}{(\sin\theta + \mu_s^2 \sin\theta)}$$

$$F_{a,min} = \frac{mg}{\sin\theta(1 + \mu_s^2)}.$$

17. $m = 985\text{ kg}, \quad R = 162\text{ m}, \quad \theta = 12°, \quad v = 24\text{ m/s}$

(a) $$F_s = \frac{mv^2}{R} = \frac{(985\text{ kg})(24\text{ m/s})^2}{162\text{ m}}$$

$$F_s = 3502.22\text{ N}$$

(b) For $v = 12\text{ m/s}$,

$$F_s = \frac{(985\text{ kg})(12\text{ m/s})^2}{162\text{ m}}$$

$$F_s = 875.6\text{ N}$$

19. With the plane moving in the x–direction, the object will deflect as shown.

$$a = g \sin \theta$$

$$a = (9.8\text{ m/s}^2) \sin 35°$$

$$a = 5.62\text{ m/s}^2$$

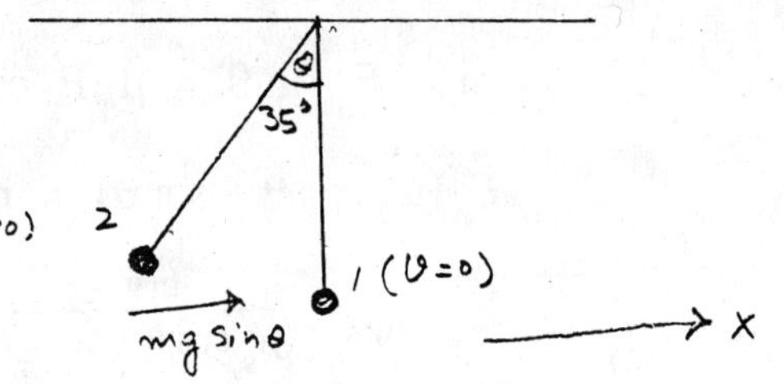

CHAPTER 7 NEWTON'S LAW OF UNIVERSAL GRAVITATION

Questions:

1. If $F = k/r^n$, for a circular orbit

$$\frac{mv^2}{r} = \frac{k}{r^n} \quad \text{or} \quad \frac{m(2\pi r)^2}{T^2 r} = \frac{k}{r^n}$$

$$\therefore \quad T^2 = 4\pi^2 m r^{n+1}$$

Since $T^2 \sim r^3$, force laws (a) and (b) are consistent with the data, but (c) is not.

3. Since $F \sim m^2$, the forces would be (a) $4F_o$, (b) $9F_o$, and (c) $16F_o$.

5. The orbiting satellite is an accelerating frame of reference. It accelerates at the local value of g. Thus, an astronaut would register zero weight on a spring scale, just as she would in a free-falling elevator.

7. The concept of mass could be explained in the context of Newton's laws. The concept of weight requires, in addition, the law of gravitation. Thus, mass is more easily explained.

9. $$m g_p = \frac{G m m_p}{r_p^2}$$

Thus, $$g_p = \frac{G m_p}{r_p^2} = \frac{G \frac{4}{3}\pi r_p^3 \rho_p}{r_p^2}$$

So $$g_p \sim \rho_p r_p$$

It follows that gravity on Egabbac would be twice that on earth.

11. $$[G] = \frac{[F][r^2]}{[m^2]} = \frac{\text{lbs ft}^2}{\text{slug}^2} = \frac{\text{slug ft}}{s^2} \cdot \frac{\text{ft}^2}{\text{slug}^2}$$

$$= \frac{\text{ft}^3}{s^2\ \text{slug}}$$

13. (a and b) The orbit becomes elliptical and the point where the burst is fired becomes the perigee.

(c) The apogee will be more distant than r_o.

(d) The period is proportional to the 3/2 power of the major axis of the ellipse and therefore increases.

15. Using $\frac{mv^2}{r} = \frac{Gmm_e}{r^2}$, we get $v^2 = \frac{Gm_e}{r}$

and with $v = \frac{2\pi r}{T}$, $T^2 = \frac{4\pi^2 r^3}{Gm_e}$

Thus, $\frac{a_C}{a_B} = \frac{r_B^2}{r_C^2} = \frac{1}{4}$

$$\frac{v_C}{v_B} = \sqrt{\frac{r_B}{r_C}} = \frac{1}{\sqrt{2}}$$

$$\frac{T_C}{T_B} = \left(\frac{r_C}{r_B}\right)^{3/2} = 2\sqrt{2}$$

17. The highest speed corresponds to the closest approach. Thus, the earth is closest to the sun in January, making the northern winter milder and it is furthest in July, making the southern winter more severe.

19. The plane of the orbit must pass through the center of the earth because the force is directed towards the center.

21. From the dynamical point of view, it is more appropriate to say that the apple accelerates because it is the less massive object and its acceleration is therefore much greater than that of the earth.

23. In a stable circular orbit, the object is accelerating towards the center of the circle. Thus, it remains in orbit while under the influence of a constant force. Without the transverse motion the only way the object can accelerate is downward toward the attracting center.

Exercises:

1. $a_C = \frac{v^2}{R} = \frac{\left(\frac{2\pi R}{T}\right)^2}{R} = \frac{4\pi^2 R}{T^2}$

	T, s	R, s	a_C, m/s^2	$a_C R^2$, m^3/s^2
Mars	5.94×10^7	2.28×10^{11}	2.55×10^{-3}	1.33×10^{20}
Neptune	522×10^7	45.0×10^{11}	6.52×10^{-6}	1.32×10^{20}

3. Using $\frac{R^3}{T^2} = \text{constant} = C$

Evaluate C from the data for Elppa

$$C = \frac{(1.22 \times 10^{11}\ m)^3}{(2.00 \times 10^7\ s)^2} = 4.54 \times 10^{18}\ m^3/s^2$$

(a) For Mulp

$$T = \sqrt{\frac{R^3}{C}} = \sqrt{\frac{(1.89 \times 10^{11}\ m)^3}{4.54 \times 10^{18}\ m^3/s^2}}$$

$$= 3.86 \times 10^7\ s$$

(b) For Healp

$$T = \sqrt{\frac{(2.45 \times 10^{11}\ m)^3}{4.54 \times 10^{18}\ m^3/s^2}} = 5.69 \times 10^7\ s$$

5. $F = \frac{G m_1 m_2}{r^2}$

$$= \frac{(6.67 \times 10^{-11}\ N \cdot m^2/kg^2)(5.2\ kg)(.15\ kg)}{(0.640\ m)^2} = 1.27 \times 10^{-10}\ N$$

7. $r^2 = \frac{G m_1 m_2}{F}$

$$r = \left(\frac{(6.67 \times 10^{-11}\ N \cdot m^2/kg^2)(5.5\ kg)(.145\ kg)}{1.3 \times 10^{-10}\ N}\right)^{1/2} = 0.640\ m$$

9. $r^2 = \frac{G m_e m}{F}$

$$r = \left(\frac{(6.67 \times 10^{-11}\ N \cdot m^2/kg^2)(5.97 \times 10^{24}\ kg)(1.00\ kg)}{1.00\ N}\right)^{1/2}$$

$$= 2.00 \times 10^7\ m$$

The initial acceleration would be

$$\frac{F}{m} = 1.00\ m/s^2$$

11. (a) When Mars is in the west at sunrise, both planets are on the same side of the sun and the distance between them is $R_{Mars} - R_{earth} = 0.78 \times 10^{11}$ m.

Thus, $$F = \frac{Gm_m m_e}{r_{me}^2} = \frac{(6.67 \times 10^{-11}\ \text{N}\cdot\text{m}^2/\text{kg}^2)(6.46 \times 10^{23}\ \text{kg})(5.97 \times 10^{24}\ \text{kg})}{(0.78 \times 10^{11}\ \text{m})^2}$$

$$= \frac{2.57 \times 10^{38}\ \text{N}\cdot\text{m}^2}{(0.78 \times 10^{11}\ \text{m})^2} = 4.2 \times 10^{16}\ \text{N}$$

(b) When Mars is in the east at sunrise,

$$r_{me} = R_{Mars} + R_{earth} = 3.78 \times 10^{11}\ \text{m},$$

and $$F = \frac{2.57 \times 10^{38}\ \text{N}\cdot\text{m}^2}{(3.78 \times 10^{11}\ \text{m})^2} = 1.80 \times 10^{15}\ \text{N}$$

(c) When Mars is overhead,

$$r_{me} = \sqrt{R_{Mars}^2 - R_{earth}^2} = 1.72 \times 10^{11}\ \text{m}$$

$$F = \frac{2.57 \times 10^{38}\ \text{N}\cdot\text{m}^2}{(1.72 \times 10^{11}\ \text{m})^2} = 8.72 \times 10^{15}\ \text{N}$$

13. $$g_m = \frac{GM_m}{r_m^2}$$

$$M_m = \frac{g_m r_m^2}{G}$$

$$= \frac{(3.7\ \text{m/s}^2)(3.4 \times 10^6\ \text{m})^2}{6.67 \times 10^{-11}\ \text{N}\cdot\text{m}^2/\text{kg}^2}$$

$$= 6.4 \times 10^{23}\ \text{kg}$$

15. (a) $$G = 1.0000\ \frac{\text{AU}^3}{\text{yr}^2\ \text{C}} \times \left(1.496 \times 10^{11}\ \frac{\text{m}}{\text{AU}}\right)^3 \times \left(\frac{\text{yr}}{365\ \text{days}} \times \frac{\text{day}}{24\ \text{hr}} \times \frac{\text{hr}}{3600\ \text{s}}\right)^2$$

$$= 3.366 \times 10^{18}\ \frac{\text{m}^3}{\text{s}^2\ \text{C}} = 6.670 \times 10^{-11}\ \frac{\text{m}^3}{\text{s}^2\ \text{kg}}$$

$$\therefore \quad 1\ \text{C} = \frac{3.366 \times 10^{18}}{6.670 \times 10^{-11}}\ \text{kg}$$

$$= 5.047 \times 10^{28}\ \text{kg}$$

(b) Using the definition of the cavendish, the acceleration of the earth towards the sun is

$$a = \frac{1.000\,M_s}{(1.000)^2}\left(\frac{AU^3}{yr^2\,C} \cdot \frac{C}{AU^2}\right)$$

$$= M_s \frac{AU}{yr^2}$$

$$= \frac{4\pi^2 R}{T^2} = 4\pi^2 \frac{AU}{yr^2}$$

Thus, $M_s = 4\pi^2 \text{ cavendish} = 39.47\ C$

17. When the forces are equal,

$$\frac{G M_e m}{r_{ep}^2} = \frac{G M_s m}{r_{sp}^2}$$

and $r_{ep} + r_{sp} = r_{es} = 1.496 \times 10^{11}$ m

$$\therefore \quad \frac{r_{ep}}{r_{sp}} = \sqrt{\frac{M_e}{M_s}} = \sqrt{\frac{5.97 \times 10^{24}}{1.99 \times 10^{30}}}$$

$$= 0.00173$$

$$\frac{r_{ep}}{r_{es}} = \frac{1}{1 + \frac{1}{.00173}} = .00173$$ i.e., 0.173% of the distance

and $r_{ep} = (.00173)(1.496 \times 10^{11}\ \text{m})$

$$= 2.59 \times 10^8\ \text{m}$$

19. Taking the earth at the origin and the moon at $x_m = 3.84 \times 10^8$ m, the force on the rocket is

$$F_x = \frac{-G M_e m}{x^2} + \frac{G M_m m}{(x_m - x)^2}$$

Let $\frac{x}{x_m} = u$

$$F_x = \frac{-G m M_e}{x_m^2}\left(\frac{1}{u^2} - \frac{M_m/M_e}{(1-u)^2}\right)$$

$$= \frac{-(6.67 \times 10^{-11}\ \mathrm{N \cdot m^2/kg^2})(5.97 \times 10^{24}\ \mathrm{kg})(6500\ \mathrm{kg})}{(3.84 \times 10^8\ \mathrm{m})^2} \left(\frac{1}{u^2} - \frac{.0123}{(1-u)^2}\right)$$

$$= -17.6\ \mathrm{N} \left(\frac{1}{u^2} - \frac{.0123}{(1-u)^2}\right)$$

Setting u = .25, .5, and .75 in this expression gives

$F_{.25} = -280$ N (towards earth)

$F_{.50} = -69.3$ N (towards earth)

$F_{.75} = -27.7$ N (towards earth)

Note: The attraction of the moon becomes greater than that of the earth only when $u > .900$.

21. When earth and Venus are nearest each other,

$$r_{ev} = 0.42 \times 10^{11}\ \mathrm{m}$$

Thus, $$\frac{F_{ev}}{F_{es}} = \frac{M_v}{r_{ev}^2} \cdot \frac{r_{es}^2}{M_s} = \frac{M_v}{M_s}\left(\frac{r_{es}}{r_{ev}}\right)^2$$

$$= \frac{4.89 \times 10^{24}\ \mathrm{kg}}{1.99 \times 10^{30}\ \mathrm{kg}} \left(\frac{1.50 \times 10^{11}\ \mathrm{m}}{.42 \times 10^{11}\ \mathrm{m}}\right)^2$$

$$= .000031$$

23. The force between each pair of particles is $F_1 = Gm^2/a^2$. The total force on each particle is

$$|F_1 i + F_1(.5i + .866j)|$$

$$= F_1\sqrt{(1.5)^2 + (.866)^2} = 1.73\ F_1$$

$$= 1.73\ Gm^2/a^2$$

25. $$g_h = g_{R_e} \frac{R_e^2}{(R_e + h)^2} = g_{R_e} \frac{1}{(1 + h/R_e)^2}$$

$$= \frac{9.80\ \mathrm{m/s^2}}{\left(1 + \frac{.02\ \mathrm{Mm}}{6.37\ \mathrm{Mm}}\right)^2} = 9.74\ \mathrm{m/s^2}$$

27. $$\frac{g_h}{g_{R_e}} = \frac{R_e^2}{(R_e + h)^2}$$

The fractional reduction in the value of g is

$$1 - \frac{g_h}{g_{R_e}} = \frac{(R_e + h)^2 - R_e^2}{(R_e + h)^2} = \frac{h(2R_e + h)}{(R_e + h)^2}$$

$$\simeq \frac{2hR_e}{R_e^2} = \frac{2h}{R_e} \quad \text{if} \quad h << R_e$$

With $h = 10^4$ m, $R_e = 6.37 \times 10^6$ m

$$1 - \frac{g_h}{g_{R_e}} = \frac{2(10^4 \text{ m})}{6.37 \times 10^6 \text{ m}} = .003$$

29. (a) $$g_m = \frac{GM_m}{R_m^2} = \frac{(6.67 \times 10^{-11} \text{ N} \cdot \text{m}^2/\text{kg}^2)(6.46 \times 10^{23} \text{ kg})}{(3.38 \times 10^6 \text{ m})^2}$$

$$= 3.77 \text{ m/s}^2$$

(b) $W = mg = 64 \text{ kg}(3.77 \text{ m/s}^2) = 240 \text{ N}$

31. $$\mathbf{g}_P = -\frac{Gm_A \mathbf{n}_{PA}}{r_{PA}^2} - \frac{Gm_B \mathbf{n}_{PB}}{r_{PB}^2}$$

$$= -G\left(\frac{(1.4 \text{ kg})\mathbf{j}}{(1.5 \text{ m})^2} + \frac{(3.1 \text{ kg})(-0.85\,\mathbf{i} + 0.53\,\mathbf{j})}{(2.4 \text{ m})^2 + (1.5 \text{ m})^2}\right)$$

$$= (6.67 \times 10^{-11} \text{ N} \cdot \text{m}^2/\text{kg}^2)\left((0.33 \text{ kg/m}^2)\mathbf{i} - (0.83 \text{ kg/m}^2)\mathbf{j}\right)$$

$$= (2.2 \times 10^{-11} \text{ N/kg})\mathbf{i} - (5.5 \times 10^{-11} \text{ N/kg})\mathbf{j}$$

33. (a) $$\frac{mv^2}{r} = \frac{GMm}{r^2} \Rightarrow v^2 = \frac{GM}{r}$$

so $$v = \sqrt{\frac{(6.67 \times 10^{-11} \text{ N} \cdot \text{m}^2/\text{kg}^2)(5.97 \times 10^{24} \text{ kg})}{7.19 \times 10^6 \text{ m}}}$$

$$= 7.44 \times 10^3 \text{ m/s}$$

(b) $v = \frac{2\pi r}{T} \Rightarrow T = \frac{2\pi r}{v}$

so $$T = \frac{2\pi(7.19 \times 10^6 \text{ m})}{7.44 \times 10^3 \text{ m/s}} = 6070 \text{ s}$$

$$= 1.69 \text{ hours}$$

35. (a) For the artificial satellite in circular orbit about the moon,

$$v_C = \sqrt{\frac{GM_m}{r}} = \sqrt{\frac{(6.67 \times 10^{-11} \text{ N} \cdot \text{m}^2/\text{kg}^2)(7.35 \times 10^{22} \text{ kg})}{1.5 \times 1.74 \times 10^6 \text{ m}}}$$

$$= 1370 \text{ m/s}$$

(b and c) To determine the orbit numerically, take the initial position to be along the x–axis and the initial velocity along the y–axis. The equations of motion are:

$$a_x = \frac{-GM_m x}{r^3}$$

$$a_y = \frac{-GM_m y}{r^3}$$

where $r^2 = x^2 + y^2$. Taking a time interval h, estimate the position at time t + h/2 as

$$\mathbf{r}(t + h/2) = \mathbf{r}(t) + \mathbf{v}(t)h/2$$

Then calculate the velocity and position at time t + h using

$$\mathbf{v}(t+h) = \mathbf{v}(t) + h\,\mathbf{a}(t + h/2)$$

and $$\mathbf{r}(t+h) = \mathbf{r}(t) + h\,\mathbf{v}(t) + \frac{1}{2}\,\mathbf{a}(t + h/2)h^2$$

This can be set up in a spreadsheet as follows:

Use the first four rows for titles and to define the constants for the problem.

Use the columns as follows:

t in column A, x (t) in B, y (t) in C, v_x (t) in D, v_y (t) in E, x (t + h/2) in F, y (t + h/2) in G, r (t + h/2) in H, a_x (t + h/2) in I, a_y (t + h/2) in J, and use the above formulas to generate the same row.

The same spreadsheet can be used for both parts (b) and (c). It can also be used for Exercises 36 and 37 by replacing the mass of the moon with that of the earth or sun, respectively.

	A	B	C	D	E	F	G	H	I	J
1	Ch.7, Exercises 35,36,37									
2	G=	6.67E-11				h=	100			
3	Mmoon=	7.35E+22				GM	4.9E+12			
4	t	x(t)	y(t)	vx(t)	vy(t)	x(t+h/2)	y(t+h/2)	r(t+h/2)	ax(t+h/2)	ay(t+h/2)
5	0	2610000	0	0	1370	2610000	68500	2610899	-0.71893	-0.01887
6	100	2606405	136905.7	-71.8925	1368.113	2602811	205311.3	2610896	-0.71695	-0.05655
7	200	2595631	273434.2	-143.587	1362.458	2588452	341557.1	2610890	-0.713	-0.09408
8	300	2577708	409209.6	-214.887	1353.05	2566963	476862.1	2610881	-0.70709	-0.13135
9	400	2552684	543857.8	-285.595	1339.914	2538404	610853.5	2610869	-0.69923	-0.16827
10	500	2520628	677007.8	-355.518	1323.088	2502852	743162.2	2610854	-0.68945	-0.20471
11	600	2481629	808293	-424.463	1302.616	2460406	873423.8	2610836	-0.67777	-0.2406
12	700	2435794	937351.6	-492.24	1278.556	2411182	1001279	2610816	-0.66422	-0.27583
13	800	2383249	1063828	-558.662	1250.973	2355316	1126377	2610792	-0.64885	-0.3103
14	900	2324138	1187374	-623.547	1219.943	2292961	1248371	2610766	-0.63169	-0.34392
15	1000	2258625	1307649	-686.717	1185.551	2224289	1366926	2610737	-0.61279	-0.37659
16	1100	2186889	1424321	-747.996	1147.892	2149490	1481715	2610706	-0.59221	-0.40823
17	1200	2109129	1537069	-807.217	1107.07	2068768	1592422	2610672	-0.56999	-0.43875
18	1300	2025557	1645582	-864.216	1063.195	1982346	1698742	2610636	-0.5462	-0.46806
19	1400	1936404	1749561	-918.836	1016.389	1890463	1800381	2610598	-0.52091	-0.49609
20	1500	1841916	1848720	-970.927	966.7801	1793370	1897059	2610557	-0.49418	-0.52275
21	1600	1742353	1942784	-1020.34	914.5051	1691335	1988509	2610514	-0.46608	-0.54798
22	1700	1637988	2031495	-1066.95	859.7073	1584640	2074480	2610470	-0.4367	-0.5717
23	1800	1529109	2114607	-1110.62	802.5376	1473578	2154734	2610423	-0.40612	-0.59385
24	1900	1416016	2191891	-1151.24	743.1529	1358454	2229049	2610375	-0.37441	-0.61436
25	2000	1299020	2263135	-1188.68	681.7168	1239586	2297221	2610325	-0.34167	-0.63319
26	2100	1178444	2328141	-1222.84	618.3981	1117302	2359061	2610274	-0.30798	-0.65027
27	2200	1054620	2386729	-1253.64	553.371	991937.9	2414398	2610222	-0.27344	-0.66556
28	2300	927888.6	2438738	-1280.99	486.8146	863839.3	2463079	2610168	-0.23814	-0.67903
29	2400	798599.2	2484025	-1304.8	418.9121	733359.2	2504970	2610113	-0.20219	-0.69062

(b)

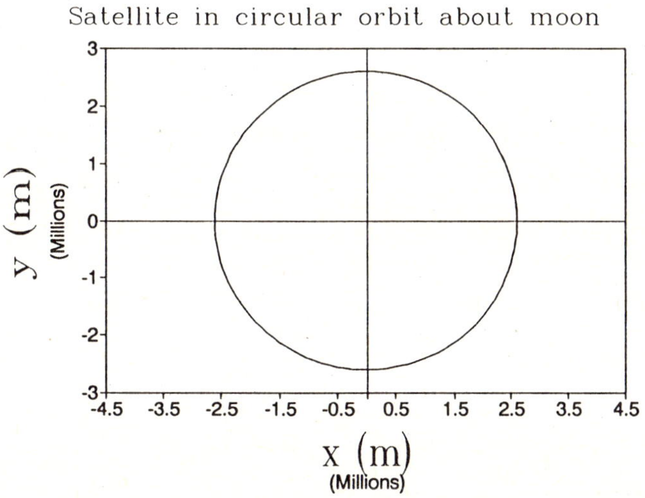

(c)

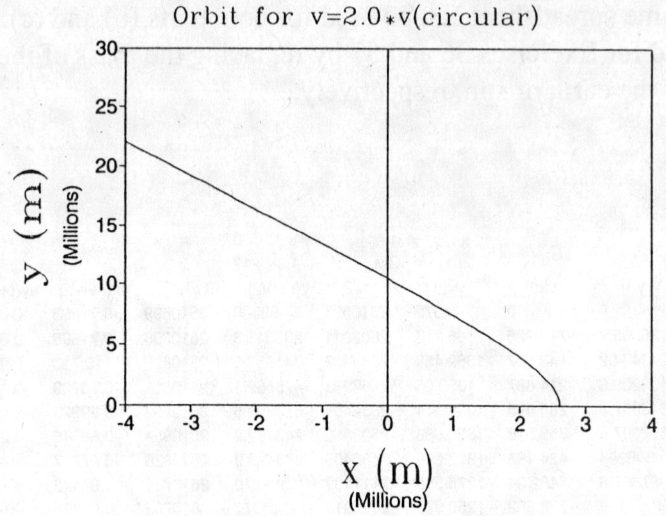

37. The same spreadsheet that was used in Exercise 35 can be used here. With the initial conditions given, the spreadsheet produces a graph like the one shown below.

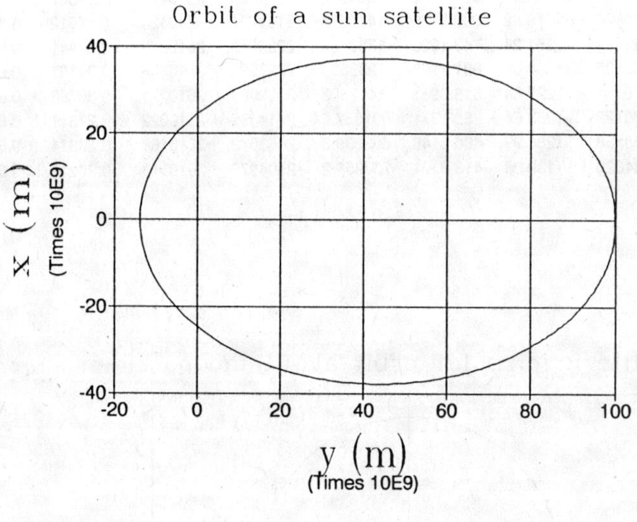

It can be seen from the graph that the initial conditions correspond to the aphelion. Reading directly from the spreadsheet (the @MIN and @MAX functions can locate these values), the extreme values can be located. This gives for the perihelion,

(b) $x = 0, \quad y = -1.39 \times 10^{10}$ m,

(d) $v_x = -129500$ m/s, $\quad v_y = 0$;

while for the aphelion,

(c) $x = 0, \quad y = 1.00 \times 10^{11}$ m,

(e) $v_x = 18000$ m/s, $\quad v_y = 0$.

39. Referring to Table 7–4 (p. 163), calculate R^3/T^2 for each satellite:

	R (Mm)	T (days)	R^3/T^2
Io	421.6	1.769	2.395×10^7
Europa	670.8	3.551	2.397×10^7
Ganymede	1070	7.155	2.393×10^7
Callisto	1882	16.689	2.393×10^7

Since R^3/T^2 is constant (within .2%), Kepler's third law is verified.

41. $R_m = 1.52\,R_e, \quad T_e = 1$ year

Using Kepler's law,

$$T_m^2 = T_e^2 \left(\frac{R_m}{R_e}\right)^3$$

or $$T_m = 1.00 \text{ yr } (1.52)^{3/2} = 1.87 \text{ years}$$

43. (a) The force on the moon, $\mathbf{F}_m$, is the sum of the force exerted on the moon by the sun, $\mathbf{F}_{sm}$, and that exerted by the earth, $\mathbf{F}_{em}$. The acceleration of the moon, $\mathbf{a}_m$, is equal to its acceleration relative to the earth, $\mathbf{a}_{me}$, and the acceleration of the earth relative to the sun, $\mathbf{a}_{es}$. Thus Newton's second law gives

$$\mathbf{F}_m = m_m \mathbf{a}_m,$$

or $$\mathbf{F}_{sm} + \mathbf{F}_{em} = m_m (\mathbf{a}_{me} + \mathbf{a}_{es})$$

Since the earth and the moon are at approximately the same distance and orientation from the sun,

$$\mathbf{F}_{sm} = \frac{-G m_s m_m}{r_{sm}^3} \mathbf{r}_{sm}$$

$$\mathbf{F}_{se} = \frac{-G m_s m_e}{r_{se}^3} \mathbf{r}_{se} = m_e \mathbf{a}_{es}$$

and $$\mathbf{r}_{sm} \simeq \mathbf{r}_{se}$$

so $$\mathbf{F}_{sm} \simeq \mathbf{F}_{se} \frac{m_m}{m_e} = m_m \mathbf{a}_{es}$$

Thus $\mathbf{F}_{em} \simeq m_m \mathbf{a}_{me}$

Now, $$a_{me} = \frac{v_{me}^2}{r_{me}} = \frac{4\pi^2 r_{me}}{T^2}$$

$$= \frac{4\pi^2(384 \times 10^6 \text{ m})}{(2.36 \times 10^6 \text{ s})^2} = 0.00272 \text{ m/s}^2$$

$$F_{em} = \frac{G m_e m_m}{r_{em}^2} = m_m a_{me}$$

$$m_e = \frac{a_{me} r_{em}^2}{G}$$

$$= \frac{(0.00272 \text{ m/s}^2)(384 \times 10^6 \text{ m})^2}{6.67 \times 10^{-11} \text{ N} \cdot \text{m}^2/\text{kg}^2}$$

$$= 6.01 \times 10^{24} \frac{\text{kg}^2 \cdot \text{m}}{\text{N} \cdot \text{s}^2}$$

$$= 6.01 \times 10^{24} \text{ kg}$$

(b) The accepted value for m_e is 5.98×10^{24} kg, so

$$\% \text{ error} = \frac{(6.01 - 5.98) \times 10^{24} \text{ kg}}{5.98 \times 10^{24} \text{ kg}} \times 100$$

$$= 0.5\%$$

45. (a) $$\mathbf{F}_a = \frac{G(4.0)(2.0)}{(4.0)^2}(-\mathbf{j}) \text{ N}$$

(b) $$\mathbf{F}_b = \frac{G(2.0)(5.0)}{(3.0)^2}(\mathbf{i}) \text{ N}$$

(c) $$\mathbf{F}_a + \mathbf{F}_b = G\left(\frac{10}{9}\mathbf{i} - \frac{1}{2}\mathbf{j}\right) \text{ N}$$

(d) $$\mathbf{g} = \frac{\mathbf{F}_a + \mathbf{F}_b}{2.0} = G\left(\frac{5}{9}\mathbf{i} - \frac{1}{4}\mathbf{j}\right) \text{ m/s}^2$$

47. (a) $$a_{es} = \frac{4\pi^2 r_{es}}{T_{orb}^2} = \frac{4\pi^2(1.496 \times 10^{11} \text{ m})}{(3.156 \times 10^7 \text{ s})^2} = 0.005929 \text{ m/s}^2$$

(b) $v_{eq} = \frac{2\pi R_e}{T_{rot}} = \frac{2\pi(6.38 \times 10^6 \text{ m})}{24\,(3600 \text{ s})} = 464 \text{ m/s}$

$$a = \frac{v^2}{R} = 0.0337 \text{ m/s}^2$$

(c) At noon the sun is overhead, so the accelerations are in opposite directions and

$$a_{net} = 0.0278 \text{ m/s}^2$$

(d) At midnight, the accelerations are in the same direction and

$$a_{net} = 0.0396 \text{ m/s}^2$$

(e) A person appears to be heavier at midnight.

49. $g = \frac{GM}{r^2} = \frac{(6.67 \times 10^{-11} \text{ N} \cdot \text{m}^2/\text{kg}^2)(2 \times 10^{30} \text{ kg})}{(1 \times 10^7 \text{ m})^2}$

$$= 1.3 \times 10^6 \text{ m/s}^2$$

Problems:

1. (a) $g = \frac{GM}{r^2}$

or $G = \frac{gr^2}{M} = \frac{gr^2}{\frac{4}{3}\pi r^3 \rho}$

$$= \frac{3g}{4\pi r \rho}$$

so $G = \frac{3\,(9.8 \text{ m/s}^2)}{4\pi(6.37 \times 10^6 \text{ m})(2.7 \times 10^3 \text{ kg/m}^3)}$

$$= 1.36 \times 10^{-10} \text{ N} \cdot \text{m}^2/\text{kg}^2$$

(b) $\rho = \frac{M}{\frac{4}{3}\pi r^3} = \frac{5.97 \times 10^{24} \text{ kg}}{\frac{4}{3}\pi(6.37 \times 10^6 \text{ m})^3} = 5.51 \times 10^3 \text{ kg/m}^3$

3. (a) From the satellite measurement,

$$\frac{GM}{r^2} = \frac{v^2}{r} = \frac{4\pi^2 r}{T^2}$$

or
$$M = \frac{4\pi^2 r^3}{GT^2}$$

Thus,
$$M = \frac{4\pi^2(1.476 \times 10^7\ \text{m})^3}{(6.67 \times 10^{-11}\ \text{N}\cdot\text{m}^2/\text{kg}^2)(8.09 \times 10^3\ \text{s})^2} = 2.91 \times 10^{25}\ \text{kg}$$

(b)
$$\rho = \frac{M}{\frac{4}{3}\pi R^3} = \frac{2.91 \times 10^{25}\ \text{kg}}{\frac{4}{3}\pi(9.54 \times 10^6\ \text{m})^3} = 8.00 \times 10^3\ \text{kg/m}^3$$

(c)
$$g = \frac{GM}{R^2} = \frac{(6.67 \times 10^{-11}\ \text{N}\cdot\text{m}^2/\text{kg}^2)(2.91 \times 10^{25}\ \text{kg})}{(9.54 \times 10^6\ \text{m})^2} = 21.3\ \text{m/s}^2$$

(d)
$$\frac{v^2}{R} = \frac{4\pi^2 R}{T^2} = \frac{4\pi^2(9.54 \times 10^6\ \text{m})}{(1.04 \times 10^4\ \text{s})^2} = 3.48\ \text{m/s}^2$$

The net acceleration at the equator is 17.8 m/s^2, so a 1 kg object weighs 17.8 N.

5. (a)
$$\frac{Gmm_B}{x^2} = \frac{Gmm_C}{(x_C - x)^2}$$

so
$$(x_C - x)^2 = \frac{m_C}{m_B}x^2$$

(b) If $x_C = 6$ m, $m_B = 2$ kg, and $m_C = 8$ kg, then

$$4x^2 = (6\ \text{m} - x)^2 = 36\ \text{m}^2 - 12\ \text{m}\,x + x^2$$

$$3x^2 + 12\ \text{m}\,x - 36\ \text{m}^2 = 0$$

$$x^2 + 4\ \text{m}\,x - 12\ \text{m}^2 = 0$$

$$(x - 2\ \text{m})(x + 6\ \text{m}) = 0$$

$$x = 2\ \text{m} \quad \text{or} \quad -6\ \text{m}$$

Of the two solutions, only x = 2 m is relevant. (At x = –6 m the two forces are equal but they are both in the same direction.)

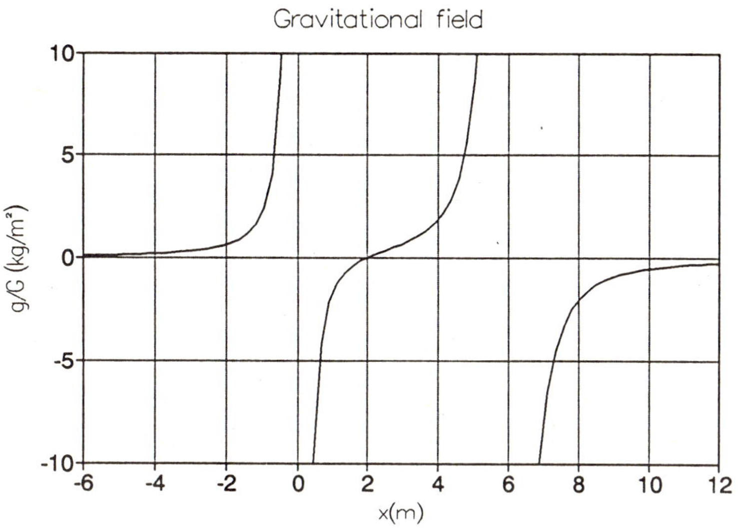

7. The period vs. radius relation plots as a straight line on log–log paper. Taking as two points on the line

$$R_1 = 10^{13}\text{ m},\ T_1 = 1.8\times10^{10}\text{ s} \quad\text{and}\quad R_2 = 10^{11}\text{ m},\ T_2 = 1.7\times10^{7}\text{ s}$$

and $\log T = \log T_o + a \log R$

then $10.255 = \log T_o + 13\,a$

$7.230 = \log T_o + 7\,a$

so $6\,a = 3.02,\ a = .504,\ \text{and}\ \log T_o = 3.701$

so $T = 5020\,R^{.504}$

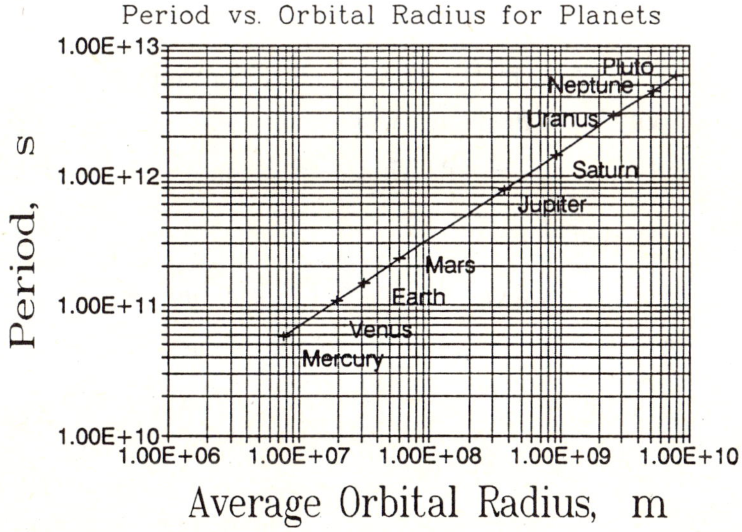

9. The spreadsheet used for Exercises 35–37 can also be used for this problem. Cells B5 and E5 have to be changed to make

$$x = 5.000 \times 10^6 \text{ m}$$

and
$$v_y = \sqrt{Gmr}$$
$$= \left((6.67 \times 10^{-11} \text{ N} \cdot \text{m}^2/\text{kg}^2)(7.35 \times 10^{22} \text{ kg})(5.00 \times 10^6 \text{ m})\right)^{1/2}$$
$$v_y = 4.951 \times 10^9 \text{ m/s}$$

11.
$$g = \frac{Gm_e}{R^2} = \frac{Gm_e}{(R_e + h)^2} = \frac{Gm_e}{R_e^2(1 + h/R_e)^2}$$
$$\simeq \frac{Gm_e}{R_e^2}\left(1 - \frac{2h}{R_e}\right) \quad \text{(using } (1+x)^{-2} \simeq 1 - 2x\text{)}$$

13. (a) From Table 7–4, for Io, $R = 421.6 \times 10^6$ m and $T = 1.769$ days.

$$v_{IJ} = \frac{2\pi R}{T} = \frac{2\pi(421.6 \times 10^6 \text{ m})}{1.769 \text{ days} \times 24 \text{ hrs/day} \times 3600 \text{ s/hr}} = 17.3 \times 10^3 \text{ m/s}$$

(b) From Table 7–1, for Jupiter, $R = 7.78 \times 10^{11}$ m and $T = 37.4 \times 10^7$ s.

$$v_{JS} = \frac{2\pi(7.78 \times 10^{11} \text{ m})}{37.4 \times 10^7 \text{ s}} = 13.1 \times 10^3 \text{ m/s}$$

(c and d) When Io is between Jupiter and the sun, their velocities are in opposite directions. The speed of Io relative to the sun is 4.2×10^3 m/s.

15. When the cavity is not present, the field is **g**. The contribution to this from the mass occupying the cavity is the field due to a point mass with a mass ρV located at the center of the cavity, $x = 0$, $y = -d$. The contribution from the rest of the mass is precisely $\mathbf{g}_{eff}$.

(a) Setting $\mathbf{g}_{eff} = \mathbf{g} + \mathbf{g}_a$, it can be seen that $\mathbf{g}_a$ is *minus* the field due to the point mass ρV, and so it is directed *away* from the center of the cavity.

(b and c) Accordingly,

$$g_{ax} = \frac{G\rho V x}{\left[x^2 + z^2 + (y+d)^2\right]^{3/2}},$$

$$g_{ay} = \frac{G\rho V (y+d)}{\left[x^2 + z^2 + (y+d)^2\right]^{3/2}},$$

$$g_{az} = \frac{G\rho V z}{\left[x^2 + z^2 + (y+d)^2\right]^{3/2}},$$

so
$$g_a = \sqrt{g_{ax}^2 + g_{ay}^2 + g_{az}^2}$$

$$= \frac{G\rho V}{x^2 + z^2 + (y+d)^2}$$

Along $z = 0$ and for $|y| << d$,

$$g_{ax} = \frac{G\rho V x}{\left[x^2 + d^2\right]^{3/2}},$$

$$g_{ay} = \frac{G\rho V d}{\left[x^2 + d^2\right]^{3/2}},$$

and
$$g_a = \frac{G\rho V}{x^2 + d^2}$$

(d) $g_{eff}^2 = \mathbf{g}_{eff} \cdot \mathbf{g}_{eff}$

$$= (\mathbf{g} + \mathbf{g}_a) \cdot (\mathbf{g} + \mathbf{g}_a)$$

$$= (-g\,\mathbf{j} + \mathbf{g}_a) \cdot (-g\,\mathbf{j} + \mathbf{g}_a)$$

$$= (-g + g_{ay})^2 + g_{ax}^2$$

so
$$g_{eff} = \left[(g - g_{ay})^2 + g_{ax}^2\right]^{1/2}$$

(e) Since $g_a << g$, the terms g_{ay}^2 and g_{ax}^2 can be neglected in the above expression,

so
$$g_{eff} \simeq \left[g^2 - 2g\,g_{ay}\right]^{1/2} \simeq g\left(1 - \frac{g_{ay}}{g}\right)$$

$$= g - g_{ay}$$

(f) When $\rho = 2.7 \times 10^3 \text{ kg/m}^3$, $V = 1.0 \times 10^6 \text{ m}^3$, $x = 0$, and $d = 100$ m,

$$\frac{g_{ay}}{g} = \frac{(2.7 \times 10^3 \text{ kg/m}^3)(1.0 \times 10^6 \text{ m}^3)}{5.97 \times 10^{24} \text{ kg}} \cdot \frac{(6.38 \times 10^6 \text{ m})^2}{(100 \text{ m})^2} = 1.84 \times 10^{-6}$$

17. (a) With the ring at $x = 0$, the distance between any point in the ring and the point P is $(R^2 + x^2)^{1/2}$. The gravitational field from a segment of mass dm will have an x–component equal to

$$dg_x = \frac{-G\,dm\,x}{(R^2 + x^2)^{3/2}}$$

Integrating, $$g = \frac{-Gmix}{(R^2 + x^2)^{3/2}}$$

(The transverse components cancel.)

(b)

$\frac{x}{R}$	$\frac{-\frac{x}{R}}{(1 + x^2/R^2)^{3/2}}$
0	0
.25	−.228
.5	−.358
.75	−.384
1.0	−.354
1.25	−.305
1.5	−.256
1.75	−.214
2.0	−.179

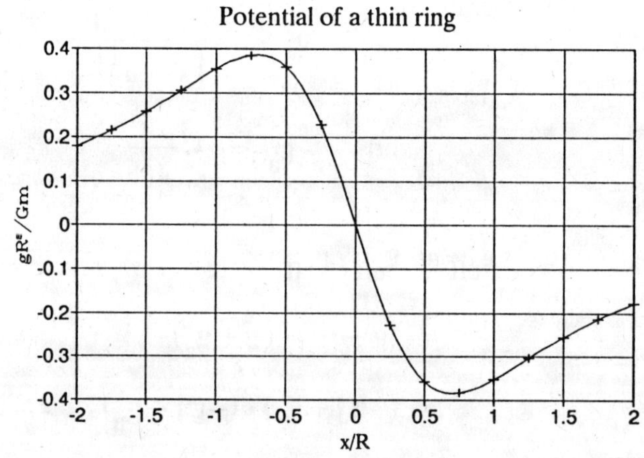

19. (a) For equilibrium, the gravitational force on a particle at the equator must be equal to or greater than mv^2/R.

$$\frac{GM}{R^2} > \frac{4\pi^2 R}{T^2}$$

or $$T^2 > \frac{4\pi^2 R^3}{GM} = \frac{4\pi^2}{\frac{4}{3}\pi\rho G} = \frac{3\pi}{\rho G}$$

(b) If $\rho = 3 \times 10^3\ \text{kg/m}^3$,

$$T_{min} = \left[\frac{3\pi}{(6.67 \times 10^{-11}\ \text{N} \cdot \text{m}^2/\text{kg}^2)(3 \times 10^3\ \text{kg/m}^3)}\right]^{1/2} = 6.9 \times 10^3\ \text{s}$$

(c) If $\rho = 3 \times 10^{17}\ \text{kg/m}^3$,

$$T_{min} = 6.9 \times 10^{-4}\ \text{s}$$

21. (a) Since $\mathbf{g} = \dfrac{-Gm\mathbf{r}}{r^3}$

and $\mathbf{r} = x\mathbf{i} - a\mathbf{j}$,

$$\mathbf{g} = \frac{-Gmx}{(x^2 + a^2)^{3/2}}\,\mathbf{i} + \frac{Gma}{(x^2 + a^2)^{3/2}}\,\mathbf{j}$$

(b) $$\frac{dg_x}{dx} = \frac{-Gm}{(x^2 + a^2)^{3/2}} + \frac{3}{2}\,\frac{Gmx(2x)}{(x^2 + a^2)^{5/2}}$$

$$= \frac{+Gm}{(x^2 + a^2)^{5/2}}\,(2x^2 - a^2)$$

Thus, the extrema occur at $x = \pm \dfrac{a}{\sqrt{2}}$

At $x = \dfrac{-a}{\sqrt{2}}$,

$$g_x = \frac{+Gm\left(\frac{a}{\sqrt{2}}\right)}{\left(\frac{3}{2}a^2\right)^{3/2}} = +0.385\,\frac{Gm}{a^2} \quad \text{(max)}$$

At $x = \dfrac{a}{\sqrt{2}}$,

$$g_x = -0.385\,\frac{Gm}{a^2} \quad \text{(min)}$$

(c and e)

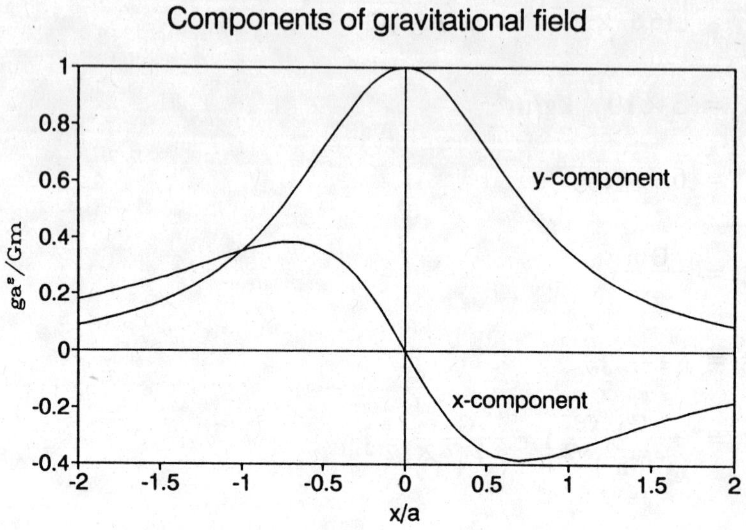

(d) $$\frac{dg_y}{dx} = \frac{-3GMax}{(x^2+a^2)^{3/2}} = 0 \text{ at } x = 0 \quad \text{(max)}$$

23. The field of a ring was found in Problem 17:

$$g_x = \frac{-Gmx}{(R^2+x^2)^{3/2}}$$

$$g_y = 0$$

In this case, $m \to \rho dV = \rho 2\pi r_o^2 \sin\theta \tau \, d\theta$,

$$R \to r_o \sin\theta, \text{ and } x \to r - r_o \cos\theta$$

Thus, $$g_x = -2\pi G\tau \int_0^{\pi} \frac{r_o^2 \sin\theta(r - r_o\cos\theta)\, d\theta}{\left[r_o^2 \sin^2\theta + (r - r_o\cos\theta)^2\right]^{3/2}}$$

$$g_x = -2\pi G\tau r_o^2 \rho \int_0^{\pi} \frac{\sin\theta(r - r_o\cos\theta)\, d\theta}{\left[r^2 + r_o^2 - 2rr_o\cos\theta\right]^{3/2}}$$

Now, $$\frac{d}{d\theta}\frac{1}{\left[r^2+r_o^2-2rr_o\cos\theta\right]^{1/2}}$$

$$=\frac{\left(-\frac{1}{2}\right)(-2rr_o)(-\sin\theta)}{\left[r^2+r_o^2-2rr_o\cos\theta\right]^{3/2}}$$

so $$g_x=\frac{2\pi G\tau r_o^2\rho}{rr_o}\int_0^\pi (r-r_o\cos\theta)\frac{d}{d\theta}\frac{1}{\left[r^2+r_o^2-2rr_o\cos\theta\right]^{1/2}}$$

$$=\frac{2\pi G\tau r_o\rho}{r}\left\{\frac{r-r_o\cos\theta}{\left[r^2+r_o^2-2rr_o\cos\theta\right]^{1/2}}\Bigg|_0^\pi-\int_0^\pi\frac{r_o\sin\theta\,d\theta}{\left[r^2+r_o^2-2rr_o\cos\theta\right]^{1/2}}\right\}$$

The first term is zero. To evaluate the second term, note that

$$\frac{d}{d\theta}\left[r^2+r_o^2-2rr_o\cos\theta\right]^{1/2}$$

$$=\frac{1}{2}\left[r^2+r_o^2-2rr_o\cos\theta\right]^{-1/2}(2rr_o\sin\theta)$$

so $$g_x=\frac{2\pi G\tau r_o\rho}{r}\left(-\frac{1}{r}\left[r^2+r_o^2-2rr_o\cos\theta\right]^{1/2}\right)_0^\pi$$

$$=\frac{2\pi G\tau r_o\rho}{r^2}\left[-(r+r_o)+(r-r_o)\right]$$

$$=\frac{-4\pi G\tau r_o^2\rho}{r^2}=\frac{-Gm}{r^2}$$

since the mass of the shell is $4\pi r_o^2\tau\rho$.

CHAPTER 8 WORK AND ENERGY

Questions:

1. The work done by the kinetic frictional force can be positive, negative or zero, depending on the angle between the force and the displacement.

3. The normal force exerted by a surface on an object will do work if there is a displacement of the object in the direction of the normal force.

5. Since $\mathbf{A} \cdot \mathbf{B} = AB \cos \theta$ and if $\mathbf{A} \cdot \mathbf{B} = 0$, it is not necessary that either $A = 0$ or $B = 0$. This dot product could also be zero if the two vectors **A** and **B** are perpendicular to each other.

7. Work is the dot product of the force and the displacement ($W = \mathbf{F} \cdot \overrightarrow{d\ell}$), and the displacement is in the direction of the velocity vector. Therefore, a force cannot do any work if it is perpendicular to the velocity, since the dot product of force and the displacement will be zero.

9. $KE = \frac{1}{2} m v^2$

   If the speed of a baseball is doubled, its kinetic energy will change by a factor of four.

11. Since the force is perpendicular to the displacement, it does not do any work.

13. Yes. Since $\mathbf{F} = m\mathbf{a}$ and if $F = 0$, the work done is $W = \mathbf{F} \cdot \overrightarrow{d\ell} = 0$, and therefore the change in kinetic energy is also zero.

    The work–energy theorem is not valid in a noninertial frame of reference because Newton's second law of motion is not valid in that frame of reference.

15. In the FPS system, the unit of work and kinetic energy is foot-pound.

17. The work is positive in one direction and negative in the opposite direction, and therefore the total work done is zero.

19. The electric company sells energy to its customers and if they don't pay the bill, it cuts off power so no energy is transmitted to its customers.

Exercises:

1. $m = 4\text{ kg}, \quad h = 2\text{ m}$

(a) $F = mg = (4\text{ kg})(9.8\text{ m/s}^2) = 39.2\text{ N}$

(b) $W = \mathbf{F} \cdot \vec{d\ell} = F\,d\ell\cos\theta$

Since the force and the displacement are in the same direction, $\theta = 0$.

Therefore, $W = (39.2\text{ N})(2\text{ m}) = 78.4\text{ J}$

3. $J = N \cdot m = kg \cdot m/s^2 \cdot m$

or $J = kg \cdot m^2 \cdot s^{-2}$

5. From Exercise 4,

$m = 16\text{ kg}, \quad d\ell = 3.2\text{ m}$

$T = 5.8\text{ N} \quad \text{and} \quad \theta = 37°$

The work done by the tension on the sled is given by

$W = \mathbf{F} \cdot \vec{d\ell} = F\,d\ell\cos\theta$

$W = (5.8\text{ N})(3.2\text{ m})\cos 37°$

$W = 14.82\text{ J}$

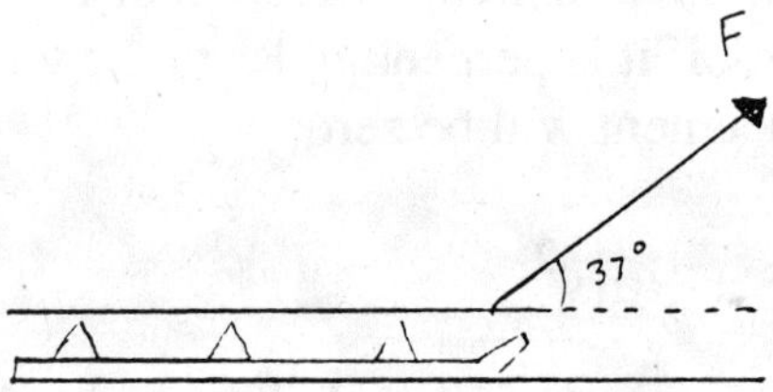

(a) Therefore, the work done by the frictional force on the sled is – 14.82 J.

(b) Since $f = \mu_k F_N$,

$$\mu_k = \frac{f}{F_N} \qquad (1)$$

$$F_N + F\sin 37° = mg$$

$$F_N = mg - F\sin 37°$$

or $$F_N = (16\text{ kg})(9.8\text{ m/s}^2) - (5.8\text{ N})\sin 37°$$

$$F_N = 156.8\text{ N} - 3.49\text{ N}$$

$$F_N = 153.31\text{ N} \qquad (2)$$

Also, $f = T\cos 37° = 4.63\text{ N}$ (frictional force) (3)

Using equations (2) and (3) in equation (1), we get

$$\mu_k = \frac{4.63\ \text{N}}{153.31\ \text{N}} = 0.03$$

7. $$\mathbf{A} \cdot \mathbf{B} = AB\cos\theta$$

or $$\cos\theta = \frac{\mathbf{A} \cdot \mathbf{B}}{AB}$$

Since $\cos\theta$ lies between -1 and $+1$,

$$-1 \le \cos\theta \le 1$$

$$-1 \le \frac{\mathbf{A} \cdot \mathbf{B}}{AB} \le 1$$

or $$-AB \le \mathbf{A} \cdot \mathbf{B} \le AB$$

9. $$\vec{d\ell} = (2\ \text{m})\mathbf{i} + (3\ \text{m})\mathbf{j} - (5\ \text{m})\mathbf{k}$$

$$\mathbf{F} = (7\ \text{N})\mathbf{i} - (7\ \text{N})\mathbf{j} - (2\ \text{N})\mathbf{k}$$

(a) $$W = \mathbf{F} \cdot \vec{d\ell} = |\mathbf{F}|\,|d\ell|\cos\theta \qquad (1)$$

$$= [(2)(7) - (3)(7) + (5)(2)]\,\text{J} = (3.0)\,\text{J}$$

(b) From (1), $$\cos\theta = \frac{W}{|\mathbf{F}|\,|d\ell|} = \frac{3.0}{[F_x^2 + F_y^2 + F_z^2]^{1/2}\,[(dx)^2 + (dy)^2 + (dz)^2]^{1/2}}$$

or $$\theta = \cos^{-1}\left[\frac{3.0}{(49 + 49 + 4)^{1/2}\,(4 + 9 + 25)^{1/2}}\right]$$

$$\theta = 87.25^\circ$$

11. If $$\mathbf{C} = \mathbf{A} + \mathbf{B},$$

then $$\mathbf{C} \cdot \mathbf{C} = (\mathbf{A} + \mathbf{B}) \cdot (\mathbf{A} + \mathbf{B})$$

or $$C^2 = \mathbf{A} \cdot \mathbf{A} + \mathbf{A} \cdot \mathbf{B} + \mathbf{B} \cdot \mathbf{A} + \mathbf{B} \cdot \mathbf{B}$$

$$C^2 = \mathbf{A} \cdot \mathbf{A} + \mathbf{B} \cdot \mathbf{B} + 2\,\mathbf{A} \cdot \mathbf{B} \qquad [\mathbf{A} \cdot \mathbf{B} = \mathbf{B} \cdot \mathbf{A}]$$

$$C^2 = A^2 + B^2 + 2AB\cos\theta$$

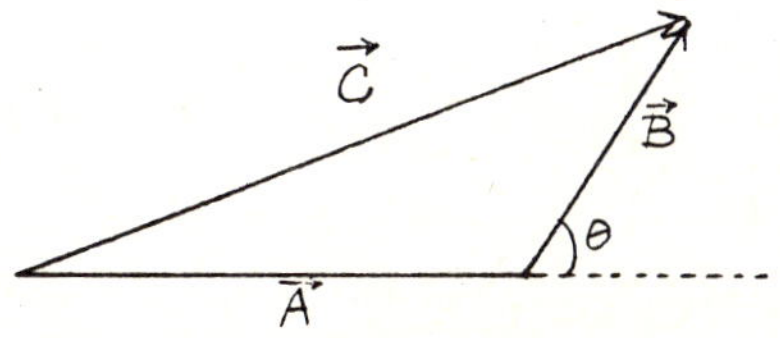

13. $k = 2100$ N/m, $x = 0.14$ m

(a) The work done by the restoring force is given by equation 8–6:

$$W = -\frac{1}{2}k\left(x_f^2 - x_i^2\right)$$

$$W = -\frac{1}{2}(2100 \text{ N/m})(0.14 \text{ m})^2 = -20.58 \text{ J}$$

(b) $F_s \text{ (min)} = 0$

$$F_s \text{ (max)} = (2100 \text{ N/m})(0.14 \text{ m}) = 294 \text{ N}$$

15. $F_y = -mg$

$$W = \int_{y_i}^{y_f} F_y \cdot dy$$

$$W = \int_{y_i}^{y_f} (-mg)\,dy = -mg\,[y]_{y_i}^{y_f}$$

$$W = -mg\,[y_f - y_i]$$

This is the same as equation 8–9.

17. $F_z(z) = -\frac{C}{z^2}$

$$W = \int_{z_i}^{z_f} \mathbf{F} \cdot \vec{dz} = -C \int_{z_i}^{z_f} \frac{dz}{z^2}$$

$$W = \left[\frac{C}{z}\right]_{z_i}^{z_f} = C\left[\frac{1}{z_f} - \frac{1}{z_i}\right]$$

19. $W = \int_i^f \mathbf{F} \cdot \vec{d\ell}$

(a) If the force F is constant,

$$W = F\ell\cos\theta = \mathbf{F} \cdot \boldsymbol{\ell}$$

(b) If the path is along the x–axis, $F = F_x$, $d\ell = dx$, $\theta = 0$, and

$$W = \int_{x_i}^{x_f} F_x \, dx \cos 0° = \int_{x_i}^{x_f} F_x \, dx$$

21. m = 1.5 kg, r = 0.75 m

(a) The work done is given by equation 8–9:

$$W = -mg(y_f - y_i) = -(1.5 \text{ kg})(9.8 \text{ m/s}^2)[0 - 1.5 \text{ m}]$$

$$W = 22.05 \text{ J}$$

(b) The work done by the tension force is zero, since the force is perpendicular to the displacement.

23. m = 1100 kg, v = 45 km/h = 12.5 m/s

(a) $KE = \frac{1}{2}mv^2 = \frac{1}{2}(1100 \text{ kg})(12.5 \text{ m/s})^2$

$$KE = 8.6 \times 10^4 \text{ J}$$

(b) 90 km/h = 25 m/s

$$KE = \frac{1}{2}(550 \text{ kg})(25 \text{ m/s})^2$$

$$KE = 1.7 \times 10^5 \text{ J}$$

25. m = 0.015 kg, k = 120 N/m, x = 0.12 m

Using the law of conservation of energy, the potential energy of the spring is equal to the kinetic energy of the ball.

$$\frac{1}{2}kx^2 = \frac{1}{2}mv^2$$

or $$v^2 = \frac{kx^2}{m}$$

$$v = \sqrt{\frac{k}{m}}\, x$$

or $$v = \sqrt{\frac{120 \text{ N/m}}{0.015 \text{ kg}}}\,(0.12 \text{ m})$$

$$v = 10.73 \text{ m/s}$$

27. $m = 0.50\text{ kg}, \quad v_o = 3.4\text{ m/s}, \quad r = 0.75\text{ m}$

(a) Work done by friction = change in kinetic energy

$$W = 0 - \frac{1}{2}mv_o^2 \quad [v = 0]$$

or $$W = -\frac{1}{2}mv^2 = -\frac{1}{2}(0.5\text{ kg})(3.4\text{ m/s})^2$$

$$W = -2.89\text{ J}$$

(b) $f = \mu_k F_N$ where $F_N = mg$

From part (a), $W = -f(s) = -2.89\text{ J}$,

where $s = 2\pi(0.75\text{ m})(2.5) = 11.78\text{ m}$

$\therefore$ $-f(11.78\text{ m}) = -2.89\text{ J}$

or $f = 0.245\text{ N}$

$$\mu_k = \frac{f}{F_N} = \frac{0.245\text{ N}}{mg} = \frac{0.245\text{ N}}{(0.50\text{ kg})(9.8\text{ m/s}^2)}$$

or $\mu_k = 0.05$

(c) The tension force provides the centripetal force:

$$T = \frac{mv^2}{r},$$

where, after one revolution,

work done by friction = change in kinetic energy

$$-2\pi rf = \frac{1}{2}mv^2 - \frac{1}{2}mv_o^2$$

or $$\frac{1}{2}mv^2 = \frac{1}{2}mv_o^2 - 2\pi rf$$

$$= \frac{1}{2}(0.5\text{ kg})(3.4\text{ m/s})^2 - 2\pi(0.75\text{ m})(0.245\text{ N})$$

$$= 1.74\text{ J}$$

or $mv^2 = 3.48\text{ J}$

Therefore, $T = \frac{3.48\text{ J}}{0.75\text{ m}} = 4.64\text{ N}$

(d) The tension force is perpendicular to the displacement and therefore does not do any work.

29. $m = 0.37$ kg, $v_o = 14$ m/s, $h = 8.4$ m

(a) As the ball is going up, the work done by air resistance is equal to the change in kinetic energy minus the work done by gravity.

$$\therefore \quad W_{\text{air resistance}} = -\frac{1}{2}mv_o^2 - (-mgh)$$

$$W_{\text{air resistance}} = -\frac{1}{2}(0.37\text{ kg})(14\text{ m/s})^2 + (0.37\text{ kg})(9.8\text{ m/s}^2)(8.4\text{ m})$$

$$W = -36.26\text{ J} + 30.46\text{ J}$$

$$W = -5.86\text{ J}$$

(b) As the ball comes down, $v_o = 0$ and the work done by gravity plus the work done by air resistance equals the change in kinetic energy.

Therefore, $$30.46\text{ J} - 5.86\text{ J} = \frac{1}{2}mv^2$$

or $$v^2 = \frac{2(24.6\text{ J})}{0.37\text{ kg}}$$

or $$v = 11.5\text{ m/s}$$

31. Work done by gravity = change in kinetic energy

$$-mgR = \frac{1}{2}mv^2 - \frac{1}{2}mv_o^2$$

$$-gR = \frac{1}{2}v^2 - \frac{1}{2}v_o^2$$

or $$v^2 = v_o^2 - 2gR \qquad (1)$$

Also, $$T + mg = \frac{mv^2}{R}$$

$$T = \frac{mv^2}{R} - mg$$

$$T = \frac{m}{R}\left[v_o^2 - 2gR\right] - mg$$

$$T = \frac{mv_o^2}{R} - 2mg - mg = \frac{mv_o^2}{R} - 3mg$$

At point C, $T \geq 0$

$$\frac{mv_o^2}{R} - 3mg \geq 0$$

or $$v_o^2 \geq 3gR = 3(9.8\text{ m/s}^2)(.8\text{ m}) = 23.52\text{ m}^2/\text{s}^2$$

or $$v_o \geq 4.8\text{ m/s}$$

33. $m_B = 15\text{ kg}, \ v_o = 0, \ a = 2.5\text{ m/s}^2$

T = truck, B = toolbox

$$x_T = \frac{1}{2}(2.5\text{ m/s}^2)t^2 = (1.25\text{ m/s}^2)t^2$$

$$m_B \frac{d^2x_B}{dt^2} = F = 0.2\,(m_B\,g)$$

$$x_B = \frac{1}{2}(0.2)(9.8\text{ m/s}^2)t^2 = (0.98\text{ m/s}^2)t^2$$

$$x_T - x_B = 2.0\text{ m} \quad \text{when} \quad 0.27\,t^2 = 2.0,$$

so $t = 2.72\text{ s}$

and $x_B = 7.26\text{ m}$

$$W = F(7.26\text{ m})$$

$$W = (0.2)(15\text{ kg})(9.8\text{ m/s}^2)(7.26\text{ m})$$

$$W = 210\text{ J}$$

35. $v = 2.3\text{ m/s}, \ P = 940\text{ W}$

Using $P = Fv$,

$$F = \frac{P}{v} = \frac{940\text{ W}}{2.3\text{ m/s}}$$

$$F = 408.7\text{ N}$$

37. $230\text{ hp} = 171580\text{ W}$

Since $P = Fv$,

$$F = \frac{P}{v} = \frac{171580\text{ W}}{0.25\text{ m/s}} = 686320\text{ N}$$

(a) $P = (686320\text{ N})(0.75\text{ m/s}) = 514740\text{ W}$

(b) $F = 686320\text{ N}$

(c) $F = 228773.3\text{ N}$

39. (a) $W = \mathbf{F}\cdot\overrightarrow{d\ell} = F\,d\ell\cos\theta,$

where $\theta = 0°$, $F = mg$ and $d\ell = 10\text{ m}$

(b) $W = mgh$

(c) Zero

41. $F = (8.0\text{ N})\mathbf{i} - (9.0\text{ N})\mathbf{j}$

$\ell = (2.0\text{ m})\mathbf{i} + (4.0\text{ m})\mathbf{j} + (4.0\text{ m})\mathbf{k}$

(a) $$W = \mathbf{F} \cdot \boldsymbol{\ell} = F_x\, dx + F_y\, dy + F_z\, dz$$

$$W = (8.0\text{ N})(2.0\text{ m}) - (9.0\text{ N})(4.0\text{ m}) + (0\text{ N})(4.0\text{ m})$$

$$W = -20\text{ J}$$

(b) $$|F|\,|\ell| \cos\theta = -20\text{ J}$$

$$\cos\theta = \frac{-20\text{ J}}{[F_x^2 + F_y^2 + F_z^2]^{1/2}[dx^2 + dy^2 + dz^2]^{1/2}}$$

$$\cos\theta = \frac{-20\text{ J}}{[64\text{ N}^2 + 81\text{ N}^2]^{1/2}[4\text{ m}^2 + 16\text{ m}^2 + 16\text{ m}^2]^{1/2}}$$

$$\cos\theta = 0.277$$

$$\theta = 73.92°$$

43. $m = 2\text{ kg}, \quad d\ell = 3\text{ m}$

(a) $$W_g = -mgh = 0$$

(b) $$W = -mgh = -(2\text{ kg})(9.8\text{ m/s}^2)(3\text{ m})$$

$$W = -58.8\text{ J}$$

(c) Work is done because of gravity. The component of the displacement in the vertical direction is responsible for work.

$$W = mgh = (2\text{ kg})(9.8\text{ m/s}^2)(3\text{ m})$$

$$W = 58.8\text{ J}$$

45. $F_x(x) = -Ax^3, \quad A = 1 \times 10^{26}\text{ N/m}^3$

(a) $$W = \int_{x_i}^{x_f} F_x(x)\, dx$$

or $$W = -A \int_{x_i}^{x_f} \frac{dx}{x^3} = \frac{A}{2} \left| x^{-2} \right|_{x_i}^{x_f}$$

$$W = \frac{A}{2}\left[\frac{1}{x_f^2} - \frac{1}{x_i^2}\right]$$

$$W = 0.5 \times 10^{26}\left[\frac{1}{x_f^2} - \frac{1}{x_i^2}\right] \text{ J}$$

(b) For $x_i = 0$ and $x_f = 2 \times 10^{-11}$ m

$$W = 0.5 \times 10^{26}\left[\frac{1}{(2 \times 10^{-11})^2}\right] = \frac{0.5 \times 10^{26}}{4 \times 10^{-22}} \text{ J}$$

$$W = 0.125 \times 10^{48} \text{ J}$$

47. $m = 0.01$ kg, $v = 0.12$ m/s

(a) $$W = mgh = (0.01 \text{ kg})(9.8 \text{ m/s}^2)h$$

$$\frac{dW}{dt} = (0.01 \text{ kg})(9.8 \text{ m/s}^2)v \qquad \left[\frac{dh}{dt} = v\right]$$

$$\frac{dW}{dt} = (0.01 \text{ kg})(9.8 \text{ m/s}^2)(0.12 \text{ m/s})$$

$$\frac{dW}{dt} = 0.012 \text{ J/s}$$

(b) Work done by the force due to the syrup is the same as that given in part (a) to attain constant speed.

49. (a) The work done on the electron as it moves from $x_i = 0.000$ m to $x_f = 0.020$ m is given by

$$W = (x_f - x_i)\left[\frac{F_1 + F_2}{2}\right],$$

where F_1 and F_2 are the values of force at x_i and x_f, respectively.

$$\therefore \quad W = (0.020 \text{ m} - 0.000 \text{ m})\left[\frac{2 \times 10^{-14} \text{ N} + 0}{2}\right]$$

$$W = 0.02 \times 10^{-14} \text{ J}$$

(b) Using the work–energy theorem,

$$W = \frac{1}{2}mv_f^2 - \frac{1}{2}mv_i^2$$

At x_i, $v_i = 0$ and at x_f the speed is v_f

Therefore, $W = \frac{1}{2} m v_f^2$

$$0.02 \times 10^{-14}\ J = \frac{1}{2}(9.11 \times 10^{-31}\ kg)\, v_f^2$$

$$v_f^2 = \frac{0.04 \times 10^{-14}\ kg \cdot m^2/s^2}{9.11 \times 10^{-31}\ kg}$$

$$v_f = \sqrt{4.39 \times 10^{14}\ m^2/s^2}$$

$$v_f = 20.9 \times 10^6\ m/s$$

Problems:

1. $m = 15\ kg,\ \theta = 30.0°,\ v_o = 4.6\ m/s,\ \mu_k = 0.34$

(a) Using the work–energy theorem, we get:

$$W = \frac{1}{2} m v_f^2 - \frac{1}{2} m v_o^2$$

or $$W = -\frac{1}{2} m v_o^2 \quad [v_f = 0]$$

$$W = -\frac{1}{2}(15\ kg)(4.6\ m/s)^2$$

$$W = -158.7\ J$$

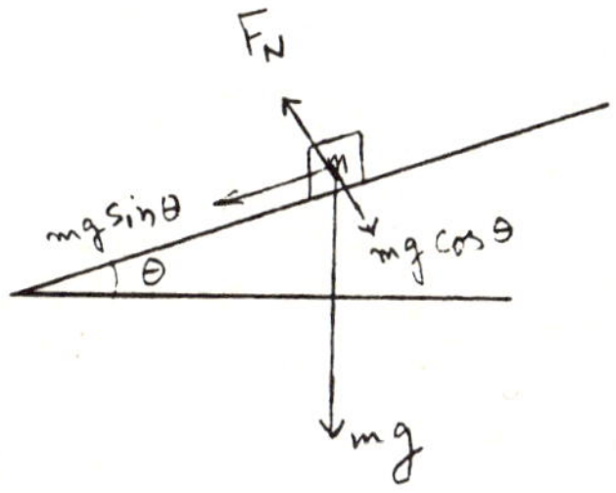

(b) The net force $F = -(mg \sin\theta + f)$, where $f = \mu_k F_N$

$$F = -mg \sin\theta - \mu_k mg \cos\theta$$

$$F = -(15\ kg)(9.8\ m/s^2) \sin 30° - 0.34(15\ kg)(9.8\ m/s^2) \cos 30°$$

$$F = -116.78\ N$$

Using $F_{Net} = ma$ gives

$$a = \frac{-116.78\ N}{15\ kg} = -7.79\ m/s^2$$

Using $v^2 - v_o^2 = 2as$, where $v = 0$, gives

$$s = \frac{0 - (4.6\ m/s)^2}{(2)(-7.79\ m/s^2)} = 1.36\ m$$

Therefore, the work done by the weight of the block is given by

$$W = -(mg \sin\theta)s = -(15\text{ kg})(9.8\text{ m/s}^2)\sin 30° \times 1.36\text{ m}$$

$$W = -99.96\text{ J}$$

(c) The work done by the normal force is zero.

(d) The work done by friction force

$$W_f = -158.7\text{ J} + 99.96\text{ J}$$

$$W_f = -58.74\text{ J}$$

(e) As calculated in part (b), $s = 1.36$ m

3. $F_a = F_e = mg$ and $ds = L\,d\theta$

For displacement ds,

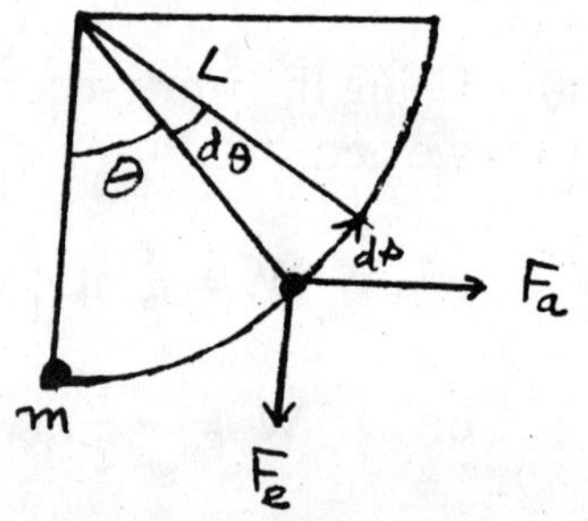

$$W = \int (F_a \cos\theta - F_e \sin\theta)\,ds$$

$$W = \int_0^{\pi/2} mg(\cos\theta - \sin\theta)\,L\,d\theta$$

$$W = mgL\left[\sin\theta + \cos\theta\right]_0^{\pi/2} = 0$$

5. (a) At $t = 0$, $v_0 = 0$ and at $t = 2$ s, $v = v_0 + at$,

or $v = (1\text{ m/s}^2)(2\text{ s}) = 2\text{ m/s}$

$$v^2 = v_0^2 + 2as_1$$

or $$s_1 = \frac{(2\text{ m/s})^2}{2(1\text{ m/s}^2)} = 2\text{ m}$$

For the next $t = 10$ s, v = constant = 2 m/s, and

$$s_2 = \bar{v}t = (2\text{ m/s})(10\text{ s}) = 20\text{ m}$$

For the next $t = 2$ s, $v_0 = 2$ m/s, $v = 0$ and $a = -1\text{ m/s}^2$

Therefore, $v^2 = v_0^2 + 2as$ gives

$$0 = (2\text{ m/s})^2 - 2(1\text{ m/s}^2)s_3$$

or $s_3 = 2\text{ m}$

$$s = s_1 + s_2 + s_3$$

$$s = 2\text{ m} + 20\text{ m} + 2\text{ m} = 24\text{ m}$$

The work done by the normal force $[F_N = mg]$ is

$$W_N = mgs$$

or $W_N = (60\text{ kg})(9.8\text{ m/s}^2)(24\text{ m})$

$$W_N = 14.11 \times 10^3\text{ J}$$

(b) The work done by the weight mg is

$$W_g = -mgs = -14.11 \times 10^3\text{ J}$$

(c) $P_{av} = \frac{\text{Work}}{\text{Time}} = \frac{14 \times 10^3\text{ J}}{14\text{ s}} = 1 \times 10^3\text{ J/s} = 1\text{ kW}$

(d) At t = 7 s, the elevator is moving with a constant velocity of 2 m/s and

$$P_{av} = (\text{Force})(\text{velocity})$$

$$P_{av} = mgv = (60\text{ kg})(9.8\text{ m/s}^2)(2\text{ m/s})$$

$$P_{av} = 1.18\text{ kW}$$

(e) At t = 13 s, the elevator is decelerating and

$$v = v_o + at, \quad a = -1\text{ m/s}^2 \text{ and } t = 1\text{ s}$$

$$v = 2\text{ m/s} - (1\text{ m/s}^2)(1\text{ s}) = 1\text{ m/s}$$

$$P_{av} = F_N v,$$

where $F_N = m(g + a) = 60\text{ kg}\,(9.8\text{ m/s}^2 - 1\text{ m/s}^2)$

$$F_N = 528\text{ N}$$

$\therefore$ $P_{av} = (528\text{ N})(1\text{ m/s}) = 528\text{ W}$

7. (a) Let v be the speed at the highest point. Using $F_{Net} = ma$ at the highest point, we get:

$$T + mg\sin\theta = \frac{mv^2}{L},$$

where L is the radius of the circular path as shown in Fig. 8–23.

Thus, $$T = \frac{mv^2}{L} - mg\sin\theta \qquad (1)$$

From the work–energy theorem,

$$\frac{1}{2}mv^2 - \frac{1}{2}mv_o^2 = -2\,mgL\sin\theta$$

$$mv^2 - mv_o^2 = -4\,mgL\sin\theta$$

$$mv^2 = mv_o^2 - 4\,mgL\sin\theta \qquad (2)$$

Using equation (2) in equation (1), we get, after simplification,

$$T = \frac{mv_o^2}{L} - 5\,mg\sin\theta$$

Since at the highest position T = 0,

$$0 = \frac{mv_o^2}{L} - 5\,mg\sin\theta$$

$$\frac{mv_o^2}{L} = 5\,mg\sin\theta$$

or $$v_o^2 = 5\,gL\sin\theta$$

$$v_o = (5\,gL\sin\theta)^{1/2} = [(5)(9.8\text{ m/s}^2)(0.75\text{ m})\sin 37°]^{1/2}$$

$$v_o = 4.7\text{ m/s}$$

(b) At the lowest point, $T = mg\sin\theta + \frac{mv_o^2}{L}$

or $$T = (1.2\text{ kg})(9.8\text{ m/s}^2)\sin 37° + \frac{(1.2\text{ kg})(4.7\text{ m/s})^2}{0.75\text{ m}}$$

$$T = 42.24\text{ N}$$

(c) If the puck has $v_o = 4.7$ m/s at the lowest point, the tension will be less at the highest point and it will leave the circular path and describe a curve.

9. (a) $W = mgh = (10{,}000\text{ kg})(9.8\text{ m/s}^2)(100\text{ m})$

$W = 9.8 \times 10^6\text{ J}$

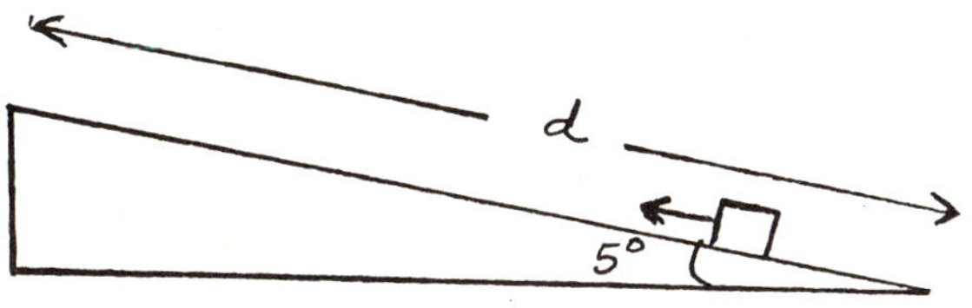

(b) $F_{Net} = ma$

$F - mg\sin\theta - f = 0$

$F = mg\sin\theta + f = mg\sin\theta + \mu_s F_N$, where $F_N = mg\cos\theta$

$F = mg\sin\theta + \mu_s\, mg\cos\theta$

or $F = (10{,}000\text{ kg})(9.8\text{ m/s}^2)\sin 5° + (0.2)(10{,}000\text{ kg})(9.8\text{ m/s}^2)\cos 5°$

$F = 2.81 \times 10^4\text{ N}$

$W = Fd = 2.81 \times 10^4 \times \dfrac{100\text{ m}}{\sin 5°}$

$W = 3.22 \times 10^7\text{ J}$

(c) $P = Fv = (2.81 \times 10^4\text{ N})(0.1\text{ m/s}) = 2.81 \times 10^3\text{ W}$

11. (a) $T - mg = ma$

$a = \dfrac{T - mg}{m} = \dfrac{600\text{ N} - (50\text{ kg})(9.8\text{ m/s}^2)}{50\text{ kg}} = 2.2\text{ m/s}^2$

$v^2 = v_o^2 + 2as$, where $v_o = 0$

$v^2 = (2)(2.2\text{ m/s}^2)(2\text{ m})$

$v = 2.97\text{ m/s}$

(b) 600 N

(c) In (a), $W = (T - mg)(s) = (110\text{ N})(2\text{ m}) = 220\text{ J}$

In (b), the tension is the same as in (a).

13. $\mathbf{F}_{air} = -Av^2\mathbf{i}$, where $A = 0.6\text{ N}\cdot\text{s}^2/\text{m}^2$

(a) $W_{air} = (F_{air})(d)$, where $d = 1\text{ km}$ and $v = 80\text{ km/h} = 22.22\text{ m/s}$

$W_{air} = -A(v^2)(d)$

or $W_{air} = -(0.6\text{ N}\cdot\text{s}^2/\text{m}^2)(22.22\text{ m/s})^2(1000\text{ m})$

$W_{air} = -2.96 \times 10^5\text{ J}$

(b) $P_{av} = (F_{air})(v) = -(Av^2)(v) = -Av^3$

$$P_{av} = -(0.6\ \mathrm{N\cdot s^2/m^2})(22.22\ \mathrm{m/s})^3 = 6.58\times 10^3\ \mathrm{W}$$

15. $\mathbf{F} = -kx\mathbf{i} - ky\mathbf{j}$

$$W = \int_i^f \mathbf{F}\cdot d\vec{\ell} = \int_i^f F_x\,dx + F_y\,dy$$

or $$W = \int_{x_i}^{x_f} F_x\,dx + \int_{y_i}^{y_f} F_y\,dy$$

$$W = \int_{x_i}^{x_f} (-kx)\,dx + \int_{y_i}^{y_f} (-ky)\,dy$$

or $$W = -k\left[\frac{x^2}{2}\right]_{x_i}^{x_f} - k\left[\frac{y^2}{2}\right]_{y_i}^{y_f}$$

$$W = -\tfrac{1}{2}k(x_f^2 - x_i^2) - \tfrac{1}{2}k(y_f^2 - y_i^2)$$

or $$W = -\tfrac{1}{2}k\left[(x_f^2 - x_i^2) + (y_f^2 - y_i^2)\right]$$

Since the gravitational force field is conservative, the work done in this force field is independent of the path.

19. $F_x(x) = -Axe^{-x^2/a^2}$

$A = 1\times 10^3\ \mathrm{N/m}$, $a = 0.25\ \mathrm{m}$, $x_i = 0.00\ \mathrm{m}$

Set up the spreadsheet as follows:

A	B	C	D	E
x	$x + \frac{\Delta x}{2}$	$F\left(x + \frac{\Delta x}{2}\right)$	$F\left(x + \frac{\Delta x}{2}\right)\Delta x$	Σ

where $A_{n+1} = A_n + \Delta x$

$$B_n = A_n + \frac{\Delta x}{2}$$

$$C_n = -(A)B_n \exp\left(-B_n^2/a^2\right)$$

$$D_n = C_n \Delta x$$

and $$E_n = D_n + E_{n-1}$$

To select Δx, the main requirement is that the force should not change too much in that distance. Choosing $\Delta x = a/5$ should be adequate.

The spreadsheet is as follows:

dx = 0.05
A = 1000
a = 0.25

x	x + dx/2	F	F*dx	dW
				0
0	0.05	−48.0395	−2.40197	−2.40197
0.05	0.1	−85.2144	−8.52144	−10.9234
0.1	0.15	−104.651	−15.6977	−26.6211
0.15	0.2	−105.458	−21.0917	−47.7128
0.2	0.25	−91.9699	−22.9925	−70.7053
0.25	0.3	−71.0783	−21.3235	−92.0288
0.3	0.35	−49.3004	−17.2552	−109.284
0.35	0.4	−30.9219	−12.3688	−121.653
0.4	0.45	−17.6238	−7.93069	−129.583
0.45	0.5	−9.15782	−4.57891	−134.162
0.5	0.55	−4.34888	−2.39188	−136.554
0.55	0.6	−1.89067	−1.1344	−137.689
0.6	0.65	−0.7535	−0.48977	−138.178
0.65	0.7	−0.27557	−0.1929	−138.371
0.7	0.75	−0.09256	−0.06942	−138.441
0.75	0.8	−0.02857	−0.02286	−138.464
0.8	0.85	−0.00811	−0.00689	−138.47
0.85	0.9	−0.00212	−0.00191	−138.472
0.9	0.95	−0.00051	−0.00048	−138.473
0.95	1	−0.00011	−0.00011	−138.473
1	1.05	−2.3E−05	−2.4E−05	−138.473
1.05	1.1	−4.3E−06	−4.7E−06	−138.473
1.1	1.15	−7.4E−07	−8.5E−07	−138.473
1.15	1.2	−1.2E−07	−1.4E−07	−138.473
1.2	1.25	−1.7E−08	−2.2E−08	−138.473
1.25	1.3	−2.3E−09	−3.1E−09	−138.473
1.3	1.35	−2.9E−10	−4E−10	−138.473
1.35	1.4	−3.4E−11	−4.7E−11	−138.473
1.4	1.45	−3.6E−12	−5.2E−12	−138.473
1.45	1.5	−3.5E−13	−5.2E−13	−138.473
1.5	1.55	−3.1E−14	−4.9E−14	−138.473

CHAPTER 9 CONSERVATION OF ENERGY

Questions:

1. The work required to move an object from one location to another is a constant, independent of the path taken for a conservative force.

3. Conservation of energy in physics means that energy remains constant. In the media, conservation refers to the preservation of certain forms of energy as opposed to depleting these forms by transforming the energy into other forms, particularly if the change is regarded as wasteful or unnecessary. Thus, leaving the refrigerator door open, for example, leads to the transformation of "useful" energy in the form of electrical power into "waste" heat.

5. When an automobile travels at constant speed on level ground, the chemical energy released from the combustion of the fuel is transformed into (a) waste heat released to the atmosphere through the radiator and by direct heat transfer from the engine or from the exhaust fumes, (b) mechanical energy which in part recharges the battery and operates the radio and the lights, but principally performs work against friction and air resistance.

7. The potential energy of the parachutist is transformed into work to overcome the frictional resistance of the air and ultimately is transformed into heat.

9. By rubbing two sticks together, mechanical energy is transformed into heat, thus raising the temperature of the sticks. The temperature may be high enough to ignite some straw.

11. Potential energy is measured relative to a given datum. For an object close to the earth's surface, the reference is a point on the surface of the earth and the energy at a point above the surface is positive. For an object at some distance from the earth, the datum is taken at infinity and the energy is lower (negative).

If the surface were used in both cases, the energy in the second case would be $Gm_em\left(\frac{1}{R} - \frac{1}{r}\right)$ which is positive for $r > R$.

13. If no forces perform work, not only mechanical energy but also kinetic energy would be conserved.

15. Elastic potential energy could be transformed directly into gravitation potential energy by allowing a compressed spring to raise an object slowly to a higher elevation.

17. (a) Riding an escalator, mechanical energy from the escalator performs work on the rider, work that is converted directly into potential energy.

(b) There are transfers of mechanical energy into kinetic energy when the elevator starts and from kinetic energy into gravitational energy and heat when the elevator stops. During the rest of the trip, mechanical energy is transformed into gravitational energy.

(c) Running up the stairs, mechanical energy is transformed into gravitational energy and heat.

19. For a satellite in a circular orbit,

$$\frac{mv^2}{r} = \frac{Gmm_e}{r^2}$$

or

$$mv^2 = \frac{Gmm_e}{r}$$

Thus, if the velocity increases, $\frac{Gmm_e}{r}$ increases also (the radius gets smaller). The total energy is

$$\frac{1}{2}mv^2 - \frac{Gmm_e}{r} = -\frac{1}{2}mv^2,$$

so the total energy decreases (it becomes more negative). This occurs because the increase in kinetic energy is half of the decrease in potential energy.

21. From Question 19, for circular orbits,

(a) the kinetic energy is $\frac{1}{2} \cdot \frac{Gmm_e}{r}$, so with $r_2 > r_1$, satellite 2 has the lesser kinetic energy;

(b) the potential energy is $-\frac{Gmm_e}{r}$, so satellite 2 has the greater potential energy;

(c) the mechanical energy is $-\frac{1}{2} \cdot \frac{Gmm_e}{r}$, so satellite 2 has the larger potential energy;

(d) the binding energy is the absolute value of the mechanical energy $\left(\text{i.e., } \frac{1}{2} \cdot \frac{Gmm_e}{r}\right)$, so satellite 2 has the lesser binding energy.

23.

Symbol	Represents	Type	SI Unit
U	potential energy (PE)	scalar	J
$U_f - U_i$	increase in PE	scalar	J
E	mechanical energy	scalar	J
$\oint \mathbf{F} \cdot d\vec{\ell}$	work in a closed path	scalar	J
mgy	gravitational PE	scalar	J
$\frac{-GMm}{r}$	gravitational PE	scalar	J
$\frac{1}{2}kx^2$	PE in a spring	scalar	J

Exercises:

1. (a) $E = \frac{1}{2}mv_i^2 = \frac{1}{2}(55\text{ kg})(14\text{ m/s})^2 = 5.4\text{ kJ}$

(b) At the highest point, $K = 0$ so $E = U$

Thus, $U = 5.4\text{ kJ}$

(c) $U = mgy$

$$\therefore \quad y_{max} = \frac{5.4 \times 10^3\text{ J}}{(55\text{ kg})(9.8\text{ m/s}^2)} = 10\text{ m}$$

3. For the block, $K = \frac{1}{2}mv^2$, $U = \frac{1}{2}kx^2$, $E = K + U = 47\text{ J}$

(a) The maximum displacement occurs when $v = 0$,

so $\frac{1}{2}(2100\text{ N/m})x^2 = 47\text{ J}$, $x = 0.21\text{ m}$

(b) The maximum velocity occurs when $x = 0$,

$$\frac{1}{2}(0.75\text{ kg})v^2 = 47\text{ J}, \quad v = 11\text{ m/s}$$

(c) When $v = 5.6$ m/s,

$$\frac{1}{2}(2100\text{ N/m})x^2 + \frac{1}{2}(0.75\text{ kg})(5.6\text{ m/s})^2 = 47\text{ J}, \quad x = 0.18\text{ m}$$

5. $E = \frac{1}{2}mv^2 = \frac{1}{2}(12000\text{ kg})(4.3\text{ m/s})^2 = 1.1 \times 10^5\text{ J}$

This energy is temporarily stored in the spring, with $x = 0.23$ m

$$\frac{1}{2}k(0.23\text{ m})^2 = 1.1 \times 10^5\text{ J}$$

$$k = 4.2 \times 10^6\text{ N/m}$$

7. (a) The normal force on the carton is

$$F_n = mg + ma = 75\text{ kg}(2.4\text{ m/s}^2 + 9.8\text{ m/s}^2)$$

$$= 915\text{ N}$$

The work done by the force is

$$W = (915\text{ N})(3.8\text{ m}) = 3.5\text{ kJ}$$

(b) The force is not conservative because the work done in a round trip is not zero.

(c) Using the work–energy theorem,

$$\frac{1}{2}mv^2 + mgy = 3.5\text{ kJ}$$

Here $y = 0$,

$$\text{so} \quad v = \sqrt{\frac{2(3.5 \times 10^3\text{ J})}{75\text{ kg}}} = 9.6\text{ m/s}$$

9. $U = \frac{1}{2}(k_1 + k_2)x^2, \quad K = \frac{1}{2}mv^2$

Calculating the energy when $v = 0$,

$$E = \frac{1}{2}(1200\text{ N/m} + 1800\text{ N/m})(0.20\text{ m})^2 = 60\text{ J}$$

(a) $K_{max} = E = 60\text{ J} \quad (\text{when } x = 0)$

$$\frac{1}{2}mv^2 + \frac{1}{2}(k_1 + k_2)x^2 = 60\text{ J}$$

$$v = \sqrt{\frac{120\text{ J} - 3000\text{ N/m } x^2}{5\text{ kg}}} = \sqrt{24\text{ m}^2/\text{s}^2 - (600\text{ N/m}\cdot\text{kg})x^2}$$

(b) When $x = 0$, $v = \sqrt{24\text{ m}^2/\text{s}^2} = 4.9\text{ m/s}$

(c) When $x = -.10$ m,

$$v = \sqrt{24\text{ m}^2/\text{s}^2 - (600\text{ N/m}\cdot\text{kg})(-.10\text{ m})^2} = \sqrt{18\text{ m}^2/\text{s}^2} = 4.2\text{ m/s}$$

11. $E = K + U$. Since $K \geq 0$, $U \leq E$

(a) The maximum height corresponds to $U = U_{max} = E$. From Fig. 9–20, for $E = 200$ J, this corresponds to $y = 2.0$ m.

(b) Since $U \geq 0$, $K_{max} = E$. This occurs when $U = 0$ (i.e., at $y = 0$).

(c) When $K = U$, each of these must be equal to E/2, or 100 J. From the graph, this occurs at $y = 1.0$ m.

(d) The force on the object is given by

$$F = -\frac{dU}{dy},$$

which is the slope of the graph. This graph has a constant slope of 100 J/m, so the force is 100 N downwards.

13. Given $\quad U = \frac{1}{4}\alpha x^4 - \frac{1}{2}kx^2,$

where $\quad \alpha = 16 \times 10^{20}\ \text{N/m}^3, \quad k = 4.0\ \text{N/m}$

x (nm)	$U(J \times 10^{-21})$
0.000	0.00
±0.020	−0.74
±0.040	−2.2
±0.060	−2.0
±0.080	3.6
±0.10	20
±0.0707	0.00

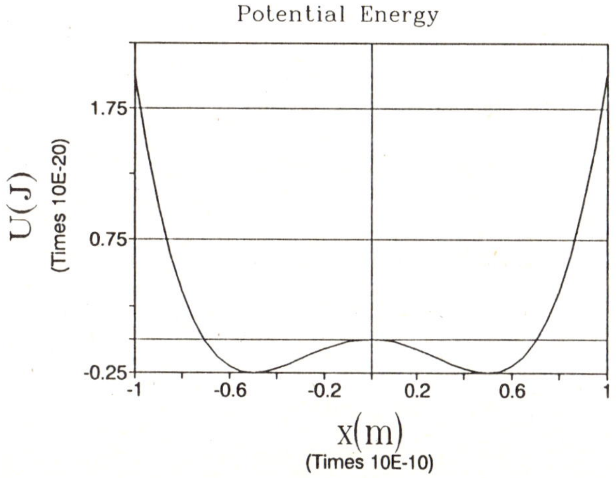

(a) If $v = 0$ when $x = -.02$ nm, then $E = -0.74 \times 10^{-21}$ J. From the graph, $U_{min} = -2.6 \times 10^{-21}$ J. Thus, $K_{max} = E - U_{min}$, so $K_{max} = 1.9 \times 10^{-21}$ J.

(b) The line U (J) = 0 on the graph corresponds to this value of E. It is seen that $K = 0$ at $x = -.065$ nm.

(c) When $E < 0$ the turning points have the same sign, so the motion is confined between two positive or two negative values of x. For $E > 0$ the turning points are positive and negative values of x with the same magnitude. One can also say that for $E < 0$ the particle never goes through $x = 0$, whereas for $E > 0$ it always does.

15. Assuming m = 70 kg

(a) The work done by gravity when the person climbs 3 m is −210 J.

(b) The work done by gravity when the person jumps down is 210 J. (No work is done by gravity in the horizontal portion of the trip.)

(c) No net work is done by gravity for the round trip. (Gravity is conservative.)

17. From Exercise 16,

$$W = 3\,\text{N}(x_f - x_i) + 4\,\text{N}(y_f - y_i)$$

regardless of the path. Thus if $x_f = x_i$ and $y_f = y_i$ (a round trip), the work is zero, so the force is conservative.

(a) $U_f - U_i = -W = -3\,\text{N}(x_f - x_i) - 4\,\text{N}(y_f - y_i)$

(b) If $x_i = 0$, $y_i = 0$, $v_i = 0$

then $U = -3\,\text{N}\,x - 4\,\text{N}\,y$

(c) $U(8\text{ m}, 6\text{ m}) = -24\text{ Nm} - 24\text{ Nm} = -48\text{ J}$

(d) $U(-8\text{ m}, 6\text{ m}) = 24\text{ Nm} - 24\text{ Nm} = 0\text{ J}$

$\left(\text{In fact, } U = 0 \text{ along } y = -\frac{3}{4}x\right)$

19. The horizontal velocity component is constant, so

$$v_x = \frac{37\text{ m}}{2.0\text{ s}} = 18.5\text{ m/s}$$

(a) At the highest elevation, $v_y = 0$, so $v = v_x$ and

$$E = \frac{1}{2}mv^2 + mgy$$

$$= \frac{1}{2}(0.25\text{ kg})(18.5\text{ m/s})^2 + (0.25\text{ kg})(9.8\text{ m/s}^2)(18\text{ m})$$

$$= 87\text{ J}$$

(b) Since energy is conserved, when $y = 0$

$$E = \frac{1}{2}mv^2,$$

so $$v = \sqrt{\frac{2E}{m}} = \sqrt{\frac{2\,(87\text{ J})}{0.25\text{ kg}}} = 26\text{ m/s}$$

(c) The maximum potential energy is obtained when the ball is at its highest elevation.

$$U_{max} = mgy_{max} = (0.25\text{ kg})(9.8\text{ m/s}^2)(18\text{ m})$$
$$= 44\text{ J}$$

21. The elevation of the ball above the lowest point is $y = L(1-\cos\theta)$.

Thus, $E = \frac{1}{2}mv^2 + mgL(1-\cos\theta)$

If $v = 0$ when $\theta = 30°$, then

$$E = m(9.8\text{ m/s}^2)(0.450\text{ m})(1-\cos 30°)$$
$$= (0.59\text{ J/kg})m$$

(a) When $y = 0$, then $\frac{1}{2}mv^2 = (0.59\text{ J/kg})m$,

so $v = \sqrt{2(.59\text{ J/kg})} = 1.09\text{ m/s}$

(b) At the lowest position,

$$T - mg = \frac{mv^2}{L}$$

so $T = m\left(9.8\text{ m/s}^2 + \frac{(1.09\text{ m/s})^2}{0.45\text{ m}}\right)$

$$= (12.4\text{ m/s}^2)m$$

23. Taking the reference for potential energy along OB,

$$E = \frac{1}{2}mv_o^2 = \frac{1}{2}mv_C^2 + mgR$$

At C, $T_C + mg = \frac{mv_C^2}{R}$

If $T_C = 0$, then $v_C^2 = gR$, and since

$$v_o^2 = v_C^2 + 2gR,$$

the minimum value of v_o is $\sqrt{3gR}$.

25. The trajectory of the particle with the origin at the edge of the counter is

$$x = v_x t, \quad y = -\frac{1}{2} g t^2,$$

or
$$y = -\frac{1}{2} \cdot \frac{g x^2}{v_x^2}$$

Now, $x = 4.2 \text{ m}$ when $y = -1.2 \text{ m}$,

so
$$v_x^2 = \frac{1}{2} \cdot \frac{(9.8 \text{ m/s}^2)(4.2 \text{ m})^2}{(1.2 \text{ m})}, \quad \text{or} \quad v_x = 8.5 \text{ m/s}$$

(a) $E = \frac{1}{2} m v_x^2 = 2.7 \text{ J}$ (since $v_y = 0$ when $y = 0$)

(b) This energy was initially in the spring, so

$$\frac{1}{2} k (0.025 \text{ m})^2 = 2.7 \text{ J}, \quad k = 8.6 \times 10^3 \text{ N/m}$$

(c) If $x = 0.037 \text{ m}$, then $E = \frac{1}{2} (8.6 \times 10^3 \text{ N/m})(.037 \text{ m})^2 = 5.9 \text{ J}$

$$v_x = \sqrt{\frac{2E}{m}} = \sqrt{\frac{2(5.9 \text{ J})}{.075 \text{ kg}}} = 12.6 \text{ m/s}$$

and
$$x = \sqrt{\frac{-2 y v_x^2}{g}} = \sqrt{\frac{2(1.2 \text{ m})(12.6 \text{ m/s})^2}{9.8 \text{ m/s}^2}} = 6.2 \text{ m}$$

27. (a) Given a crate of $m = 50$ kg sliding on rough surface with $\mu_k = 0.3$,

$$F_F = \mu N = (0.3)(50 \text{ kg})(9.8 \text{ m/s}^2)$$
$$= 150 \text{ N}$$

In going around a full circle of radius 4.5 m, the crate is moved a total distance of $2\pi(4.5 \text{ m}) = 28$ m. The force and displacement are parallel, so the total work done by the frictional force is $-(150 \text{ N})(28 \text{ m}) = -4.2$ kJ.

(b) This calculation shows that the force is not conservative because the work done in going around a closed path is not zero.

29. (a) Taking $m = 70$ kg, the only change in the mechanical energy is

$$m g y = 70 \text{ kg} (9.8 \text{ m/s}^2)(10 \text{ m}) = 6.9 \times 10^3 \text{ J}$$

(b) The normal force on the shoes does no work because the steps do not move,

so $\mathbf{F} \cdot d\vec{\ell} = 0$

(c) The mechanical energy increases because mechanical work is done by the leg muscles (i) when the forward leg is raised to the next step and (ii) when the forward leg straightens, bringing the rest of the body up one step.

31. $v = \frac{36\text{ km}}{\text{h}} \times \frac{1000\text{ m}}{\text{km}} \times \frac{\text{h}}{3600\text{ s}} = 10\text{ m/s}$

(a) Since the automobile moves with a constant velocity, the net force is zero.

Thus, $mg\sin\theta - F_{ret} = 0$

so $$\frac{F_{ret}}{m} = 9.8\text{ m/s}^2(.05) = 0.49\text{ N/kg}$$

For a 1000 kg automobile, $F_{ret} = 490\text{ N}$.

(b) If the car is moving uphill at the same speed, it can be assumed that the retarding force will remain the same, so the force required to move the car is

$$F = F_{ret} + mg\sin\theta = 980\text{ N}$$

and the power is $Fv = 980\text{ kW}$,

or $$\frac{Fv}{m} = 0.98\text{ kW/kg}$$

33. (a) In this problem, the earth can be regarded as being unaffected by the motion of the projectile, so the energy transformations are as follows:

Initially,

$$K = \frac{1}{2}mv^2 = \frac{1}{2}(0.33\text{ kg})(23\text{ m/s})^2 = 87\text{ J}$$

$$U = 0$$

At $y = 14\text{ m}$,

$$K = \frac{1}{2}(0.33\text{ kg})(13\text{ m/s})^2 = 28\text{ J}$$

$$\Delta K = 28\text{ J} - 87\text{ J} = -59\text{ J}$$

$$U = mgy = (0.33\text{ kg})(9.8\text{ m/s}^2)(14\text{ m}) = 45\text{ J}$$

$$\Delta U = 45\text{ J} - 0\text{ J} = 45\text{ J}$$

$$E = K + U = 73\text{ J}$$

$$\Delta E = 73\text{ J} - 87\text{ J} = -14\text{ J}$$

The projectile has lost 14 J of energy. This energy has gone to overcome frictional losses and eventually becomes heat added to the atmosphere.

(b) There will be additional losses of energy as the projectile reaches its maximum height and returns to the 14 m level. Therefore, the kinetic energy will be less than 28 J and the speed of the projectile will be less than 13 m/s.

35. For a circular orbit,

$$\frac{mv^2}{r} = \frac{GmM}{r^2},$$

$$K = \frac{1}{2}mv^2, \qquad U = -\frac{GmM}{r}$$

so $$K = -\frac{1}{2}U \quad \text{and} \quad E = \frac{1}{2}U$$

Given $m = 30 \times 10^3$ kg, $r = 70 \times 10^6$ m,

(a) $$U = \frac{-(6.67 \times 10^{-11}\ \text{N} \cdot \text{m}^2/\text{kg}^2)(30 \times 10^3\ \text{kg})(5.98 \times 10^{24}\ \text{kg})}{(70 \times 10^6\ \text{m})}$$

$$= -1.7 \times 10^{11}\ \text{J}$$

(b) $K = -\frac{1}{2}U = 8.5 \times 10^{10}$ J

(c) $E = \frac{1}{2}U = -8.5 \times 10^{10}$ J

(d) If r is made smaller, then the kinetic energy would be the only one of these quantities to increase.

37. From Exercise 35,

$$E = \frac{1}{2}U = -\frac{GmM}{2r}$$

(a) $$E = \frac{-(6.67 \times 10^{-11}\ \text{N} \cdot \text{m}^2/\text{kg}^2)(3.3 \times 10^{23}\ \text{kg})(2.0 \times 10^{30}\ \text{kg})}{2\,(5.8 \times 10^{10}\ \text{m})}$$

$$= -3.8 \times 10^{32}\ \text{J}$$

(b) $E(r = 15 \times 10^{10}\ \text{m}) = -1.5 \times 10^{32}$ J, $\quad \Delta E = 2.3 \times 10^{32}$ J

39. $r_p = 7.2 \times 10^6$ m, $\quad v_p = 8.0 \times 10^3$ m/s

$r_a = 9.9 \times 10^6$ m

$$E = \frac{1}{2}mv^2 - \frac{GMm}{r}$$

$$GM = (6.67 \times 10^{-11}\ \text{N} \cdot \text{m}^2/\text{kg}^2)(5.98 \times 10^{24}\ \text{kg})$$

$$= 3.99 \times 10^{14}\ \text{N} \cdot \text{m}^2/\text{kg}$$

$$v^2 = \frac{2E}{m} + \frac{2GM}{r}$$

Evaluate $\frac{2E}{m}$ at the perigee:

$$\frac{2E}{m} = v_p^2 - \frac{2GM}{r_p} = 64 \times 10^6\ \text{m}^2/\text{s}^2 - \frac{2\,(3.99 \times 10^{14}\ \text{N} \cdot \text{m}^2/\text{kg})}{7.2 \times 10^6\ \text{m}}$$

$$= -46.8 \times 10^6\ \text{J/kg}$$

(a) $$v_a^2 = \frac{2E}{m} + \frac{2GM}{r} = -46.8 \times 10^6\ \text{m}^2/\text{s}^2 + \frac{2\,(3.99 \times 10^{14}\ \text{N} \cdot \text{m}^2/\text{kg})}{9.9 \times 10^6\ \text{m}}$$

$$v_a = 5.8 \times 10^3\ \text{m/s}$$

(b) $$v_{8.4}^2 = -46.8 \times 10^6\ \text{m}^2/\text{s}^2 + \frac{2\,(3.99 \times 10^{14}\ \text{N} \cdot \text{m}^2/\text{kg})}{8.4 \times 10^6\ \text{m}}$$

$$v_{8.4} = 6.9 \times 10^3\ \text{m/s}$$

41. Assuming there is no air resistance,

$$E_i = \frac{1}{2} m v^2 = E_f = mgy$$

$$\therefore \quad E = (200 \times 10^{-6}\ \text{kg})(9.8\ \text{m/s}^2)(0.20\ \text{m})$$

$$= 3.9 \times 10^{-4}\ \text{J}$$

43. (a) $$\Delta F = 4\,(50\ \text{kg})(9.8\ \text{m/s}^2) = 1960\ \text{N}$$

$$= 4k\Delta x = 4k\,(0.01\ \text{m})$$

$$\therefore \quad k = 49 \times 10^3\ \text{N/m}$$

(b) Each spring supports 250 kg to start with

$$kx = (250\ \text{kg})(9.8\ \text{m/s}^2), \qquad x = 0.05\ \text{m}$$

$$U_i = \frac{1}{2} kx^2 = \frac{1}{2}\,(49 \times 10^3\ \text{N/m})(0.05\ \text{m})^2 = 61\ \text{J}$$

$$U_f = \frac{1}{2}\,(49 \times 10^3\ \text{N/m})(0.06\ \text{m})^2 = 88\ \text{J}$$

45. (a) $\Delta U_{elevator} = mg\Delta y = (800\text{ kg})(9.8\text{ m/s}^2)(-10\text{ m})$

$$= -78 \times 10^3\text{ J}$$

(b) $\Delta U_{weight} = (700\text{ kg})(9.8\text{ m/s}^2)(10\text{ m}) = 69 \times 10^3\text{ J}$

(c) The net loss in potential energy is equal to the gain in kinetic energy:

$$\frac{1}{2}(700\text{ kg} + 800\text{ kg})v^2 = +9.8 \times 10^3\text{ J}$$

$$\therefore \quad v = 3.6\text{ m/s}$$

47. (a) The positions of stable equilibrium occur at the minima of U. From Fig. 9–29, these are at

$$x = \pm 3 \times 10^{-11}\text{ m}$$

(b) The value of U when $x = -4 \times 10^{-11}$ m is approximately -0.4×10^{-19} J. Drawing a horizontal line through this point, it is found that U has the same value at $x = -1.8 \times 10^{-11}$ m. This is the second turning point.

(c) Following the same procedure,

if $x_1 = 4 \times 10^{-11}$ m then $x_2 = 1.8 \times 10^{-11}$ m, and

(d) if $x_1 = 4.5 \times 10^{-11}$ m then $x_2 = -4.5 \times 10^{-11}$ m

Note that in (a), (b), and (c) the turning points are on the same side of the origin. That is, the origin is not a point on the trajectory, whereas in (d) the turning points are on opposite sides of the origin.

Problems:

1. In this problem, energy is conserved. The initial energy is

$$E = mgy = (7.3\text{ kg})(9.8\text{ m/s}^2)(3.0\text{ m})$$

$$= 215\text{ J}$$

(a) If the spring is compressed a distance x before the block comes to rest, it will be at a vertical position $y = -\frac{3}{5}x$ at that point. Setting

$$E = \frac{1}{2}kx^2 + mgy$$

$$215\text{ J} = \frac{1}{2}(210\text{ N/m})x^2 + (7.3\text{ kg})(9.8\text{ m/s}^2)\left(-\frac{3}{5}x\right)$$

or $$(105\text{ N/m})x^2 - (43\text{ N})x - (215\text{ J}) = 0,$$

so $$x = \frac{(43\text{ N}) \pm \sqrt{(43\text{ N})^2 + 4(105\text{ N/m})(215\text{ J})}}{2(105\text{ N/m})}$$

$$= 1.65\text{ m} \quad \text{or} \quad -1.24\text{ m}$$

The negative root is not relevant to this situation, so $x = 1.65\text{ m}$

(b) When the block just reaches the spring, $y = 0$ and $E = \frac{1}{2}mv^2$,

so $$v = \sqrt{\frac{2E}{m}} = \sqrt{\frac{2(215\text{ J})}{7.3\text{ kg}}} = 7.7\text{ m/s}$$

3. (a) $U = \frac{1}{2}ky^2 + mgy$

(b) $F = -ky - mg$

(c) $F = 0$ when $y = -\frac{mg}{k} = y_1$

(d) $$U = \frac{1}{2}k\left[(y - y_1) + y_1\right]^2 - ky_1\left[(y - y_1) + y_1\right]$$

$$= \frac{1}{2}k\left[(y - y_1)^2 + 2y_1(y - y_1) + y_1^2\right] - ky_1\left[(y - y_1) + y_1\right]$$

$$= \frac{1}{2}k(y - y_1)^2 - \frac{1}{2}ky_1^2$$

(e) If $U' = U + \frac{1}{2}ky_1^2$ and $y' = y - y_1$, then $U' = \frac{1}{2}ky'^2$

5. Since $$F = F(r),$$

then $$W = \int_{r_1}^{r} F(r)\,dr$$

and no work is done in moving at a constant radius. Thus the work depends only on the initial and final positions and $U = -W$.

7. (a) Since the inertia of the pulley and friction are neglected, the tension is the same in both sections of the string. Both masses move equal distances, m_2 in the direction of the tension, m_1 in the direction opposite the tension. Thus the work done by m_1 on the tension cancels the work done by the tension on m_2.

(b) When m_1 has fallen from rest a distance h,

$$\frac{1}{2}(m_1 + m_2)v^2 = m_1 g h$$

or $$v = \sqrt{\frac{2 m_1 g h}{m_1 + m_2}}$$

9. The force is minus the slope of the potential energy curve. Thus, from x = 0 to x = 0.1 the slope is $\frac{2-0}{0.1} = 20$ N. Proceeding in this way,

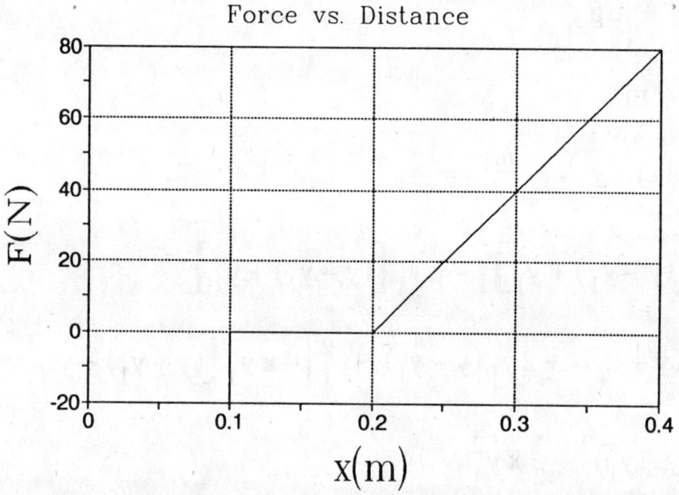

11. The same spreadsheet as in Chapter 7, Exercise 35, can be used for this problem. The mass of the earth has to be put into cell B3 and, in order to calculate K, U, and E, four columns have been introduced: D (for r(t)) and G, H, and I (for K(t), U(t), E(t)), thus moving the other columns to the right. The output with a time interval h = 240 s (4 minutes) shows that the apogee is at 9.9×10^6 m and is reached in approximately 67 minutes. The velocity at the apogee is 5.8×10^3 m/s and the velocity when $r = 8.4 \times 10^6$ m (at $t \simeq 1600$ s) is 6.9×10^3 m/s. The energy per unit mass remains constant at -2.3×10^7 J/kg.

	A	B	C	D	E	F	G	H	I	J	K	L	M	N
1	Ch. 9, Prob. 11													
2	G =	6.67E-11								h =	240			
3	Mearth =	5.98E+24								GM =	3.99E+14			
4	t	x(t)	y(t)	r(t)	vx(t)	vy(t)	K(t)	U(t)	E(t)	x(t+h/2)	y(t+h/2)	r(t+h/2)	ax(t+h/2)	ay(t+h/2)
5	0	0	7200000	7200000	8000	0	32000000	-5.5E+07	-2.3E+07	960000	7200000	7263718	-0.99913	-7.49346
6	240	1891225	6984188	7235718	7760.209	-1798.43	31727601	-5.5E+07	-2.3E+07	2822450	6768377	7333290	-2.85468	-6.84565
7	480	3671461	6355410	7339677	7075.087	-3441.39	30950005	-5.4E+07	-2.3E+07	4520471	5942443	7466411	-4.33187	-5.69451
8	720	5244724	5365475	7503029	6035.439	-4808.07	29772034	-5.3E+07	-2.3E+07	5968977	4788506	7652351	-5.31303	-4.26228
9	960	6540214	4088784	7713142	4760.312	-5831.02	28330676	-5.2E+07	-2.3E+07	7111451	3389062	7877721	-5.80208	-2.76506
10	1200	7515589	2609706	7955793	3367.813	-6494.63	26761217	-5E+07	-2.3E+07	7919727	1830350	8128484	-5.88178	-1.35935
11	1440	8154469	1011844	8217006	1956.187	-6820.88	25175525	-4.9E+07	-2.3E+07	8389212	193339.1	8391439	-5.6629	-0.13051
12	1680	8460863	-628925	8484205	597.0912	-6852.2	23654584	-4.7E+07	-2.3E+07	8532513	-1451189	8655041	-5.24924	0.892778
13	1920	8452986	-2247741	8746732	-662.727	-6637.93	22250685	-4.6E+07	-2.3E+07	8373459	-3044293	8909688	-4.7222	1.716825
14	2160	8157932	-3791401	8995920	-1796.06	-6225.9	20993795	-4.4E+07	-2.3E+07	7942406	-4538508	9147670	-4.13854	2.364878
15	2400	7607689	-5217507	9224929	-2789.31	-5658.33	19898434	-4.3E+07	-2.3E+07	7272972	-5896506	9362954	-3.53427	2.865386
16	2640	6836469	-6492982	9428474	-3637.53	-4970.63	18969412	-4.2E+07	-2.3E+07	6399965	-7089458	9550914	-2.93001	3.245676
17	2880	5879077	-7592458	9602550	-4340.74	-4191.67	18206040	-4.2E+07	-2.3E+07	5358188	-8095459	9708071	-2.33586	3.529147
18	3120	4770027	-8496820	9744183	-4901.34	-3344.67	17604996	-4.1E+07	-2.3E+07	4181867	-8898181	9831868	-1.75505	3.734394
19	3360	3543160	-9191991	9851228	-5322.55	-2448.42	17162163	-4E+07	-2.3E+07	2904454	-9485802	9920498	-1.18656	3.875258
20	3600	2231575	-9668005	9922209	-5607.33	-1518.36	16873768	-4E+07	-2.3E+07	1558695	-9850208	9972769	-0.62682	3.961185
21	3840	867763.7	-9918328	9956217	-5757.76	-567.674	16737049	-4E+07	-2.3E+07	176832	-9986449	9988015	-0.07079	3.997611
22	4080	-516138	-9939439	9952831	-5774.75	391.7527	16750619	-4E+07	-2.3E+07	-1209109	-9892429	9966047	0.487218	3.986219
23	4320	-1888047	-9730615	9912093	-5657.82	1348.445	16914617	-4E+07	-2.3E+07	-2566985	-9568802	9907138	1.052945	3.925003
24	4560	-3215599	-9293948	9834508	-5405.11	2290.446	17230696	-4.1E+07	-2.3E+07	-3864213	-9019095	9812044	1.631584	3.808128
25	4800	-4465837	-8634567	9721083	-5013.53	3204.396	17701835	-4.1E+07	-2.3E+07	-5067461	-8250040	9682061	2.226968	3.625597

Cell Formulas

```
A1: 'Ch. 9, Prob. 11
A2: 'G =
B2: 6.67E-11
J2: 'h =
K2: 240
A3: 'Mearth =
B3: 5.98E+24
J3: 'GM =
K3: +$B$2*$B$3
A4: ^t
B4: ^x(t)
C4: ^y(t)
D4: ^r(t)
E4: ^vx(t)
F4: ^vy(t)
G4: ^K(t)
H4: ^U(t)
I4: ^E(t)
J4: 'x(t+h/2)
K4: 'y(t+h/2)
L4: 'r(t+h/2)
M4: 'ax(t+h/2)
N4: 'ay(t+h/2)
A5: 0
B5: 0
C5: 7200000
D5: @SQRT(B5^2+C5^2)
E5: 8000
F5: 0
G5: (E5^2+F5^2)/2
H5: -$K$3/D5
I5: +G5+H5
J5: +B5+0.5*E5*$K$2
K5: +C5+0.5*F5*$K$2
L5: (J5^2+K5^2)^0.5
M5: -$K$3*J5/L5^3
N5: -$K$3*K5/L5^3
A6: +A5+$K$2
B6: +B5+E5*$K$2+M5*$K$2^2/2
C6: +C5+F5*$K$2+N5*$K$2^2/2
D6: @SQRT(B6^2+C6^2)
E6: +E5+M5*$K$2
F6: +F5+N5*$K$2
G6: (E6^2+F6^2)/2
H6: -$K$3/D6
I6: +G6+H6
J6: +B6+0.5*E6*$K$2
K6: +C6+0.5*F6*$K$2
L6: (J6^2+K6^2)^0.5
M6: -$K$3*J6/L6^3
N6: -$K$3*K6/L6^3
```

13. $F_x = -\dfrac{2Ax}{(x^2+a^2)^2}$

(a) $U = -\int_{\infty}^{x} F_x\, dx = \int_{\infty}^{x} \dfrac{A\, d(x^2)}{(x^2+a^2)^2}$

$= -\dfrac{A}{x^2+a^2}$

(b)

$\frac{x}{a}$	$\frac{Ua^2}{A}$
0	− 1.00
0.5	− 0.80
1.0	− 0.50
1.5	− 0.31
2.0	− 0.20

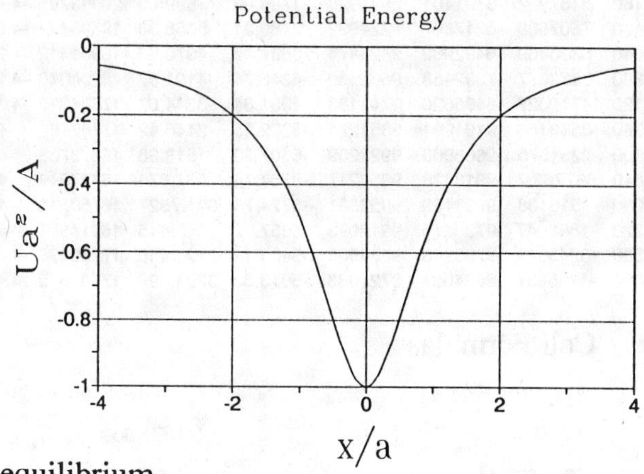

x = 0 is a position of stable equilibrium.

15. (a) In Chapter 7, Problems 22 and 23, it was shown that the potential of a uniform shell is zero for a point P inside the shell and is equal to that of a point mass at the center for a point P outside the shell. Thus, if we assume that the galaxy can be approximated by a uniform distribution of mass, then

$$U = -\frac{GM_G m_s}{r} = \frac{-(6.67\times10^{-11}\ \mathrm{N\cdot m^2/kg^2})(2\times10^{41}\ \mathrm{kg})(2\times10^{30}\ \mathrm{kg})}{3\times10^{20}\ \mathrm{m}}$$

$$= -9\times10^{40}\ \mathrm{J}$$

(b) Assuming a circular orbit,

$$\frac{m_s v_s^2}{r} = \frac{G m_s M_G}{r^2}$$

so $$v_s = \sqrt{\frac{GM_G}{r}} = \sqrt{\frac{(6.67\times10^{-11}\ \mathrm{N\cdot m^2/kg^2})(2\times10^{41}\ \mathrm{kg})}{3\times10^{20}\ \mathrm{m}}}$$

$$= 2.1\times10^5\ \mathrm{m/s}$$

17. Referring to Problem 1, the initial energy is

$$E_i = E = \tfrac{1}{2}ks^2 - mg\left(\tfrac{3}{5}s\right)$$

where $k = 210$ N/m, and $m = 7.3$ kg

The final energy, given that $v = 3.4$ m/s when $y = 3$ m, is

$$E_f = \tfrac{1}{2}m(3.4\text{ m/s})^2 + mg(3\text{ m})$$

Setting $E_i = E_f$,

$$s^2 - \frac{1.2\,g}{k}s = \frac{m}{k}\left((3.4\text{ m/s})^2 + (3\text{ m})g\right)$$

or $$s^2 - (0.056\text{ m})s - 1.42\text{ m}^2 = 0$$

so $$s = 1.22\text{ m}$$

19. (a) $$\text{Power} = \frac{\text{work}}{\text{time}}$$

$$= \frac{\text{weight} \times \text{distance}}{\text{time}}$$

Thus, $$\frac{\text{weight}}{\text{time}} = \frac{100 \times 10^6\text{ W}}{100\text{ m}} = 10^6\text{ N/s}$$

or $$\frac{\text{mass}}{\text{time}} = \frac{10^6\text{ N/s}}{9.8\text{ N/kg}} = 1.02 \times 10^5\text{ kg/s}$$

or $$\frac{\text{volume}}{\text{time}} = \frac{1.02 \times 10^5\text{ kg/s}}{10^3\text{ kg/m}^3} = 102\text{ m}^3\text{/s}$$

(b) If the area of the reservoir is $(10^3)^2\text{ m}^2 = 10^6\text{ m}^2$, the water level drops at the rate of

$$\frac{dh}{dt} = \frac{\frac{V}{A}}{\Delta t} = \frac{102\text{ m}^3\text{/s}}{10^6\text{ m}^2} = 1.02 \times 10^{-4}\text{ m/s}$$

21. If $F = -\frac{k}{r^2}$, then

(a) $U = -\frac{k}{r}$

In a circular orbit,

$$\frac{mv^2}{r} = \frac{k}{r^2}, \text{ so}$$

(b) $K = \frac{1}{2} mv^2 = \frac{1}{2} \cdot \frac{k}{r}$

Thus, $K = -\frac{1}{2} U$

(c) $E = \frac{1}{2} U$

(d) If $k = 2.3 \times 10^{-28}\ N \cdot m^2$ and $r = 0.53 \times 10^{-10}$ m,

$$U = \frac{-2.3 \times 10^{-28}\ N \cdot m^2}{0.53 \times 10^{-10}\ m} = -4.3 \times 10^{-8}\ J$$

and $K = 2.2 \times 10^{-8}$ J, $E = -2.2 \times 10^{-8}$ J

23. Since the rest mass energy has decreased by 8×10^{-14} J and the initial kinetic energy is zero, the final kinetic energy must be 8×10^{-14} J. Of this, the kinetic energy of the other two products is given as 6×10^{-14} J. The remaining 2×10^{-14} J must correspond to the energy of the antineutrino.

CHAPTER 10 MOMENTUM AND THE MOTION OF SYSTEMS

Questions:

3. Not necessarily. There can be objects in the shape of a ring or a hoop which have a center of mass at a point where there is no mass.

5. For a spherical basketball, the center of mass is at the center of the ball. For a hula hoop and for a doughnut it is also at the center. The center of mass of this book is approximately at the geometric center, and for a horseshoe it is somewhere along the line of symmetry.

7. The center of mass of a uniform hemisphere is on the line that is normal through the center of the circular plane base of the hemisphere. The center of mass of a D-shaped cylinder is in the plane normal to the cylindrical axis which bisects the cylinder, on a line which is the normal bisector of the straight edge of the cylindrical section. The center of mass of an isosceles triangle is the point where the three perpendicular bisectors meet.

9. For an object in uniform circular motion, the kinetic energy is conserved but the momentum is not conserved. On the other hand, for objects having inelastic collisions, the momentum is conserved but the kinetic energy is not conserved.

11. It is not necessary that the forces that do work during an elastic collision be conservative. The forces can be nonconservative, but the sum of the work done by these forces should be zero. On the other hand, it is sufficient that the forces are conservative for the collision to be elastic.

13. Air pressure turns a fan, which in turn applies torque on the drill. When the torque applied exceeds a given value, it disengages a mechanical device and no more torque is applied on the drill. At that point, the fan just turns without applying a torque on the drill.

15. You need to apply the law of conservation of momentum. You need to remove a piece of cloth from your body and throw it horizontally and keep doing this until you get to the shore.

17. Typically, it goes about three miles before coming to a complete stop and the turbulence in water provides the external force to stop the tanker.

21. This is an impulse problem. The car that has the shorter collision time will experience the greater force and will have more damage.

23. Yes. The change in the rocket's speed is directly proportional to the speed of the exhaust gases and depends logarithmically on the fractional reduction of the mass.

Exercises:

1. Using equation 10–2, we get:

$$x_{cm} = \frac{\sum m_i x_i}{M}; \quad y_{cm} = \frac{\sum m_i y_i}{M}$$

$$x_{cm} = \frac{(3\text{ kg})(1\text{ m}) - (5\text{ kg})(2\text{ m}) - (4\text{ kg})(3\text{ m}) + (2\text{ kg})(2\text{ m})}{14\text{ kg}}$$

$$x_{cm} = -1.071\text{ m}$$

$$y_{cm} = \frac{(3\text{ kg})(2\text{ m}) + (5\text{ kg})(3\text{ m}) - (4\text{ kg})(2\text{ m}) - (2\text{ kg})(1\text{ m})}{14\text{ kg}}$$

$$y_{cm} = 0.786\text{ m}$$

3. Using equation 10–2, we get:

$$x_{cm} = \frac{\sum m_i x_i}{M}; \quad y_{cm} = \frac{\sum m_i y_i}{M}$$

$$x_{cm} = \frac{0.041\text{ kg}[2(0.1\text{ m}) + 2(0.15\text{ m}) + 2(0.05\text{ m}) - 2(0.1\text{ m})]}{8\,(0.041\text{ kg})}$$

$$x_{cm} = 0.05\text{ m}$$

Similarly,

$$y_{cm} = 0.05\text{ m}$$

5. Using example 10–3,

$$x_{cm} = \frac{1}{M}\int_0^{\ell} x\left(\frac{M}{\ell A}\right) A\,dx$$

$$x_{cm} = \frac{1}{M}\cdot\frac{M}{\ell}\int_0^{\ell} x\,dx$$

$$= \frac{1}{\ell}\cdot\frac{x^2}{2}\Bigg|_0^{\ell} = \frac{\ell}{2}$$

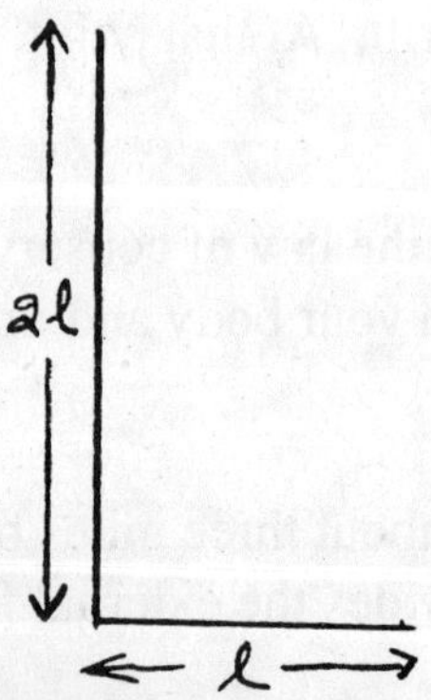

Similarly,

$$y_{cm} = \frac{1}{M}\int_0^{\ell} y\left(\frac{M}{2\ell A}\right) A\, dy$$

$$y_{cm} = \frac{1}{M}\cdot\frac{M}{2\ell}\int_0^{2\ell} y\left(\frac{M}{2\ell A}\right) A\, dy$$

$$y_{cm} = \frac{1}{2\ell}\left|\frac{y^2}{2}\right|_0^{2\ell} = \ell$$

7. $M_G = 42\ \text{kg}$, $M_B = 21\ \text{kg}$, $v = 0.75\ \text{m/s}$

Using equation 10–6,

$$v_{cm} = \frac{1}{M}\sum m_i v_i$$

$$v_{cm} = \frac{1}{63\ \text{kg}}\left[(42\ \text{kg})(0.75\ \text{m/s}) + (21\ \text{kg})(0\ \text{m/s})\right]$$

$$v_{cm} = 0.50\ \text{m/s}$$

9. Using equation 10–7, we get:

$$a_{cm} = \frac{1}{M}\sum m_i a_i$$

(a) $$\mathbf{a}_{cm} = \frac{1}{M}\left[m_A a_y(-\mathbf{j}) + m_B a_y(\mathbf{j})\right]$$

$$\mathbf{a}_{cm} = \frac{1}{M}\left[m_A(3.4\ \text{m/s}^2)(-\mathbf{j}) + m_B(1.7\ \text{m/s}^2)(\mathbf{j})\right]$$

Let $m_A = m_B = m$

$$\mathbf{a}_{cm} = \frac{1.7}{M}\, m\left[2\ \text{m/s}^2(-\mathbf{j}) + 1\ \text{m/s}^2(\mathbf{j})\right]$$

$$\mathbf{a}_{cm} = \frac{1.7}{2}\left[1\ \text{m/s}^2\right](-\mathbf{j})$$

$$\mathbf{a}_{cm} = 0.85\ \text{m/s}^2(-\mathbf{j})$$

(b) $$m_A = \frac{1}{2}\, m_B$$

$$\mathbf{a}_{cm} = \frac{1}{M}\left[m_A(3.4\ \text{m/s}^2)(-\mathbf{j}) + 2m_A(1.7\ \text{m/s}^2)(\mathbf{j})\right]$$

$$\mathbf{a}_{cm} = \frac{3.4\ m_A}{M}\left[1\ \text{m/s}^2(-\mathbf{j}) + 1\ \text{m/s}^2(\mathbf{j})\right]$$

$$\mathbf{a}_{cm} = 0$$

11. (a) The acceleration of the center of mass is given by:

$$a_{cm} = \frac{1}{M}\sum m_i a_i$$

$$a_{cm} = \frac{1}{M}\left[m_1 a - m_2 a\right]$$

$$a_{cm} = \frac{a}{M}\left[m_1 - m_2\right]$$

where a is given by $a = \dfrac{(m_2 - m_1)g}{(m_2 + m_1)}$

Therefore, $$a_{cm} = \frac{(m_2 - m_1)(m_1 - m_2)g}{M(m_2 + m_1)} \qquad (1)$$

(b) Using equation 10–9, we get:

$$\sum F_{ext} = M a_{cm}, \qquad (2)$$

where a_{cm} is given by equation (1).

(c) Gravity and tension force are responsible for this net external force.

13. Use $P_e = M_e v_e$,

where M_e = mass of earth = 5.97×10^{24} kg

$$v_e = R\omega_e = R\frac{2\pi}{T}$$

R = radius of earth, T = time period = 1 year

$$P_e = \frac{(5.97 \times 10^{24}\ \text{kg})(6.38 \times 10^{6}\ \text{m})\, 2\pi}{3.15 \times 10^{7}\ \text{s}}$$

$$P_e = 7.6 \times 10^{24}\ \text{kg m/s}$$

15. (a) Momentum before collision:

$$\mathbf{P}_1 = m_1 v_1$$

$$\mathbf{P}_1 = (3 \times 10^{-26}\ \text{kg})(350\ \text{m/s})(\mathbf{j})$$

$$\mathbf{P}_1 = 1.05 \times 10^{-23}\ \text{kg m/s}\ (\mathbf{j})$$

(b) Momentum after collision:

$$\mathbf{P}_2 = 1.05 \times 10^{-23}\ \text{kg m/s}\ (-\mathbf{j})$$

(c) $$\mathbf{P}_1 - \mathbf{P}_2 = 2.1 \times 10^{-23}\ \text{kg m/s}$$

17. Using equation 10–14, we get:

(a) $J = \Delta P$ where $J = \bar{F}\Delta t$

Therefore, $\bar{F}\Delta t = \Delta P$

Using $v^2 - v_o^2 = 2gs$, where $s = 1.4$ m, $v = 0$

$$v_o^2 = 2(9.8\ \text{m/s}^2)(1.4\ \text{m})$$

$$v_o = 5.24\ \text{m/s}$$

Therefore, $\Delta P = (413\ \text{kg})(5.24\ \text{m/s}) = 2163\ \text{kg m/s}$

(b) $\bar{F}\Delta t = \Delta P$

$$\bar{F} = \frac{\Delta P}{\Delta t}$$

$$\bar{F} = \frac{2163\ \text{kg m/s}}{43 \times 10^{-3}\ \text{s}}$$

$$F = 5.03 \times 10^4\ \text{N}$$

19. $m = 0.071$ kg, $v_i = 0$

From equation 10–14,

$J = \Delta P$, where J = impulse

$$J = mv_f - mv_i \qquad (1)$$

Using the principle of conservation of energy, we find v_f:

$$\not{m}gh_1 = -\not{m}gh_2 + \frac{1}{2}\not{m}v_f^2$$

$$gh_1 = -gh_2 + \frac{1}{2}v_f^2$$

$$(9.8\ \text{m/s}^2)(1\ \text{m}) = -(9.8\ \text{m/s}^2)(0.48\ \text{m}) + \frac{1}{2}v_f^2$$

$$v_f^2 = 29.01\ \text{m}^2/\text{s}^2$$

$$v_f = 5.39\ \text{m/s} \qquad (2)$$

Using equation (2) in equation (1), we get:

$$J = (0.071\text{ kg})(5.39\text{ m/s}) - 0$$

$$J = 0.383\text{ kg m/s}$$

This is the magnitude of the impulse exerted by the floor on the ball and it will act vertically up, or in the + **j** direction.

The magnitude of the impulse exerted by the ball on the floor is the same but in the opposite direction (−**j**).

21. $m_A = 4.52\text{ kg}, \quad m_B = 2.37\text{ kg}$

(a) Momentum before = momentum after

$$0 = -m_A v_A + m_B v_B$$

$$v_B = \frac{m_A v_A}{m_B} = \frac{(4.52\text{ kg})(2.11\text{ m/s})}{2.37\text{ kg}}$$

$$v_B = 4.02\text{ m/s}$$

(b) The energy stored initially is the same as the final kinetic energy of the system, which is given by

$$\frac{1}{2} m_A v_A^2 + \frac{1}{2} m_B v_B^2$$

$$= \frac{1}{2}(4.52\text{ kg})(-2.11\text{ m/s})^2 + \frac{1}{2}(2.37\text{ kg})(4.02\text{ m/s})^2$$

$$= 29.2\text{ J}$$

23. Momentum before = momentum after

Let v_1 and V_1 be the velocities of the girl and the boat before the girl leaves and let v_2 and V_2 be the velocities after the girl leaves the boat.

$$m v_1 + M V_1 = m v_2 + M V_2$$

$$(45\text{ kg})(5.2\text{ m/s}) + 0 = 0 - 1000\text{ kg } V_2$$

$$V_2 = -0.23\text{ m/s}$$

25. $m_1 = 0.050\text{ kg}, \quad m_2 = 0.100\text{ kg}, \quad v_1 = 0.40\text{ m/s}$

(a) Momentum is conserved, thus

$$m_1 v_1 + m_2 v_2 = 0$$

$$(0.050\text{ kg})(0.40\text{ m/s}) + (0.100\text{ kg})(v_2) = 0$$

$$v_2 = -0.2\text{ m/s}$$

(b) $v_{cm} = \frac{1}{M}\sum m_i v_i$

$$= \frac{1}{(0.05\text{ kg}) + (0.1\text{ kg})}[(0.05\text{ kg})(0.4\text{ m/s}) - (0.1\text{ kg})(0.2\text{ m/s})]$$

or $v_{cm} = 0$

(c) No

27. $m_1 = 14.2\text{ Mg}, \quad v_1 = 1.8\text{ m/s}, \quad m_2 = 23.5\text{ Mg}$

(a) Newton's second law of motion states that

$$\frac{d\mathbf{P}}{dt} = \sum \mathbf{F}_{ext}$$

If for a system $\sum \mathbf{F}_{ext} = 0$,

$$\frac{d\mathbf{P}}{dt} = 0 \quad \text{or} \quad \mathbf{P} = \text{constant}$$

(b) Momentum before collision = momentum after collision

$$m_1 v_1 = (m_1 + m_2)V$$

$$V = \frac{m_1 v_1}{m_1 + m_2} = \frac{(14.2\text{ Mg})(1.8\text{ m/s})}{(37.7\text{ Mg})}$$

$$V = 0.68\text{ m/s}$$

29. $v = 2.5\text{ m/s}, \quad M = 34\text{ kg}$ (mass of the sled)

Let m = mass of the girl

Momentum is conserved, so

momentum before = momentum after

$$m(2.5\text{ m/s}) = (m + M)\,1.5\text{ m/s}$$

$$(2.5\text{ m}) = (1.5\text{ m}) + (1.5\text{ M})$$

$$(1.0\text{ m}) = (1.5)(34\text{ kg})$$

$$m = 51\text{ kg}$$

31. Let $m_1 = 50\ \text{kg}$, $v_1 = 1\ \text{m/s}$, and $m_2 = 500\ \text{kg}$

Momentum is conserved, so

$$m_1 v_1 + m_2 v_2 = 0$$

$$(50\ \text{kg})(1\ \text{m/s}) + (500\ \text{kg})(v_2) = 0$$

$$v_2 = -0.1\ \text{m/s}$$

Therefore, Ann's velocity relative to the spaceship is 1.1 m/s. This is also Ann's speed relative to Bob as she drifts across the spaceship.

33. Momentum is conserved, so

$$\text{momentum before} = \text{momentum after}$$

$$mv = (m + M)V$$

$$(3.3\ \text{kg})v = (51.3\ \text{kg})(0.32\ \text{m/s})$$

$$v = 4.97\ \text{m/s}$$

35. $m_u = 59.5\, m_\alpha$

m_u = mass of uranium nucleus, m_α = mass of alpha particle

Momentum is conserved, so

$$\text{momentum before collision} = \text{momentum after collision}$$

$$m_\alpha v_{\alpha_1} = m_\alpha v_{\alpha_2} + m_u v_{u_2} \qquad (1)$$

Since the collision is elastic, kinetic energy is also conserved:

$$\frac{1}{2} m_\alpha v_{\alpha_1}^2 = \frac{1}{2} m_\alpha v_{\alpha_2}^2 + \frac{1}{2} m_u v_{u_2}^2 \qquad (2)$$

Solve equations (1) and (2) for v_{u_2} (see Exercise 45)

37. From conservation of momentum:

$$mv = (m + M)V \qquad (1)$$

From conservation of total energy:

$$\text{KE right after collision} = \text{potential energy}$$

$$\frac{1}{2} \cancel{(m+M)} V = \cancel{(m+M)} g h$$

or $$V = \sqrt{2gh} \qquad (2)$$

Using equation (2) in equation (1), we get:

$$mv = (m+M)\sqrt{2gh}$$

or $$v = \left(\frac{m+M}{m}\right)\sqrt{2gh}$$

$$\therefore \quad v = \frac{(4.5\times10^{-3}\text{ kg})+(1.5\text{ kg})}{(4.5\times10^{-3}\text{ kg})}\sqrt{2(9.8\text{ m/s}^2)(0.08\text{ m})} = 420\text{ m/s}$$

39. Let m_1 be the mass of the cart with unknown mass and m_2 be the mass of the cart with known mass.

From conservation of momentum:

$$m_1v_1 = m_1v_1' + m_2v_2' \qquad (1)$$

From conservation of KE:

$$\frac{1}{2}m_1v_1^2 = \frac{1}{2}m_1v_1'^2 + \frac{1}{2}m_2v_2'^2 \qquad (2)$$

Rewriting equation (2):

$$m_1(v_1^2 - v_1'^2) = m_2v_2'^2$$

or $$m_1(v_1 - v_1')(v_1 + v_1') = m_2v_2'^2 \qquad (3)$$

Rewriting equation (1):

$$m_1(v_1 - v_1') = m_2v_2' \qquad (4)$$

Equation (3) divided by equation (4) gives:

$$v_1 + v_1' = v_2' \quad \text{or} \quad v_1' = v_2' - v_1$$

$$\therefore \quad v_1' = 1.4\text{ m/s }\mathbf{i} - 2.8\text{ m/s }\mathbf{i} = -1.4\text{ m/s }\mathbf{i}$$

Since the cart with the known mass is travelling with a smaller velocity than the cart with the unknown mass, and since the cart with the unknown mass is deflected back, then the cart with known mass has the larger mass.

41. $m_1 = 1\text{ Mg}, \quad v_1 = 30\text{ km/h}$ (eastward)

$m_2 = 8\text{ Mg}, \quad v_2 = 20\text{ km/h}$ (southward)

This is an inelastic collision, and since momentum is conserved,

momentum before collision = momentum after collision

$$m_1 v_1 = M v_x \quad \text{conservation of P in x–direction}$$

$$(1\ \text{Mg})(30\ \text{km/h}) = (9\ \text{Mg}) v_x \tag{1}$$

$$m_2 v_2 = M v_y \quad \text{conservation of P in y–direction}$$

$$(8\ \text{Mg})(20\ \text{km/h}) = (9\ \text{Mg}) v_y \tag{2}$$

From equation (1), $v_x = \frac{1}{9}(30\ \text{km/h})$ (3)

From equation (2), $v_y = \frac{1}{9}(160\ \text{km/h})$ (4)

From equations (3) and (4), $v = \sqrt{v_x^2 + v_y^2}$

or $v = \sqrt{\frac{1}{81}\left[(30\ \text{km/h})^2 + (160\ \text{km/h})^2\right]} = 18.1\ \text{km/h}$

$$\theta = \tan^{-1}\frac{v_y}{v_x} = \tan^{-1}\left[\frac{160\ \text{km/h}}{30\ \text{km/h}}\right] = 79.4° \text{ south of east}$$

43. $m_A = 2\ \text{kg}, \quad m_B = 3\ \text{kg}, \quad m_C = 1\ \text{kg}$

$v'_A = 3\ \text{m/s}\ \mathbf{i}, \quad v'_B = -1.0\ \text{m/s}\ \mathbf{j}$

Momentum is conserved, so

momentum before explosion = momentum after explosion

x–direction: $0 = m_A v'_A + m_C v'_C$

$$0 = (2\ \text{kg})(3\ \text{m/s}) + (1\ \text{kg})\, v'_C \cos\theta \tag{1}$$

y–direction: $0 = m_B v'_B + m_C v'_C$

$$0 = (3\ \text{kg})(-1.0\ \text{m/s}) + (1\ \text{kg})\, v'_C \sin\theta \tag{2}$$

where θ is the angle of v'_C above the x–axis.

From equation (1), $v'_C \cos\theta = -6\ \text{m/s}$ (3)

From equation (2), $v'_C \sin\theta = 3\ \text{m/s}$ (4)

Dividing equation (4) by equation (3), we get:

$$\tan\theta = -0.5$$

$$\theta = 26.6° \text{ north of west}$$

Using this value of θ in equation (4), we get:

$$v'_C = 6.71\ \text{m/s}$$

45. Let m_1 be the mass of the cue ball and m_2 be the mass of the numbered ball $[m_1 = m_2]$.

Momentum before collision = momentum after collision

(a) $$m_1 v_1 = m_1 v_1' + m_2 v_2' \quad (1)$$

Apply conservation of kinetic energy:

$$K_i = K_f$$

$$\frac{1}{2} m_1 v_1^2 = \frac{1}{2} m_1 v_1'^2 + \frac{1}{2} m_2 v_2'^2 \quad (2)$$

Equation (2) can be rearranged to give

$$m_1 (v_1^2 - v_1'^2) = m_2 v_2'^2$$

or $$m_1 (v_1 - v_1')(v_1 + v_1') = m_2 v_2'^2 \quad (3)$$

Equation (1) can be rearranged to give

$$m_1 (v_1 - v_1') = m_2 v_2' \quad (4)$$

Dividing equation (3) by equation (4), we get:

$$v_1 + v_1' = v_2'$$

or $$v_2' = (0.88 \text{ m/s}) + (0.23 \text{ m/s}) = 1.11 \text{ m/s},$$

where v_1 and v_1' are the velocities of the cue ball before and after collision and v_2' is the velocity of the numbered ball after collision.

(b) From equation (3), since $m_1 = m_2$,

$$v_1^2 = v_1'^2 + v_2'^2$$

When applied to the velocity vector triangle in Fig. 10–26, this expression is the Pythagorean theorem.

(c) Thus the angle between the sides of the triangle labelled v_1' and v_2' must be 90°.

47. $$v\left|\frac{dM}{dt}\right| = \text{m/s}\left|\frac{\text{kg}}{\text{s}}\right| = \text{kg m/s}^2 = \text{N}$$

49. $F = v_e \left| \frac{dM}{dt} \right| = M \frac{dv}{dt}$

$$\therefore \quad v_f - v_i = -v_e \int_{M_i}^{M_f} \frac{dM}{M}$$

$$= v_e \ln\left(\frac{M_i}{M_f}\right)$$

In this case, $v_f = v_e$ and $v_i = 0.$

Therefore, $\ln\left(\frac{M_i}{M_f}\right) = 1$

or $M_i = M_f(e)$

$$M_f = \frac{2000 \text{ kg}}{e} = 736 \text{ kg}$$

51. $\frac{dM}{dt} = 0.01$ kg/s

$v_e = 2 \times 10^3$ m/s

(a) The engine's thrust is

$$F_t = v_e \frac{dM}{dt} = (2 \times 10^3 \text{ m/s})(0.01 \text{ kg/s})$$

$$F_t = 20 \text{ N}$$

(b) $v = v_o + at,$

where $v_o = 0$ and $a = \frac{F_t}{M} = \frac{20 \text{ N}}{10{,}000 \text{ kg}} = 2 \times 10^{-3} \text{ m/s}^2$

Therefore, $2.0 \text{ m/s} = 0 + (2 \times 10^{-3} \text{ m/s}^2)t$

or $t = 1000$ s

(c) $\Delta M = \left(\frac{dM}{dt}\right)\Delta t$

or $\Delta M = (0.01 \text{ kg/s})(1000 \text{ s}) = 10 \text{ kg}$

53. Impulse = change in momentum

$$F\Delta t = \Delta(mv)$$

$$F = \frac{\Delta(m\,v)}{\Delta t}$$

$$200\text{ N} = \frac{n\,(0.03\text{ kg})(300\text{ m/s})}{\Delta t},$$

where $\frac{n}{\Delta t}$ = number of bullets per second

From the above equation,

$$\frac{n}{\Delta t} = \frac{200\text{ N}}{9\text{ kg m/s}} \approx 22\text{ bullets/s}$$

55. Momentum is conserved, so

$$m_1 v_1 = (m_1 + m_2)V$$

$$(6000\text{ kg})(2.8\text{ m/s}) = [(6000\text{ kg}) + (10{,}000\text{ kg})]\,V$$

$$V = 1.05\text{ m/s}$$

57. Velocity of cannon relative to ship = $V\mathbf{i}$

Velocity of cannonball relative to ship = $-v\,\mathbf{i}$

Muzzle velocity $\mathbf{v}_m = -(V + v)\,\mathbf{i}$

so $v_m = v + V$

Using conservation of momentum,

$$MV - mv = 0$$

or $$M(v_m - v) = mv$$

(a) $$v = \frac{Mv_m}{M + m}$$

(b) $$V = \frac{m\,v_m}{M + m}$$

When $M = 100\text{ kg}$ and $v_m = 100\text{ m/s}$,

(c) $$v = \frac{100\text{ kg}}{101\text{ kg}}(100\text{ m/s}) = 99\text{ m/s}$$

(d) $$V = \frac{1\text{ kg}}{101\text{ kg}}(100\text{ m/s}) = 0.99\text{ m/s}$$

59. $\frac{dM}{dt} = 10\text{ kg/s}, \quad v_w = 20\text{ m/s}$

The force exerted on the truck by the streaming water is

$$F_t = v_w \left|\frac{dM}{dt}\right|$$

$$F_t = (20\text{ m/s})(10\text{ kg/s}) = 200\text{ N}$$

Problems:

1. (a) The equation of a parabola with its vertex at the origin is $y = kx^2$.

When $x = \frac{w}{2}$, $y = h$,

so $\quad h = k\left(\frac{w}{2}\right)^2 \quad$ or $\quad k = \frac{4h}{w^2}$

Thus, $\quad y = \frac{4hx^2}{w^2}, \qquad x = \pm\frac{w}{2}\sqrt{\frac{y}{h}}$

(b) To find the volume

$$V = \tau\int_0^h w\sqrt{\frac{y}{h}}\,dy = \frac{\tau w}{\sqrt{h}}\cdot\frac{2}{3}y^{3/2}\Big|_0^h$$

$$= \frac{2}{3}hw\tau$$

(c) $$y_{cm}V = \tau\int_0^h y\left(w\sqrt{\frac{y}{h}}\right)dy = \frac{\tau w}{\sqrt{h}}\cdot\frac{2}{5}y^{5/2}\Big|_0^h$$

$$y_{cm}V = \frac{2}{5}h^2 w\tau$$

$$y_{cm} = \frac{\frac{2}{5}h^2 w\tau}{\frac{2}{3}hw\tau} = \frac{3}{5}h$$

3. $$dV = \pi R^2\,dy = \frac{\pi R_o^2}{h^2}y^2\,dy$$

(a) $R = ky$ and $R = R_o$ when $y = h$

Thus, $\quad R = \frac{R_o y}{h}$

(b) $$V = \int_0^h dV = \frac{\pi R_o^2}{h^2}\int_0^h y^2\,dy = \frac{1}{3}\pi R_o^2 h$$

$$\bar{y}V = \int_0^h y\,dV = \frac{\pi R_o^2}{h^2}\int_0^h y^3\,dy = \frac{1}{4}\pi R_o^2 h^2$$

(c) $$\bar{y} = \frac{\frac{1}{4}\pi R_o^2 h^2}{\frac{1}{3}\pi R_o^2 h} = \frac{3}{4}h$$

5. $$dV = AR_o\,d\theta, \qquad y = R_o \sin\theta$$

$$V = \int_0^\pi AR_o\,d\theta = \pi AR_o$$

$$y_{cm}V = \int_0^\pi y\,dV = AR_o^2\int_0^\pi \sin\theta\,d\theta$$

$$= AR_o^2(-\cos\theta)\Big|_0^\pi = 2AR_o^2$$

$$y_{cm} = \frac{2AR_o^2}{\pi AR_o} = \frac{2R_o}{\pi}$$

7. (a) The velocity of the block relative to the cart is directed along the slope,

so $$\mathbf{v}_{B/C} = v_{B/C}(\mathbf{i}\cos\phi - \mathbf{j}\sin\phi)$$

The velocity of the cart is $-\mathbf{i}V$. There are no external forces in the x–direction, so the total momentum in the x–direction is conserved:

$$-MV + m(v_{B/C}\cos\phi - V) = 0$$

so $$v_{B/C} = \frac{(M+m)V}{m\cos\phi}$$

Now $$v_B^2 = \mathbf{v}_B \cdot \mathbf{v}_B = (\mathbf{v}_{B/C} + \mathbf{V})\cdot(\mathbf{v}_{B/C} + \mathbf{V})$$

$$= (v_{B/C}\cos\phi - V)^2 + (v_{B/C}\sin\phi)^2$$

$$= v_{B/C}^2 - 2Vv_{B/C}\cos\phi + V^2$$

so
$$v_B^2 = V^2 - \frac{2(M+m)}{m} V^2 + \left(\frac{M+m}{m\cos\phi}\right)^2 V^2$$

$$= \frac{V^2}{m^2\cos^2\phi}\left(m^2\cos^2\phi - 2m(M+m)\cos^2\phi + (M+m)^2\right)$$

Conservation of energy gives

$$\frac{1}{2}MV^2 + \frac{1}{2}mv_B^2 = mgh$$

$$V^2\left\{\frac{1}{2}M + \frac{1}{2}\cdot\frac{\left(m^2\cos^2\phi - 2m(M+m)\cos^2\phi + (M+m)^2\right)}{m\cos^2\phi}\right\} = mgh$$

$$V^2 = \frac{2mgh}{M + \frac{1}{m\cos^2\phi}\left((m^2 - 2m(M+m))\cos^2\phi + (M+m)^2\right)}$$

$$= \frac{2m^2gh\cos^2\phi}{-(m^2 + Mm)\cos^2\phi + (M+m)^2}$$

$$= \frac{2m^2gh\cos^2\phi}{(M+m)(-m\cos^2\phi + M + m)}$$

$$= \frac{2m^2gh\cos^2\phi}{(M+m)(M + m\sin^2\phi)}$$

and

(b)
$$v_B^2 = \frac{2gh\left((M+m)^2 - (m^2 + 2mM)\cos^2\phi\right)}{(M+m)(M + m\sin^2\phi)}$$

$$= \frac{2gh\left(M^2 + 2Mm\sin^2\phi + m^2\sin^2\phi\right)}{(M+m)(M + m\sin^2\phi)}$$

9. Momentum before collision = momentum after collision

$$m_1v_{1i} + m_2v_{2i} = m_1v_{1f} + m_2v_{2f} \qquad (1)$$

KE before collision = KE after collision

$$\frac{1}{2}m_1v_{1i}^2 + \frac{1}{2}m_2v_{2i}^2 = \frac{1}{2}m_1v_{1f}^2 + \frac{1}{2}m_2v_{2f}^2 \qquad (2)$$

Equation (2) can be rearranged as follows:

$$m_1(v_{1i}^2 - v_{1f}^2) = m_2(v_{2f}^2 - v_{2i}^2)$$

$$m_1(v_{1i} - v_{1f})(v_{1i} + v_{1f}) = m_2(v_{2f} - v_{2i})(v_{2f} + v_{2i}) \qquad (3)$$

Equation (1) can be rearranged as follows:

$$m_1(v_{1i}-v_{1f}) = m_2(v_{2f}-v_{2i}) \qquad (4)$$

Dividing equation (3) by equation (4), we get:

$$v_{1i}+v_{1f} = v_{2f}+v_{2i}$$

or $$v_{1i}-v_{2i} = v_{2f}-v_{1f}$$

or $$v_{1i}-v_{2i} = -(v_{1f}-v_{2f})$$

11. $v_{1i} = (2.0\ \text{m/s})\mathbf{i}, \quad v_{2i} = (-1.0\ \text{m/s})\mathbf{i}$

(a) For $m_1 = m_2$, using equations given in Problem 10, we get:

$$v_{1f} = \frac{m_1-m_2}{m_1+m_2}v_{1i}+\frac{2m_2}{m_1+m_2}v_{2i}$$

or $$v_{1f} = \left(0+\frac{2m_2}{2m_2}\right)(-1.0\ \text{m/s})\mathbf{i} = -1.0\ \text{m/s}\ \mathbf{i}$$

$$v_{2f} = \frac{m_2-m_1}{m_2+m_1}v_{2i}+\frac{2m_1}{m_2+m_1}v_{1i}$$

or $$v_{2f} = \left(0+\frac{2m_1}{2m_1}\right)(2.0\ \text{m/s})\mathbf{i} = 2.0\ \text{m/s}\ \mathbf{i}$$

(b) $$m_1 = 2m_2$$

$$v_{1f} = \frac{m_2}{3m_2}v_{1i}+\frac{2m_2}{3m_2}v_{2i}$$

$$v_{1f} = \frac{2.0\ \text{m/s}}{3}\mathbf{i}+\frac{2}{3}(-1.0\ \text{m/s})\mathbf{i}$$

or $$v_{1f} = 0$$

$$v_{2f} = \frac{-m_2}{3m_2}(-1.0\ \text{m/s})\mathbf{i}+\frac{4m_2}{3m_2}(2.0\ \text{m/s})\mathbf{i}$$

$$v_{2f} = \frac{1}{3}\ \text{m/s}\ \mathbf{i}+\frac{8}{3}\ \text{m/s}\ \mathbf{i} = 3.0\ \text{m/s}\ \mathbf{i}$$

(c) $m_2 = 2m_1$

$$v_{1f} = \frac{-m_1}{3m_1}(2.0\ \text{m/s})\mathbf{i}+\frac{4m_1}{3m_1}(-1.0\ \text{m/s})\mathbf{i}$$

$$v_{1f} = -\frac{2}{3}\ \text{m/s}\ \mathbf{i}-\frac{4}{3}\ \text{m/s}\ \mathbf{i} = -2.0\ \text{m/s}\ \mathbf{i}$$

$$v_{2f} = \frac{m_1}{3m_1}(-1.0\text{ m/s})\mathbf{i} + \frac{2m_1}{3m_1}(2.0\text{ m/s})\mathbf{i}$$

$$v_{2f} = -\tfrac{1}{3}\text{ m/s }\mathbf{i} + \tfrac{4}{3}\text{ m/s }\mathbf{i} = 1.0\text{ m/s }\mathbf{i}$$

13. Average velocity $= \dfrac{\text{Distance}}{\text{Time}}$

or $\bar{v} = \dfrac{s}{t}$

$$\frac{v_o + v_f}{2} = \frac{s}{t}$$

$$\frac{200\text{ m/s} + 0}{2} = \frac{0.01\text{ m}}{t}$$

or $t = \dfrac{0.01\text{ m}}{100\text{ m/s}} = 1 \times 10^{-4}\text{ s}$

To find the block's vertical rise, we use the law of conservation of energy:

$$\text{KE (at the end of t)} = \text{PE at the rise}$$

$$\tfrac{1}{2}(m + M)V^2 = (m + M)gh$$

or $V^2 = 2gh$

$$h = \frac{V^2}{2g} = \frac{(1\text{ m/s})^2}{2(9.8\text{ m/s}^2)} = 0.051\text{ m}$$

15. (a) In the single-stage rocket,

$$M = 11000\text{ kg}$$

$$M_p = 200\text{ kg}$$

$$M_f = 9700\text{ kg}$$

$$M_e = 1100\text{ kg}$$

$$\Delta v = v_f - v_i = v_e \ln\left(\frac{M_i}{M_{final}}\right)$$

$$= v_e \ln\left(\frac{11000\text{ kg}}{1300\text{ kg}}\right)$$

$$= 2.14\, v_e$$

(b) In the two-stage rocket,

$$M_{f1} = 9000 \text{ kg} \qquad M_{f2} = 700 \text{ kg}$$

$$M_{e1} = 1000 \text{ kg} \qquad M_{e2} = 100 \text{ kg}$$

$$\Delta v_1 = v_e \ln\left(\frac{11000 \text{ kg}}{2000 \text{ kg}}\right)$$

$$= 1.71\, v_e$$

$$\Delta v_2 = v_e \ln\left(\frac{1000 \text{ kg}}{300 \text{ kg}}\right)$$

$$= 1.20\, v_e$$

$$\Delta v_{TOTAL} = 2.91\, v_e$$

So the two-stage rocket has the larger velocity when the fuel is exhausted.

17. The initial kinetic energy of object 1 is:

$$KE_{1i} = \frac{1}{2} m_1 v_{1i}^2$$

The final kinetic energy of object 2 is:

$$KE_{2f} = \frac{1}{2} m_2 v_{2f}^2$$

$$\frac{KE_{2f}}{KE_{1i}} = \frac{\frac{1}{2} m_2 v_{2f}^2}{\frac{1}{2} m_1 v_{1i}^2} \qquad (1)$$

From conservation of momentum, we get:

$$m_1 v_{1i} = m_1 v_{1f} + m_2 v_{2f} \qquad (2)$$

since $\quad v_{1i} = v_{2f} - v_{1f} \qquad (3)$

From equations (2) and (3), we get:

$$v_{1f} = \left(\frac{m_1 - m_2}{m_1 + m_2}\right) v_{1i}$$

$$v_{2f} = \left(\frac{2 m_2}{m_1 + m_2}\right) v_{1i} \qquad (4)$$

Substituting equation (4) in equation (1), we get:

$$\frac{KE_{2f}}{KE_{1i}} = \frac{\frac{1}{2} m_2 \left[\left(\frac{2 m_2}{m_1 + m_2} \right) v_{1i} \right]^2}{\frac{1}{2} m_1 v_{1i}^2}$$

$$\frac{KE_{2f}}{KE_{1i}} = \frac{4 m_1 m_2}{(m_1 + m_2)^2}$$

19. $v = 4.8\ \text{m/s}, \quad \frac{dm}{dt} = 12\ \text{kg/s}$

$$F_t = v \left(\frac{dm}{dt} \right) = (4.8\ \text{m/s})(12\ \text{kg/s}) = 57.6\ \text{N}$$

CHAPTER 11 STATIC EQUILIBRIUM OF A RIGID BODY

Questions:

1. No body is perfectly rigid, but some are more rigid than others. Examples of "rigid" bodies are rocks, steel beams, and concrete floors. Examples of nonrigid bodies are rubber bands, string, or plastic rulers.

3. This post can be considered rigid for most applications.

5. An unequal arm balance can be used to compare weights because it would be in balance when the moments are equal; that is, when $W_1L_1 = W_2L_2$. Thus if W_2 is unknown, it can be compared to a known weight W_1, with $W_2 = \frac{W_1L_1}{L_2}$.

7. Taking distances from the midpoint, $W_1L_1 = W_2L_2$. Thus,

$$L_2 = \frac{W_1L_1}{W_2} = \frac{(20\text{ kg})(9.8\text{ m/s}^2)(1.5\text{ m})}{W_2} = \frac{290\text{ N m}}{W_2}$$

Uncle could press with a force of 290 N at a distance 1.0 m away from the midpoint, or with 580 N at 0.50 m, or any other combination of force and distance that satisfies this relation.

9. This procedure does give an accurate measure of the truck's weight: Let the distance between the front and rear wheels be L and let the center of gravity of the truck be at a distance x_0 from the front wheels. When the rear wheels are on the scale, the truck is supported by the front wheels and the moments about the front wheels balance when

$$W(L - x_0) = W_RL$$

where W_R is the force exerted by the scale on the rear wheels. When the front wheels are on the scale,

$$Wx_0 = W_FL,$$

where W_F is the force exerted by the scale on the front wheels. Adding,

$$WL = (W_R + W_F)L$$

or $$W = W_R + W_F$$

11. If the wires are vertical, the sum of the tensions will equal the weight of the object. If the wires are not vertical, the tensions will be pulling against each other in the horizontal direction, while the sum of their vertical components is still equal to the weight of the object being supported. Thus the tensions will be greater in this case.

13. It is possible for **A** × **B** to be zero without either vector being zero. This will happen if **A** and **B** are parallel to each other.

15. If a body is in translational equilibrium, the sum of the forces acting on the body is zero, and the sum of the torques acting on the body is also zero. From the first condition, the center of mass of the body must be at rest or in uniform motion (thus there is an inertial frame in which the center of mass is at rest). If the net torque is zero, it will be seen in Chapter 13 that the angular momentum of the body is constant or zero. Thus, it is possible for a body to be in translational equilibrium but not to be in static equilibrium because, at the very least, it could be rotating about its center of mass with constant angular momentum.

17. A planet whose diameter is not negligible when compared to the distance between the planet and the star that it is orbiting is an example of a situation where the center of gravity and the center of mass do not coincide.

19. The condition $\sum \tau = 0$ can be satisfied for any choice of axes as follows:

If $\sum x_i F_i = 0$, then shifting the origin gives

$$\sum (x_i' + d)F_i = 0$$

so

$$\sum x_i' F_i = -d \sum F_i$$

which is zero if $\sum F_i = 0$; that is, if the body is in static equilibrium.

21.

Symbol	Represents	Type	SI units
τ_x	torque or moment	component	Nm
$\vec{\tau}$	torque	vector	Nm
$\sum \vec{\tau}_{ext}$	net external torque	vector	Nm
A × **B**	cross product	vector	—
$r_\perp$	perpendicular distance	component	m

Exercises:

1. The torques are:

(a) $-(65\ \text{N})(0.120\ \text{m}) = -7.8\ \text{Nm}$

(b) 0

(c) $+(65\ \text{N})(0.120\ \text{m}) = 7.8\ \text{Nm}$

(d) 0

3. (a) The force is $(2.0\ \text{kg})(9.8\ \text{m/s}^2) = 19.6\ \text{N}$

(b) Assuming the distance between the hand and the nose when the arm is extended horizontally to be 1.0 m, then the torque is $(1.0\ \text{m})(19.6\ \text{N}) \simeq 20\ \text{Nm}$.

5. (a) $W_1 d_1 = W_2 d_2$

$$\therefore \quad d_2 = \frac{W_1 d_1}{W_2} = \frac{m_1 d_1}{m_2} \quad \text{(provided } g \neq 0\text{)}$$

$$= \left(\frac{478\ \text{g}}{642\ \text{g}}\right)(89.1\ \text{cm} - 50\ \text{cm}) = 29.1\ \text{cm from midpoint,}$$

which is at the 20.9 cm mark.

(b) The force exerted by the knife edge is

$$(0.642\ \text{kg} + 0.478\ \text{kg} + 0.325\ \text{kg})\, 9.8\ \text{m/s}^2$$

$$= 14.2\ \text{N}$$

(c) On the moon, the 642 g mass would be suspended from the same point, but the force exerted by the knife edge would change because of the different value of g.

In interstellar space, g = 0, so none of the masses have weight, so they would not be "suspended" from the meter stick nor would the meter stick be "supported" by the knife edge.

7.

$$F_1 d_1 = F_2 d_2$$

$$F(1.25\ \text{m}) = 2000\ \text{N}(0.25\ \text{m})$$

$$F = 400\ \text{N}$$

9. (a) The cable can only sustain a tension along its length. Thus, a free-body diagram of the boom is as follows:

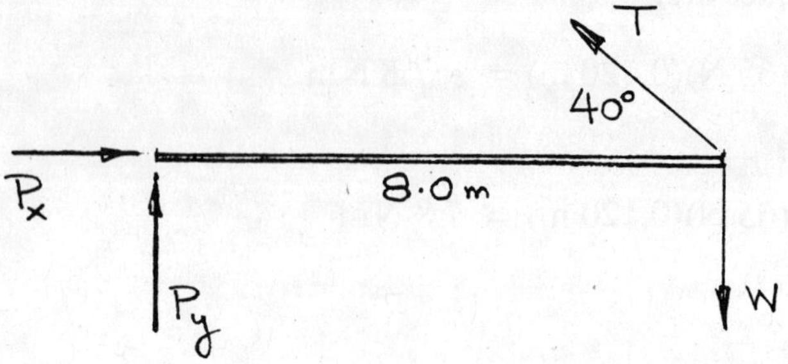

Taking horizontal and vertical components of the forces and setting $\sum F_x = 0$ and $\sum F_y = 0$,

$$P_x = T\cos 40°$$

$$P_y = W - T\sin 40°$$

and setting the moment about P equal to zero,

$$8\,T\sin 40° - 8\,W = 0$$

Therefore, $$T = \frac{W}{\sin 40°} = \frac{20.0\text{ kN}}{\sin 40°} = 31.1\text{ kN}$$

$$P_x = (31.1\text{ kN})\cos 40° = 23.8\text{ kN}$$

$$P_y = 20.0\text{ kN} - (31.1\text{ kN})\sin 40° = 0\text{ kN}$$

(b) The direction of the pin force is horizontal.

11. A free-body diagram of the truck is as follows:

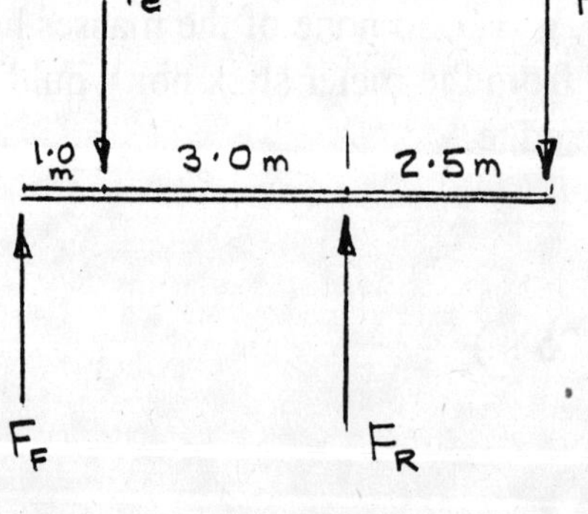

with $F_e = 30\text{ kN}$ and $F_L = 20\text{ kN}$

(a) Taking moments about the front wheel,

$$F_R(4.0\text{ m}) = F_e(1.0\text{ m}) + F_L(6.5\text{ m})$$

$$F_R = \frac{1}{4.0\text{ m}}(30\text{ kN}\cdot\text{m} + 130\text{ kN}\cdot\text{m}) = 40\text{ kN}$$

Now, $F_e + F_L = F_F + F_R$, so $F_F = 10\text{ kN}$

(b) Suppose that F_L is increased until $F_F = 0$. Then, taking moments about the rear wheel,

$$F_L(2.5\text{ m}) = F_e(3.0\text{ m})$$

$$F_L = \frac{30\text{ kN}(3.0\text{ m})}{2.5\text{ m}} = 36\text{ kN}$$

Since F_F cannot be negative, any larger value of F_L will cause the front wheels to leave the ground.

13. A free-body diagram of the boom is as follows:

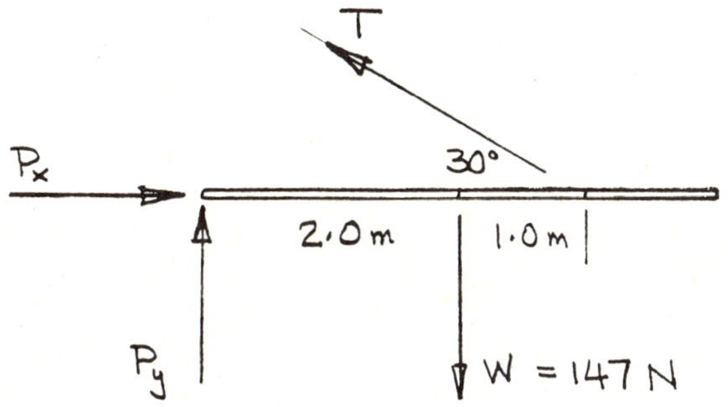

The equilibrium equations are:

$$P_x = T\cos 30°$$

$$P_y + T\sin 30° = W$$

and (taking moments about P),

$$(2.0\text{ m})W = (3.0\text{ m})T\sin 30°$$

Thus,

(a) $$T = \frac{(2.0\text{ m})(147\text{ N})}{(3.0\text{ m})\sin 30°} = 196\text{ N}$$

(b) $$P_x = (196\text{ N})\cos 30° = 170\text{ N}$$

(c) $$P_y = 147\text{ N} - (196\text{ N})\sin 30° = 49\text{ N}$$

15.

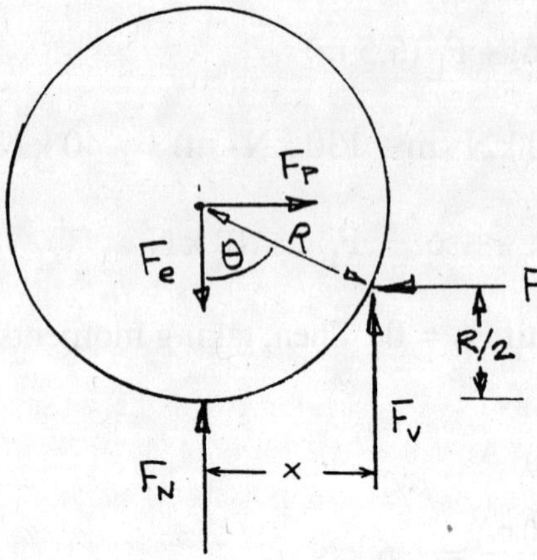

The free-body diagram is as shown. From the diagram,

$$R\cos\theta = \frac{R}{2} \Rightarrow \cos\theta = \frac{1}{2}, \quad \theta = 60^\circ$$

$$x = R\sin\theta = 0.866\,R$$

The equilibrium equations are:

$$\sum F_x = 0 \Rightarrow F_P = F_H$$

$$\sum F_y = 0 \Rightarrow F_e = F_N + F_v$$

and, taking moments about the corner,

$$\sum \tau = 0 \Rightarrow F_N(x) - F_e(x) + F_P\left(\frac{R}{2}\right) = 0$$

Thus,

(a) $F_N = F_e - F_P\left(\frac{R}{2x}\right)$

$$= F_e - \frac{F_e}{3}\left(\frac{1}{1.73}\right) = 0.808\,F_e$$

(b) $F_H = F_P = \frac{F_e}{3} = 0.333\,F_e$

$$F_v = F_e - F_N = F_e - 0.808\,F_e = 0.192\,F_e$$

(c) The force exerted by the corner is

$$\mathbf{F} = F_H\,\mathbf{i} + F_v\,\mathbf{j} = (-0.333\,\mathbf{i} + 0.192\,\mathbf{j})F_e$$

$$= 0.385\ F_e \quad \text{at an angle of } 150°$$

This shows that the force exerted by the corner is directed toward the center of the disk.

17. The free-body diagram is as shown:

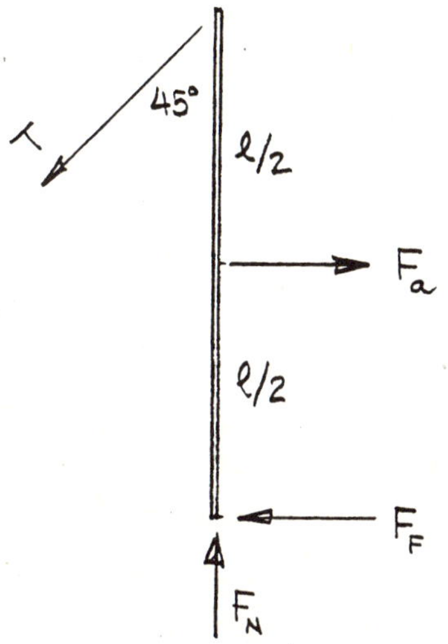

(a) The equilibrium equations are

$$F_N = T\cos 45°$$

$$F_F + T\sin 45° = F_a$$

$$F_a\left(\frac{\ell}{2}\right) = T\sin 45°\ (\ell)$$

Thus,

$$T = \frac{F_a}{2\sin 45°} = 0.707\ F_a$$

$$F_F = F_a - (0.707\ F_a)\sin 45° = 0.5\ F_a$$

$$F_N = (0.707\ F_a)\cos 45° = 0.5\ F_a$$

(b) Since $F_F \le \mu_s F_N$, $\mu_s \ge 1$

19. Breaking the structure into two pieces, with the origin at the base,

Piece	Weight W (N)	Center x (m)	Center y (m)	W (x) (Nm)	W (y) (Nm)
Vertical	500	0	2.5	0	1250
Horizontal	300	1.5	5.0	450	1500
Totals	800			450	2750

Thus,
$$x_{cm} = \frac{450\ \text{N m}}{800\ \text{N}} = 0.56\ \text{m}$$

$$y_{cm} = \frac{2750\ \text{N m}}{800\ \text{N}} = 3.44\ \text{m}$$

21. A free-body diagram for the block is as shown:

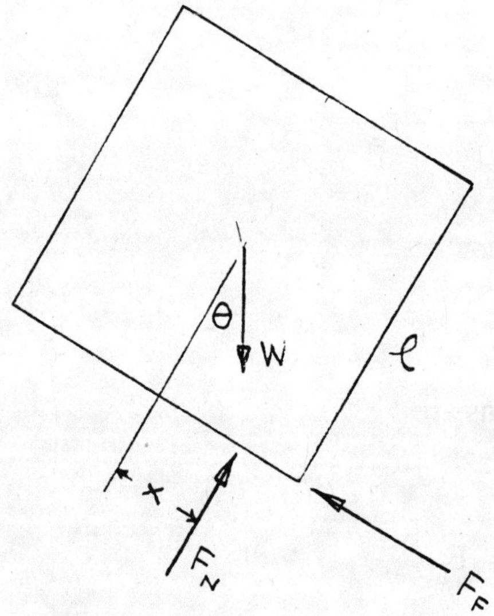

The equilibrium equations are:

$$F_F = W \sin\theta$$

$$F_N = W \cos\theta$$

and, taking moments about the center,

$$x\, F_N = \frac{\ell}{2} F_F$$

Here x is the distance along the bottom from the center to the point of application of F_N.

Thus, $\tan\theta = \frac{F_F}{F_N} = \frac{2x}{\ell}$

But $x \le \frac{\ell}{2}$, so $\tan\theta \le 1$ and $\theta \le 45°$,

so the block tips over if $\theta > 45°$.

23. $\vec{\tau} = \mathbf{r} \times \mathbf{F}$

$= (1.35\text{ m}\,\mathbf{i} - 2.22\text{ m}\,\mathbf{j}) \times (174\text{ N}\,\mathbf{i} + 203\text{ N}\,\mathbf{j} - 166\text{ N}\,\mathbf{k})$

$= (1.35\text{ m})(203\text{ N})\,\mathbf{k} - (1.35\text{ m})(-166\text{ N})\,\mathbf{j}$

$- (-2.22\text{ m})(174\text{ N})\,\mathbf{k} + (-2.22\text{ m})(-166\text{ N})\,\mathbf{i}$

$= (369\,\mathbf{i} + 224\,\mathbf{j} + 660\,\mathbf{k})\text{Nm}$

25. The corners of the tetrahedron are:

A	(0, 0, 0)
B	(1, 0, 0)
C	(1, 1, 0)
D	(1, 1, 1)

where all dimensions are in meters.

Identical tetrahedra can be formed with corners at A, B, D, and E (1, 0, 1), and with corners at A, D, E, and F (0, 0, 1).

ABDE shares triangle ABD with ABCD and ED = BC, BE = CD, AE = CA so they are equal.

Likewise, ADEF shares triangle ADE with ABDE and BE = AF, BA = EF, and EF = BA, so they are equal.

The three tetrahedra form half of the parallelepiped formed by AB, AC, and AD. From Exercise 24, the volume of the parallelepiped is

$$\vec{AB} \cdot (\vec{AC} \times \vec{AD})$$

and thus the volume of the tetrahedron is

$$V = \frac{1}{6}\mathbf{i} \cdot \left[(\mathbf{i} + \mathbf{j}) \times (\mathbf{i} + \mathbf{j} + \mathbf{k})\right]$$

$$= \frac{1}{6}\mathbf{i} \cdot (\mathbf{i} - \mathbf{j}) = \frac{1}{6}\text{ m}^3$$

27. (a) Expanding the determinant,

$$\mathbf{D} = \begin{vmatrix} \mathbf{i} & \mathbf{j} & \mathbf{k} \\ A_x & A_y & A_z \\ B_x & B_y & B_z \end{vmatrix}$$

$$= \mathbf{i}\,(A_yB_z - A_zB_y) + \mathbf{j}\,(A_zB_x - A_xB_z) + \mathbf{k}\,(A_xB_y - A_yB_x)$$

$$= \mathbf{A} \times \mathbf{B}$$

(b)

$$\begin{vmatrix} \mathbf{i} & \mathbf{j} & \mathbf{k} \\ 1 & 2 & 3 \\ 2 & -1 & -1 \end{vmatrix} = \mathbf{i}\,(-2+3) + \mathbf{j}\,(6+1) + \mathbf{k}\,(-1-4)$$

$$= \mathbf{i} + 7\,\mathbf{j} - 5\,\mathbf{k}$$

29. Let $\mathbf{D} = a\mathbf{A} + b\mathbf{B}, \quad \mathbf{C} = \mathbf{A} \times \mathbf{B}$

If **D** is perpendicular to **C**, then $\mathbf{D} \cdot \mathbf{C} = 0$

$$\mathbf{D} \cdot \mathbf{C} = (a\mathbf{A} + b\mathbf{B}) \cdot (\mathbf{A} \times \mathbf{B})$$

$$= a\mathbf{A} \cdot (\mathbf{A} \times \mathbf{B}) + b\mathbf{B} \cdot (\mathbf{A} \times \mathbf{B})$$

$$= a\mathbf{B} \cdot (\mathbf{A} \times \mathbf{A}) + b\mathbf{A} \cdot (\mathbf{B} \times \mathbf{B})$$

but $\mathbf{A} \times \mathbf{A} = 0$ and $\mathbf{B} \times \mathbf{B} = 0$, so $\mathbf{D} \cdot \mathbf{C} = 0$ as desired.

Geometrically, **D** is a vector in the plane defined by **A** and **B**, while **C** is perpendicular to that plane.

31. (a) With the origin at the midpoint,

$$\tau_{z1} = \left(-\tfrac{d}{2}\right)(-F_1) = 5.0\ \text{Nm}$$

$$\tau_{z2} = \left(\tfrac{d}{2}\right)(F_2) = 5.0\ \text{Nm}$$

$$\tau_z = 10.0\ \text{Nm}$$

(b) With the origin at point 1,

$$\tau_{z1} = 0$$

$$\tau_{z2} = dF_2 = 10.0\ \text{Nm}$$

$$\tau_z = 10.0\ \text{Nm}$$

(c) With the origin at point 2,

$$\tau_{z1} = (-d)(-F_1) = 10.0\ \text{Nm}$$

$$\tau_{z2} = 0$$

$$\tau_z = 10.0\ \text{Nm}$$

33. A free-body diagram for this case is:

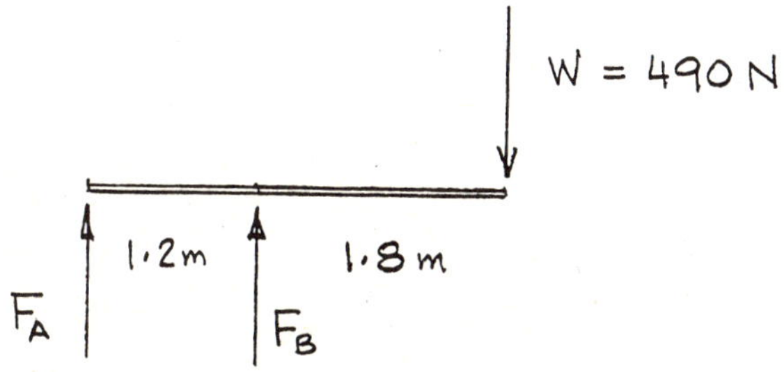

(a) The moment exerted by the diver about A is

$$(-3.0\ \text{m})\,W = -1470\ \text{Nm}$$

(b) Taking moments about A,

$$(1.2\ \text{m})\,F_B - (3.0\ \text{m})\,W = 0$$

$$\therefore \quad F_B = \frac{(3.0\ \text{m})(490\ \text{N})}{1.2\ \text{m}} = 1225\ \text{N}$$

(c) Since $F_A + F_B = W$, $\quad F_A = 490\ \text{N} - 1225\ \text{N} = -735\ \text{N}$

35. A free-body diagram in this case is:

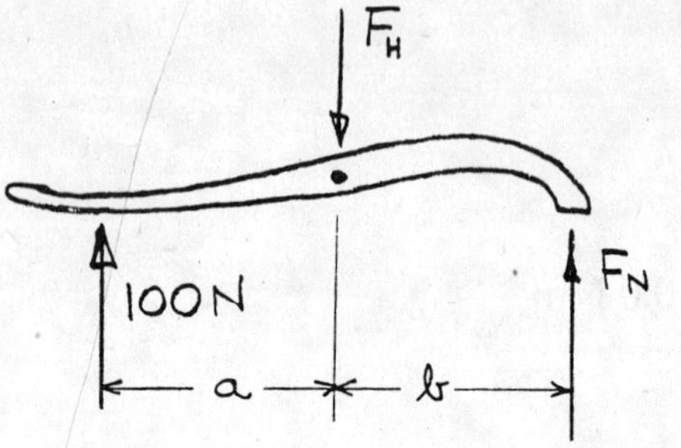

In the drawing,

$$a = 18.5 \text{ mm}$$

$$b = 9.0 \text{ mm}$$

Taking moments about the hinge,

$$100\,a = F_N(b)$$

$$\therefore \quad F_N = 100 \text{ N}\left(\frac{18.5 \text{ mm}}{9.0 \text{ mm}}\right) \simeq 200 \text{ N}$$

37. When the crate is about to tip over, the reaction is located at the lower right-hand corner. Taking moments about the corner,

$$F_a H = W(0.3 \text{ m})$$

$$H = \frac{(40 \text{ kg})(9.8 \text{ m/s}^2)(0.3 \text{ m})}{300 \text{ N}} = 0.39 \text{ m}$$

39. From the equilibrium conditions,

$$\sum F_x = 0 \quad \Rightarrow \quad F_{Px} = F_2 - F_1 = 0$$

$$\sum F_y = 0 \quad \Rightarrow \quad F_{Py} = F_3 = 2F$$

Thus, $\mathbf{F_P} = 0\,\mathbf{i} + 2F\,\mathbf{j}$

(a) $F_P = |\mathbf{F_P}| = 2F$

(b) $\mathbf{F_P}$ is along the positive y–axis

(c) Taking moments about the center,

$$x\,F_{Py} = F_1 a$$

$$\therefore \quad x = \frac{aF_1}{F_{Py}} = \frac{aF}{2F} = \frac{a}{2}$$

41. Let $\mathbf{r} = x\,\mathbf{i} + y\,\mathbf{j}$

$$\vec{\tau} = \mathbf{r} \times \mathbf{F} = (x\,\mathbf{i} + y\,\mathbf{j}) \times (60\text{ N}\,\mathbf{i} + 80\text{ N}\,\mathbf{j})$$

$$= ((80\text{ N})x - (60\text{ N})y)\mathbf{k} = 24\text{ Nm}\,\mathbf{k}$$

Thus any combination of x and y such that $(80\text{ N})x - (60\text{ N})y = 24\text{ Nm}$ will satisfy this condition. For example,

$$x = \left(\frac{24}{80}\right)\text{m}, \quad y = 0 \quad \text{or} \quad x = 0, \quad y = \left(\frac{24}{60}\right)\text{m}$$

are possible answers.

Problems:

1. $\vec{\tau} = \mathbf{r}_1 \times \mathbf{F}_1 + \mathbf{r}_2 \times \mathbf{F}_2,$ but $\mathbf{F}_2 = -\mathbf{F}_1,$

thus $\vec{\tau} = (\mathbf{r}_1 - \mathbf{r}_2) \times \mathbf{F}_1 = (\mathbf{r}_1 - \mathbf{r}_2)_\perp \times \mathbf{F}_1,$

where $(\mathbf{r}_1 - \mathbf{r}_2)_\perp$ indicates the component perpendicular to $\mathbf{F}_1$. Thus the magnitude of τ is $d\,F_1$, regardless of the choice of origin.

3. When the block is about to tip over, F_N acts through the lower right-hand corner of the block. Taking moments about that point,

$$F_e\left(\frac{L}{2}\right) = F_P(L)$$

Also, $F_N = F_e$ and $F_{fs} = F_P$

Thus, $F_{fs} = \frac{F_e}{2}$, and since

$$\frac{F_{fs}}{F_N} \le \mu_s, \quad \mu_s \ge \frac{1}{2}$$

If $\mu_s < \frac{1}{2}$, the block will slide rather than tip.

5. The free-body diagrams for this problem are:

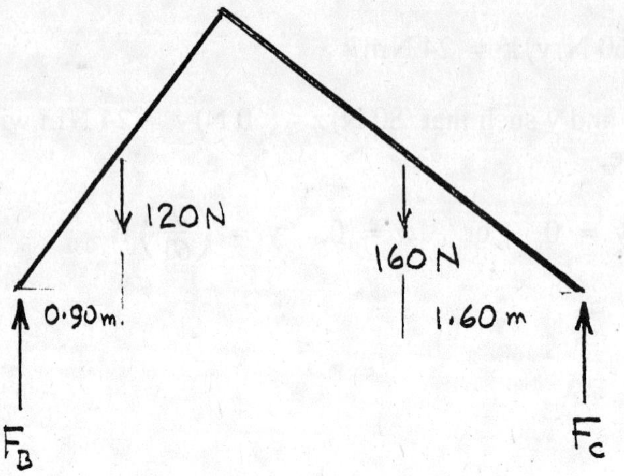

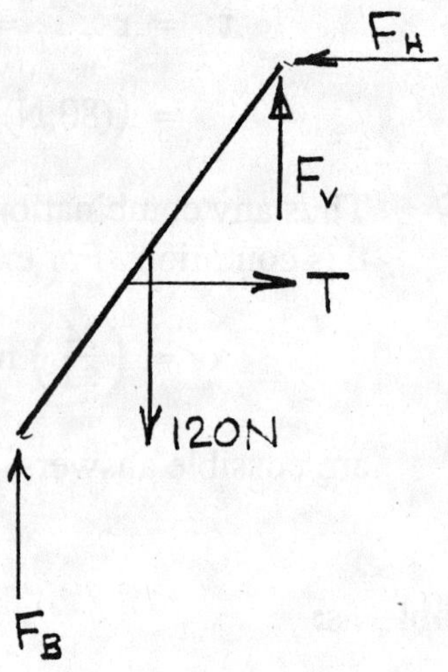

The first drawing is for the overall system, the second for the left member. The third drawing shows the geometry. All of the triangles are similar, with the corresponding sides in ratios of 3:4:5.

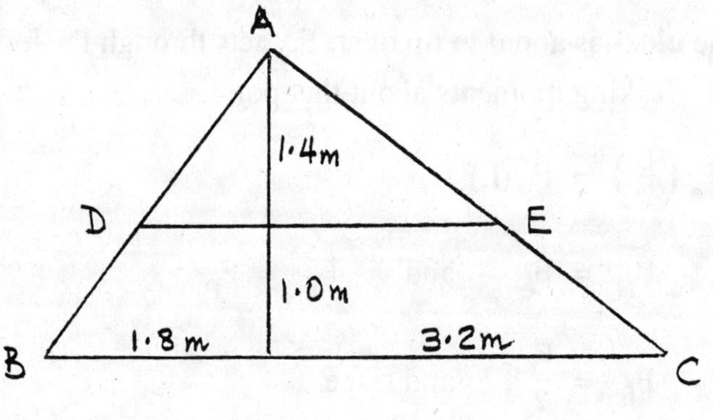

(a) Taking moments about B for the whole structure,

$$(5\text{ m})F_C = 120\text{ N}(0.90\text{ m}) + 160\text{ N}(1.8\text{ m} + 1.6\text{ m})$$

so $$F_C = 130\text{ N}$$

and since $F_B + F_C = 280\text{ N}$, $F_B = 150\text{ N}$

(b) Taking moments about A in the second free-body diagram,

$$T(1.4\text{ m}) + 120\text{ N}(0.90\text{ m}) = F_B(1.8\text{ m})$$

so $$T = \frac{(150\text{ N})(1.8\text{ m}) - (120\text{ N})(0.90\text{ m})}{1.4\text{ m}}$$

$$= 115\text{ N}$$

(c) The horizontal and vertical force equations are

$$F_H = T \quad \text{and} \quad F_V = 120\text{ N} - F_B,$$

so $$F_H = 115\text{ N} \quad \text{and} \quad F_V = -30\text{ N}$$

Thus, the force exerted by the 4 m board on the 3 m board is

$$\mathbf{F} = (115\,\mathbf{i} - 30\,\mathbf{j})\text{N}$$

or $$F = 119\text{ N} \quad \text{at an angle} \quad \theta = -14.6°$$

7. The geometry of the crane is as follows:

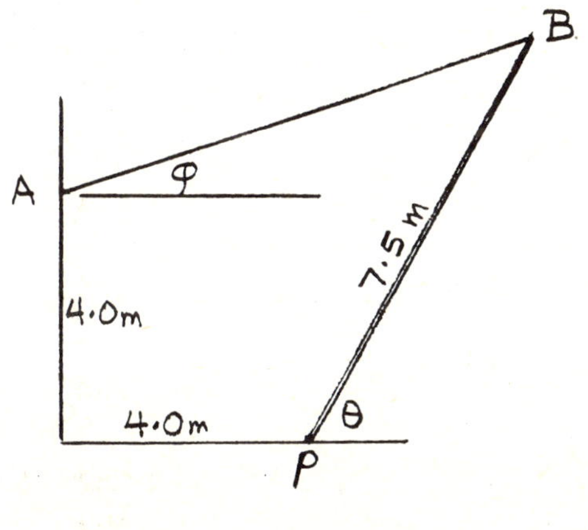

$$AB\cos\phi = 4\text{ m} + (7.5\text{ m})\cos\theta$$

$$AB\sin\phi = (7.5\text{ m})\sin\theta - 4\text{ m}$$

Thus, $$\tan\phi = \frac{(7.5\text{ m})\sin\theta - 4\text{ m}}{(7.5\text{ m})\cos\theta + 4\text{ m}}$$

The free-body diagram for the boom is:

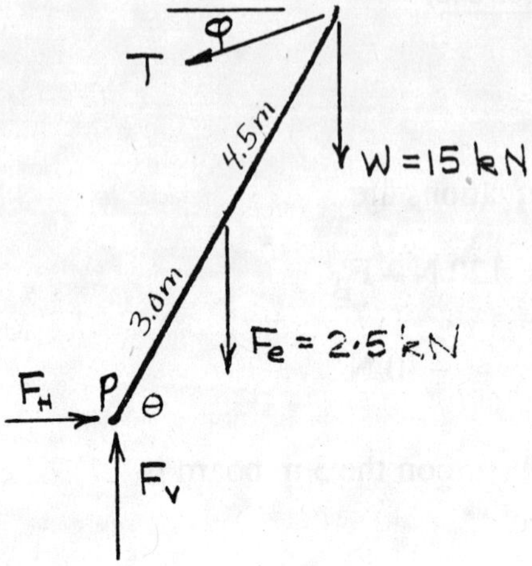

Note that T ≠ W because of friction at the pulley.

Taking moments about P:

$$(T\cos\phi)(7.5\text{ m}\sin\theta) - (W + T\sin\phi)(7.5\text{ m}\cos\theta)$$

$$= F_e(3.0\text{ m})\cos\theta$$

Thus, $T(7.5\text{ m})\sin(\theta - \phi) = W(7.5\text{ m})\cos\theta + F_e(3.0\text{ m})\cos\theta$

or $T = \dfrac{[(2.5\times10^3\text{ N})(3.0\text{ m}) + (15\times10^3\text{ N})(7.5\text{ m})]\cos\theta}{(7.5\text{ m})\sin(\theta-\phi)}$

Summing the horizontal forces,

$$F_H = T\cos\phi$$

and summing the vertical forces,

$$F_V = F_e + W + T\sin\phi$$

(a) When $\theta = 30°$, $\tan\phi = -0.0238$, so $\phi = -1.36°$

Then $T = 26.6\text{ kN}$, $F_H = 26.6\text{ kN}$, and $F_V = 16.9\text{ kN}$

(b) When $\theta = 60°$, $\tan\phi = 0.322$, so $\phi = 17.8°$

Then $T = 38.0\text{ kN}$, $F_H = 36.2\text{ kN}$, and $F_V = 29.1\text{ kN}$

In (a) the cable is practically horizontal and the vertical force at the pin balances the two weights.

In (b) the cable is pulling down, so the pin needs to support the vertical component of the tension as well.

9. Geometrically (see Exercise 24), both $\mathbf{A} \cdot (\mathbf{B} \times \mathbf{C})$ and $\mathbf{C} \cdot (\mathbf{A} \times \mathbf{B})$ represent the volume of the same parallelepiped.

Thus, $$\mathbf{A} \cdot (\mathbf{B} \times \mathbf{C}) = \mathbf{C} \cdot (\mathbf{A} \times \mathbf{B}) = (\mathbf{A} \times \mathbf{B}) \cdot \mathbf{C}$$

Alternatively, using determinants, the first product is represented by the determinant

$$\begin{vmatrix} A_x & A_y & A_z \\ B_x & B_y & B_z \\ C_x & C_y & C_z \end{vmatrix}$$

By two row interchanges, the third row is brought to the top and the other two are pushed down, giving the determinant representation of the second product. By the rules for determinants, the value of the two determinants is the same. Thus, the two products are the same.

11. Since Cavendish's experiment would not work if the spheres are physically touching, assume that the distance between adjacent spheres is 5.6 cm, allowing for a 1 mm gap. The force between them is

$$F_{near} = \frac{G m_1 m_2}{r^2} = \frac{(6.67 \times 10^{-11}\ \mathrm{N\,m^2/kg^2})(0.25\ \mathrm{kg})(2.0\ \mathrm{kg})}{(0.056\ \mathrm{m})^2}$$

$$= 1.06 \times 10^{-8}\ \mathrm{N}$$

The equation for the attraction to the more distant sphere has $r = 15$ cm,

so $$F_{distant} = 1.48 \times 10^{-9}\ \mathrm{N}$$

This force, while not negligible in comparison, is directed nearly parallel to the rod, while the other is perpendicular to the rod. The net component perpendicular to the rod, taking into account the 5.6 cm separation, is

$$F_{perp} = F_{near} - \frac{5.6\ \mathrm{cm}}{((5.6\ \mathrm{cm})^2 + (15\ \mathrm{cm})^2)^{1/2}} F_{distant}$$

$$= 1.01 \times 10^{-8}\ \mathrm{N}$$

and $$\tau = dF_{perp} = (0.15\ \mathrm{m})(1.01 \times 10^{-8}\ \mathrm{N})$$

$$= 1.5 \times 10^{-9}\ \mathrm{Nm}$$

13. Weight of door $= 25\text{ kg}(9.8\text{ m/s}^2) = 245\text{ N}$

Weight of child $= 5\text{ kg}(9.8\text{ m/s}^2) = 49\text{ N}$

Total weight $= 294\text{ N}$

Each of the two hinges supports half of this weight, namely 147 N.

Assume that the hinges cannot support a torque perpendicular to the plane of the door. (This is not a trivial assumption. A hinge is designed to yield to a torque along its axis, that is, along the hinge pin. Here we are assuming that it cannot support a torque at right angles to the hinge pin.)

(a) Taking moments about the lowest hinge,

$$F_{M_x}(1.0\text{ m}) + (245\text{ N})(1.0\text{ m}) + (49\text{ N})(1.9\text{ m}) = 0$$

$$\therefore \quad F_{M_x} = -338\text{ N}$$

Also, since $F_{M_x} + F_{L_x} = 0$,

$$F_{L_x} = +338\text{ N}$$

Also $F_{M_y} = F_{L_y} = 147\text{ N}$

(b) Again taking moments about the lowest hinge,

$$F_{U_x}(2.0\text{ m}) + 338\text{ N m} = 0$$

$$\therefore \quad F_{U_x} = -169\text{ N}$$

and $F_{L_x} = 169\text{ N}$

Also $F_{U_y} = F_{L_y} = 147\text{ N}$

15. The free-body diagram is as shown. Taking moments about P,

$$W(3.0\text{ m}) + (3.92\text{ kN})(1.5\text{ m}) = (16\text{ kN})(2.0\text{ m})$$

$$W = 8.7\text{ kN}$$

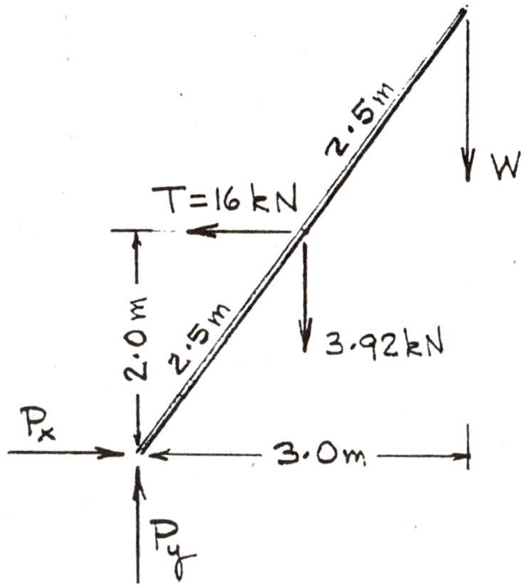

17.

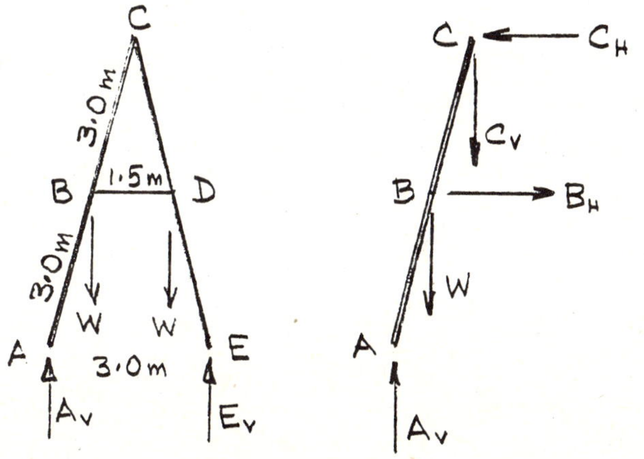

The free-body diagrams are as shown.

(a) From the diagram for the entire system, taking moments about E,

$$A_V(3.0\text{ m}) = W(.75\text{ m}) + W(2.25\text{ m})$$

$$\therefore \quad A_V = W = 150\text{ N}$$

The floor is smooth, so the force exerted by the floor on the ladder has no horizontal component and the vertical component is A_V, 150 N upward.

(b) From the free-body diagram for the left side of the ladder, taking moments about point C,

$$B_H(2.91\text{ m}) + W(.75\text{ m}) = A_V(1.5\text{ m})$$

so $$B_H = \frac{(150\text{ N})(1.5\text{ m}) - (150\text{ N})(.75\text{ m})}{2.91\text{ m}}$$

$$= 38.7\text{ N}$$

The strap can only exert a force along its length, so the force exerted by the strap on the ladder at B has a horizontal component of 38.7 N towards the right and zero vertical component.

(c) Applying the equilibrium conditions to member ABC,

$$\sum F_x = 0 \quad \Rightarrow \quad B_H = C_H, \quad \text{so} \quad C_H = 38.7\text{ N}$$

and

$$\sum F_y = 0 \quad \Rightarrow \quad C_V = A_V - W = 0$$

so the force on ABC applied by CDE at C is horizontal and it is 38.7 N towards the left.

(d) The tension on the strap pulls on both B and D equally so the force due to the strap at D is horizontal and equal to 38.7 N towards the left.

(e) From the diagram for the entire structure, taking moments about A,

$$E_V(3.0\text{ m}) = W(.75\text{ m}) + W(2.25\text{ m})$$

$$\therefore \quad E_V = \frac{W(.75\text{ m} + 2.25\text{ m})}{3.0\text{ m}} = W$$

$$= 150\text{ N}$$

The floor at E exerts a force with no horizontal component and an upward component of 150 N.

19.

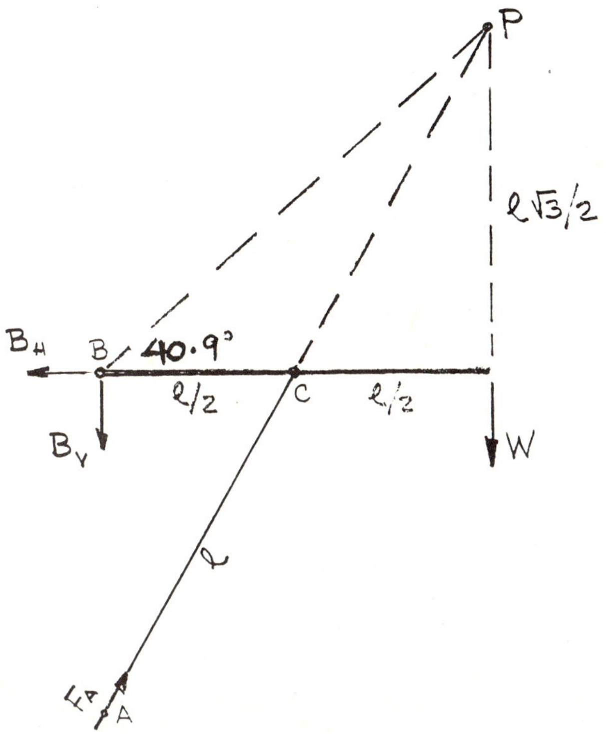

Assuming that beam AC is pinned to the wall at A and to the other boom at C, then the force exerted by the wall on AC must be along its length. Likewise, assume that B is also a pin connection so that no torque is exerted by the wall at B (as might be the case if BC cantilevered out). Then the free-body diagram is as shown.

Extend the lines of $\mathbf{F}_A$ and $\mathbf{W}$ so they meet at P. Taking moments about P, it is clear that neither $\mathbf{F}_A$ nor $\mathbf{W}$ have moments about that point, so $\mathbf{B}$ must also pass through that point. The forces then form the triangle shown. Using the sine rule,

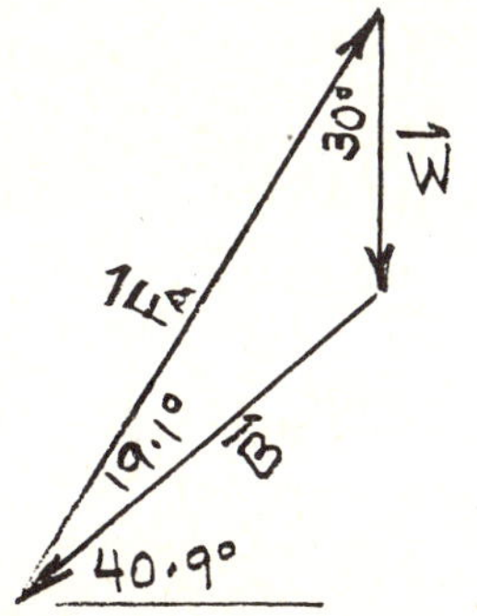

$$\frac{W}{\sin 19.1^\circ} = \frac{B}{\sin 30^\circ} = \frac{F_A}{\sin 49.1^\circ}$$

so $\quad F_A = 2.31\ W$ at an angle of 60°

$\quad B = 1.53\ W$ at an angle of 229°

21. Assume that the hinges do not exert torques (see discussion in Problem 13). Then the free-body diagram is as shown.

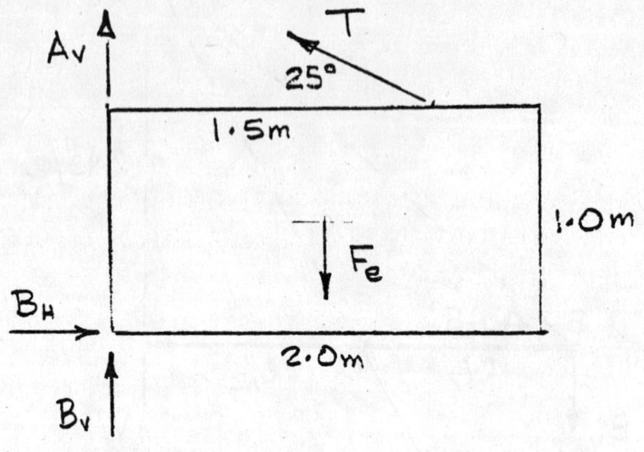

Taking moments about B,

$$(T \sin 25°)(1.5\text{ m}) + (T \cos 25°)(1.0\text{ m}) = F_e(1.0\text{ m})$$

so
$$T = \frac{(480\text{ N})(1.0\text{ m})}{(1.5\text{ m})\sin 25° + (1.0\text{ m})\cos 25°} = 312\text{ N}$$

Now,
$$B_H = T\cos 25° = 282\text{ N}$$

and
$$A_V + B_V = F_e - T\sin 25°$$
$$= 348\text{ N}$$

CHAPTER 12 ROTATION I

Questions:

1. Radians should be used when it is necessary to determine arc length from an angle and also when taking derivatives of trigonometric functions. Degrees or radians can be used for geometry or trigonometry.

3. The path is a circle or an arc of a circle.

5. If a rigid body is in purely translational motion, then every point in the body has the same velocity.

7. All of the points along a line passing through the center of mass and parallel to the axis of rotation would have the same velocity as the center of mass.

9. I would hold the dime and move it until it just covers the moon and measure the distance from my eyes to the dime and the diameter of the dime. The diameter of the dime divided by the distance is the angle (in radians) subtended by the moon at the earth. Radian measure requires less arithmetic and is easier to use in this application.

11. The direction of the angular velocity is along the axis of rotation and following the right-hand rule. The velocity of a point in the rotating body is perpendicular to both the axis and the line joining the point to the axis.

13. The angular velocities point into the wall and the angular accelerations point out of the wall.

15. The angular velocity is toward the driver's left; whereas the angular acceleration is toward the driver's right.

17. The hinges should be placed on the heavier side. That way the larger mass gets multiplied by the square of the smaller distance and the moment of inertia is less.

19. By the parallel axis theorem,

$$I_A = I_C + Md^2$$

Since M and d^2 are positive, $I_A > I_C$. Thus, it is not possible to find an axis A for which the moment of inertia is smaller than that about an axis parallel to A but passing through the center of mass C.

21. Since it is desired to maximize the translational kinetic energy, the moment of inertia of the wheels should be as small as possible. If the moment of inertia is fixed, then increasing the mass of the car will mean that the initial energy is larger. For a given forward speed the energy in the wheels is fixed; thus the larger potential energy will have to increase the translational energy.

23. The speed is the same for all three, since the same increase in total kinetic energy occurs in each case. For a rolling ball, there is a fixed ratio between the rotational energy and translational energy, so the rotation of the ball does not affect the answer.

25.

Symbol	Represents	Type	SI units
θ	angle	scalar	radians
$\vec{\omega}$	angular velocity	vector	radians/s
ω_x	angular velocity component	component	rad/s
α_x	angular acceleration component	component	rad/s^2
v_t	tangential velocity	component	m/s
a_t	tangential acceleration	component	m/s^2
a_R	radial acceleration	component	m/s^2
I	moment of inertia	tensor	kg • m^2

Exercises:

1. (a) $\frac{2}{3}\pi$ rad (b) 120° (c) $\frac{1}{3}$ rev

3. (a) $s = r\theta = (0.84\text{ m})\left(\frac{35}{180}\pi\text{ rad}\right) = 0.51\text{ m}$

(b) $s = r\theta = (0.84\text{ m})(0.61\text{ rad}) = 0.51\text{ m}$

5. $s = r\theta = (1.5\times10^{11}\text{ m})\left(2\pi\frac{1}{365}\text{ rad}\right)$

$= 2.6\times10^{9}\text{ m}$

7. $\omega = 78\ \frac{\text{rev}}{\text{min}} \times \frac{1\ \text{min}}{60\ \text{s}} \times 2\pi\ \frac{\text{rad}}{\text{rev}}$

$= 8.17\ \text{rad/s}$

9. $\theta(t) = -(1.4\ \text{rad/s}^3)t^3 + (6.8\ \text{rad})$

(a) $\omega_x(t) = \frac{d\theta}{dt} = -(4.2\ \text{rad/s}^3)t^2$

(b) $\omega_x(2.1\ \text{s}) = -(4.2\ \text{rad/s}^3)(2.1\ \text{s})^2 = -19\ \text{rad/s}$

(c) $\omega = 19\ \text{rad/s}$.

(d) $\theta(2.1\ \text{s}) = -(1.4\ \text{rad/s}^3)(2.1\ \text{s})^3 + (6.8\ \text{rad})$

$= -6.2\ \text{rad}$

11. Referring to Example 12–2, a turntable rotates clockwise, so with the coordinate system chosen, the angular velocity is in the –**k** direction.

$$\omega = 33\ \frac{\text{rev}}{\text{min}} \times \frac{1\ \text{min}}{60\ \text{s}} \times 2\pi\ \frac{\text{rad}}{\text{rev}} = 3.5\ \text{rad/s}$$

$$\theta(t) = \omega_z t + \theta_o = (-3.5\ \text{rad/s})\,t$$

13. $\omega_z(t) = (5.8\ \text{rad/s}) - (2.2\ \text{rad/s}^2)t$

(a) $\alpha_z = \frac{d\omega_z}{dt} = -2.2\ \text{rad/s}$

(b) $\theta(t) = \theta_o + \int_0^t \omega\ dt$

$= (5.8\ \text{rad/s})t - (1.1\ \text{rad/s}^2)t^2$

(c) $\omega_z = 0\ \text{rad/s}$ when $t = \frac{5.8\ \text{rad/s}}{2.2\ \text{rad/s}^2} = 2.6\ \text{s}$

(d) For $t < t_q$, v_{top} is towards the north and for $t > t_q$ it is towards the south.

(e) $\omega_x^2 = \omega_o^2 + 2\alpha(\theta - \theta_o)$

$\omega_x^2 = (5.8\ \text{rad/s})^2 - (4.4\ \text{rad/s}^2)\theta$

15. If one faces toward the north when looking at the door from the inside, then the hinge is on the left and an inward opening door is opening clockwise as viewed from above.

Thus, $\omega_z = -1.4 \text{ rad/s}$ when $\theta = 0.72 \text{ rad}$

and $\omega_z = 0$ and $\theta = 0$ when $t = 0$

Since α_z is constant,

$$\omega_z^2 - \omega_{zo}^2 = 2\alpha_z(\theta - \theta_o)$$

$$\alpha_z = \frac{\omega_z^2}{2\theta} = \frac{(1.4 \text{ rad/s})^2}{2(-0.72 \text{ rad})} = -1.4 \text{ rad/s}^2$$

(a) $$\omega_z(t) = \omega_{zo} + \alpha_z t$$

$$= -(1.4 \text{ rad/s}^2)\, t$$

(b) $$\theta(t) = \theta_o + \omega_{zo} t + \frac{1}{2}\alpha_z t^2$$

$$= -(0.72 \text{ rad/s}^2)\, t^2$$

(c) $$\omega_z^2 = \omega_{zo}^2 + 2\alpha_z(\theta - \theta_o)$$

$$= -(2.8 \text{ rad/s}^2)\, \theta$$

17. $$v = r\omega = (0.093 \text{ m})\left(\frac{1 \text{ rev}}{\text{min}} \times \frac{1 \text{ min}}{60 \text{ s}} \times \frac{2\pi \text{ rad}}{\text{rev}}\right)$$

$$= 0.0097 \text{ m/s}$$

19. (a) $$a_t = \alpha r = (0.18 \text{ rad/s}^2)(1.4 \text{ m}) = 0.25 \text{ m/s}^2$$

$$a_R = \omega^2 r = (0.34 \text{ rad/s})^2(1.4 \text{ m}) = 0.16 \text{ m/s}^2$$

(b) $$v = \omega r = (0.34 \text{ rad/s})(1.4 \text{ m}) = 0.48 \text{ m/s}$$

$$a = \sqrt{a_t^2 + a_R^2} = 0.30 \text{ m/s}^2$$

21. (a) $\left[\frac{1}{2} I\omega^2\right] = [ML^2]\left[\frac{1}{T^2}\right] = \left[\frac{ML^2}{T^2}\right]$ = energy

(b) In SI units, $[I] = kg \cdot m^2$, $[\omega]$ = rad/s

$\therefore \quad \left[\frac{1}{2} I\omega^2\right] = kg \cdot m^2/s^2$ = joules

23. Given $K = 29$ J when $\omega = 13$ rad/s,

$$K = \frac{1}{2} I\omega^2, \quad \text{or} \quad I = \frac{2K}{\omega^2}$$

$$I = \frac{2\,(29\text{ J})}{(13\text{ rad/s})^2} = 0.34\text{ kg} \cdot \text{m}^2$$

25. (a) $I_x = \sum m_i y_i^2$

$$= 3.0\text{ kg}(1.5\text{ m})^2 + 2.0\text{ kg}(1.5\text{ m})^2 + 2.5\text{ kg}(1.5\text{ m})^2 + 1.5\text{ kg}(1.5\text{ m})^2$$

$$= 20\text{ kg} \cdot \text{m}^2$$

(b) $I_y = \sum m_i x_i^2$

$$= 3.0\text{ kg}(1.0\text{ m})^2 + 2.0\text{ kg}(1.0\text{ m})^2 + 2.5\text{ kg}(1.0\text{ m})^2 + 1.5\text{ kg}(1.0\text{ m})^2$$

$$= 9.0\text{ kg} \cdot \text{m}^2$$

(c) $r = \sqrt{x^2 + y^2} = \sqrt{(1.0\text{ m})^2 + (1.5\text{ m})^2} = 1.80\text{ m}$

$$I_z = \sum m_i r_i^2 = 9.0\text{ kg}(1.80\text{ m})^2$$

$$= 29\text{ kg} \cdot \text{m}^2 = I_x + I_y$$

27. $I = \frac{1}{3} ML^2 = \frac{1}{3}(0.88 \times 10^3\text{ kg/m}^3)(0.042\text{ m})(2.1\text{ m})(1.1\text{ m})^3$

$$= 34\text{ kg} \cdot \text{m}^2$$

29. The triangle is contained within the lines y = 0, z = 0, and $z = b\left(1 - \frac{y}{a}\right)$. Let the thickness be t. Then

$$I = \rho t \int_0^a y^2 b\left(1 - \frac{y}{a}\right) dy$$

$$= \rho b t\left[\frac{1}{3} y^3 - \frac{1}{4} \cdot \frac{y^4}{a}\right]_0^a$$

$$= \frac{1}{12} \rho b t a^3$$

Now, $M = \frac{1}{2} \rho b t a,$ so $I = \frac{1}{6} M a^2$

31. From Table 12–2, $I_C = \frac{2}{5} M r_o^2$.

Using the parallel axis theorem,

$$I = I_C + M\left(\frac{r_o}{2}\right)^2 = \left(\frac{2}{5} + \frac{1}{4}\right) M r_o^2$$

$$= \frac{13}{20} M r_o^2$$

33. (a) $v = \omega r, \quad \omega = \frac{v}{r} = \frac{1.4 \text{ m/s}}{0.340 \text{ m}} = 4.1 \text{ rad/s}$

(b) $v = \omega(2r) = 2.8 \text{ m/s}$

35. (a) Using conservation of energy,

$$m g y_o = \frac{1}{2} m v^2 + \frac{1}{2} I \omega^2$$

$$= \frac{1}{2}\left(m + \frac{I}{r_2^2}\right) v^2$$

From Table 12–2,

$$I = \frac{1}{2} m\left(r_1^2 + r_2^2\right)$$

so

$$v^2 = \frac{2 g y_o}{1 + \frac{I}{m r_2^2}}$$

$$= \frac{2 g y_0}{1 + \frac{1}{2}\left(1 + \frac{r_1^2}{r_2^2}\right)}$$

$$= \frac{2\,(9.8\ \text{m/s}^2)(6.7\ \text{m} \sin 12^\circ)}{1 + \frac{1}{2}\left(1 + \frac{(75\ \text{m})^2}{(96\ \text{m})^2}\right)}$$

$$= 2.26\ \text{m}^2/\text{s}^2$$

$$v = \sqrt{2.26\ \text{m}^2/\text{s}^2} = 1.50\ \text{m/s}$$

(b) $\omega = \frac{v}{r_2} = \frac{1.50\ \text{m/s}}{0.096\ \text{m}} = 15.7\ \text{rad/s}$

Note that the mass of the object does not affect these results.

37. Estimate the mass of the frisbee as 0.2 kg, the radius as 0.1 m, its initial speed as 10 m/s and its angular velocity at 100 rpm (~ 10 rad/s). Since most of the mass is around the rim, estimate the moment of inertia as $0.002\ \text{kg} \cdot \text{m}^2$. Thus the kinetic energy of translation is

$$K_T = \frac{1}{2} m v^2 = \frac{1}{2}(.2\ \text{kg})(10\ \text{m/s})^2 = 10\ \text{J}$$

and that of rotation is

$$K_R = \frac{1}{2} I \omega^2 = \frac{1}{2}(.002\ \text{kg} \cdot \text{m}^2)(10\ \text{rad/s})^2 = 0.1\ \text{J}$$

Thus, in this estimate, only .01 of its total energy is due to rotation.

39. Given $\omega_0 = 39\ \text{rad/s}, \quad \alpha = -0.63\ \text{rad/s}^2$

(a) $\omega(t) = \omega_0 + \alpha t, \quad$ so $\omega = 0$

when $t = -\frac{\omega_0}{\alpha} = \frac{39\ \text{rad/s}}{0.63\ \text{rad/s}^2} = 62\ \text{s}$

(b) $\theta(t) = \theta_0 + \omega_0 t + \frac{1}{2}\alpha t^2$

$$\theta(62\ \text{s}) = (39\ \text{rad/s})(62\ \text{s}) + \frac{1}{2}(-0.63\ \text{rad/s}^2)(62\ \text{s})^2$$

$$= 1200\ \text{rad}$$

41. $\theta(t) = 4.2\text{ rad} - (2.9\text{ rad/s})t + (0.31\text{ rad/s}^3)t^3$

(a) $\omega_z(t) = \dfrac{d\theta}{dt} = -2.9\text{ rad/s} + (0.93\text{ rad/s}^3)t^2$

(b) $\alpha_z(t) = \dfrac{d\omega_z}{dt} = (1.86\text{ rad/s}^3)t$

43. (a) If we regard the earth as a particle and ignore the rotation of the earth about its own axis, then the angular speed of the earth is its angular speed around the sun, with a period of 1 year; that is,

$$\omega = \frac{2\pi\text{ rad}}{1\text{ yr}} \times \frac{1\text{ yr}}{365\text{ days}} \times \frac{1\text{ day}}{24\text{ h}} \times \frac{1\text{ h}}{3600\text{ s}} = 2.0\times 10^{-7}\text{ rad/s}$$

(b) $v = \omega R = (2.0\times 10^{-7}\text{ rad/s})(1.5\times 10^{11}\text{ m}) = 30\times 10^3\text{ m/s}$

(c) $a = \dfrac{v^2}{R} = \dfrac{(30\times 10^3\text{ m/s})^2}{1.5\times 10^{11}\text{ m}} = 0.0060\text{ m/s}^2$

45. Approximating the propeller as a uniform rod,

$$I = \frac{1}{12} m\ell^2,$$

so $$K = \frac{1}{24} m\ell^2\omega^2$$

$$= \frac{1}{24}(35\text{ kg})(3.2\text{ m})^2\left(2\pi \times \frac{1000}{60\text{ s}}\right)^2$$

$$= 164\text{ kJ}$$

47. $\omega = \dfrac{v}{r} = \dfrac{2.0\text{ m/s}}{r}$

(a) $\mathbf{v}_T = \vec{\omega} \times \mathbf{r}_T = \dfrac{-(2.0\text{ m/s})}{r}\,\mathbf{k} \times 2r\,\mathbf{j}$

$= 4.0\,\mathbf{i}\text{ m/s}$

(b) $\mathbf{v}_F = \vec{\omega} \times \mathbf{r}_F = \dfrac{-(2.0\text{ m/s})}{r}\,\mathbf{k} \times (r\,\mathbf{i} + r\,\mathbf{j})$

$= 2.0\text{ m/s}\,\mathbf{i} - 2.0\text{ m/s}\,\mathbf{j}$

49. Using conservation of energy,

$$mgh = \frac{1}{2}mv^2 + \frac{1}{2}I\omega^2$$

Since the cylinder rolls without slipping, $v = \omega r$.

From Table 12–2,

$$I = \frac{1}{2}\,m\left(r_1^2 + r_2^2\right) = \frac{1}{2}\,m\left(\left(\frac{r}{2}\right)^2 + r^2\right)$$

$$= \frac{5}{8}\,mr^2$$

Thus, $$mgh = \frac{1}{2}\,mv^2 + \frac{5}{16}\,mv^2 = \frac{13}{16}\,mv^2$$

(a) $$v = \sqrt{\frac{16\,gh}{13}}$$

(b) $K_T = \frac{1}{2}\,mv^2$, $K_R = \frac{5}{16}\,mv^2$,

so $\frac{5}{13}$ of the total kinetic energy is rotational.

Problems:

1. (a) For rotation with constant angular acceleration,

$$\theta(t) = \theta_o + \omega_o t + \frac{1}{2}\,\alpha t^2$$

Now, $\theta(0\text{ s}) = 0.0\text{ rad}$, $\theta(1.0\text{ s}) = -3.8\text{ rad}$,

and $\theta(2.0\text{ s}) = -5.0\text{ rad}$

$$\therefore \quad 0 = \theta_o$$

$$-3.8\text{ rad} = \omega_o(1.0\text{ s}) + \frac{1}{2}\,\alpha(1.0\text{ s})^2$$

$$-5.0\text{ rad} = \omega_o(2.0\text{ s}) + \frac{1}{2}\,\alpha(2.0\text{ s})^2$$

Thus, $$\omega_o = \frac{-3.8\text{ rad}}{1.0\text{ s}} - \frac{1}{2}\,\alpha(1.0\text{ s})$$

and $$\omega_o = \frac{-5.0\text{ rad}}{2.0\text{ s}} - \frac{1}{2}\,\alpha(2.0\text{ s})$$

$$\therefore \quad \frac{1}{2}\,\alpha(2.0\text{ s}^2 - 1.0\text{ s}^2) = -2.5\text{ rad/s} + 3.8\text{ rad/s}$$

$$\alpha = \frac{2\,(1.3\text{ rad/s})}{1.0\text{ s}} = 2.6\text{ rad/s}^2$$

$$\omega_o = -3.8\text{ rad/s} - \frac{1}{2}\,(2.6\text{ rad/s}^2)(1.0\text{ s})$$

$$= -5.1\text{ rad/s}$$

(b) Hence, $\omega_z(t) = -5.1 \text{ rad/s} + (2.6 \text{ rad/s}^2)\,t$

and $\theta(t) = (-5.1 \text{ rad/s})\,t + (1.3 \text{ rad/s}^2)\,t^2$

3. The earth rotates about the sun once a year,

so $$\omega_{sun} = \frac{2\pi}{T} = \frac{2\pi}{1 \text{ year}} \times \frac{\text{year}}{365 \text{ days}} \times \frac{\text{day}}{24 \text{ h}} \times \frac{\text{h}}{3600 \text{ s}}$$

$$= 1.99 \times 10^{-7} \text{ rad/s}$$

The earth rotates about its own axis with a period of 24 h

$$\omega_{rot} = \frac{2\pi}{24 \text{ h}} \times \frac{\text{h}}{3600 \text{ s}} = 7.27 \times 10^{-5} \text{ rad/s}$$

Thus, $$K_{orb} = \frac{1}{2} M_e v^2 = \frac{1}{2} M_e \omega_{sun}^2 R_{orbit}^2$$

$$= \frac{1}{2}(6.0 \times 10^{24} \text{ kg})(1.99 \times 10^{-7} \text{ rad/s})^2 (1.5 \times 10^{11} \text{ m})^2$$

$$= 2.7 \times 10^{33} \text{ J}$$

$$K_{rot} = \frac{1}{2} I \omega_{rot}^2 \quad \text{where} \quad I = \frac{2}{5} M_e R_e^2$$

so $$K_{rot} = \frac{1}{5} M_e R_e^2 \omega_{rot}^2$$

$$= \frac{1}{5}(6.0 \times 10^{24} \text{ kg})(6.4 \times 10^{6} \text{ m})^2 (7.27 \times 10^{-5} \text{ rad/s})^2$$

$$= 2.6 \times 10^{29} \text{ J}$$

Thus, $$\frac{K_{rot}}{K_{orb}} \sim 10^{-4}$$

5. (a) Reproducing I from Table 12–2,

Object	**I**	$k = \sqrt{\frac{I}{M}}$
Thin cylinder	MR_o^2	R_o
Thin cylinder	$\frac{1}{2} MR_o^2 + \frac{1}{12} ML^2$	$\sqrt{\frac{1}{2} R_o^2 + \frac{1}{12} L^2}$
Hollow sphere	$\frac{2}{3} MR_o^2$	$R_o \sqrt{\frac{2}{3}}$
Solid cylinder	$\frac{1}{2} MR_o^2$	$\frac{R_o}{\sqrt{2}}$
Solid cylinder	$\frac{1}{4} MR_o^2 + \frac{1}{12} ML^2$	$\sqrt{\frac{1}{4} R_o^2 + \frac{1}{12} L^2}$
Solid sphere	$\frac{2}{5} MR_o^2$	$R_o \sqrt{\frac{2}{5}}$
Thick-walled cylinder	$\frac{1}{2} M\left(R_1^2 + R_2^2\right)$	$\sqrt{\frac{1}{2} \left(R_1^2 + R_2^2\right)}$
Rod	$\frac{1}{12} ML^2$	$L \sqrt{\frac{1}{12}}$
Rectangular plate	$\frac{1}{12} M(a^2 + b^2)$	$\sqrt{\frac{a^2 + b^2}{12}}$

(b) $v = \sqrt{\dfrac{2gh}{1 + \dfrac{k^2}{R^2}}}$

7. (a) $I_x = \int \rho x^2 \, dV = \rho t \int_0^{2\pi} \int_0^R (r \cos \theta)^2 r \, dr \, d\theta$

$$= \rho t \left(\frac{1}{4} R^4\right)(\pi) = \frac{1}{4} MR^2$$

(b) By symmetry, $I_x = I_y$

and by the plane figure theorem,

$$I_z = I_x + I_y = 2 I_x$$

Thus, $I_x = \frac{1}{2} I_z = \frac{1}{2}\left(\frac{1}{2} MR^2\right) = \frac{1}{4} MR^2$

9. From example 12–9,

$$Mgh = \frac{1}{2}Mv^2 + \frac{1}{2}I\frac{v^2}{R^2}$$

Thus,

$$I = \frac{(2Mgh - Mv^2)R^2}{v^2}$$

$$= MR^2\left(\frac{2gh}{v^2} - 1\right)$$

$$= (0.175\text{ kg})(.026\text{ m})^2\left(\frac{2(9.8\text{ m/s}^2)(.130\text{ m})}{(1.3\text{ m/s})^2} - 1\right)$$

$$= 6.01\times 10^{-5}\text{ kg}\bullet\text{m}^2$$

11. (a) Using conservation of energy,

$$mgh = \frac{1}{2}mv^2 + \frac{1}{2}I\omega^2,$$

where $\omega = \frac{v}{R_o}$ and $I = \frac{1}{2}MR_o^2$

Thus,

$$mgh = \frac{1}{2}mv^2 + \frac{1}{2}I\frac{v^2}{R_o^2}$$

$$v^2 = \frac{2gh}{1 + \frac{M}{2m}}$$

so

$$v = \sqrt{\frac{2gh}{1 + \frac{M}{2m}}}$$

and

(b) $\omega = \frac{1}{R_o}\sqrt{\frac{2gh}{1 + \frac{M}{2m}}}$

13. (a) $\omega = \frac{v}{R} = \frac{13\text{ m/s}}{5.3\text{ m}} = 2.5\text{ rad/s}$

$$\vec{\omega} = \omega\mathbf{k} = (2.5\text{ rad/s})\mathbf{k}$$

(b) The tangential acceleration of the car is entirely due to gravity since the normal force is horizontal at that point.

$$\alpha = \frac{a_t}{R} = \frac{9.8\text{ m/s}^2}{5.3\text{ m}} = 1.8\text{ rad/s}^2$$

$$\vec{\alpha} = -\alpha\mathbf{k} = (-1.8\text{ rad/s}^2)\mathbf{k}$$

(c) The vertical acceleration is g downward, and in addition there is an acceleration $\frac{v^2}{R}$ towards the center of the loop

$$a_n = \frac{v^2}{R} = \frac{(13\ \text{m/s})^2}{5.3\ \text{m}} = 32\ \text{m/s}^2$$

Thus, $$\mathbf{a} = (-32\ \text{m/s}^2)\mathbf{i} - (9.8\ \text{m/s}^2)\mathbf{j}$$

15. The forces on the marble are the force of gravity and the normal force exerted by the bowl, **F**. Let the z–axis point vertically up and consider the marble to be located on the x–z plane at the instant in question. Then

$$m\omega^2 R = -F_x$$

$$mg = F_z$$

and by geometry, $\frac{F_x}{F} = \frac{-R}{r_o}$.

Thus, $$\frac{F_z}{F} = \sqrt{1 - \left(\frac{F_x}{F}\right)^2} = \sqrt{1 - \left(\frac{R}{r_o}\right)^2}$$

and $$\frac{F_z}{F_x} = \frac{-r_o}{R}\sqrt{1 - \left(\frac{R}{r_o}\right)^2} = -\sqrt{\left(\frac{r_o}{R}\right)^2 - 1}$$

Equating this to $\frac{-g}{\omega^2 R}$,

$$\frac{g^2}{\omega^4 R^2} = \left(\frac{r_o}{R}\right)^2 - 1$$

or $$\frac{g^2}{\omega^4} = r_o^2 - R^2$$

Thus, $$R = \sqrt{r_o^2 - \frac{g^2}{\omega^4}}$$

17. To find the perpendicular distance from, say, particle 1 to the axis passing through particles 2 and 8, we can use the fact that the magnitude of $(\mathbf{r}_2 - \mathbf{r}_1) \times (\mathbf{r}_2 - \mathbf{r}_8)$ is equal to the length of $(\mathbf{r}_2 - \mathbf{r}_8)$ multiplied by the desired distance.

Thus, $$d = \frac{|(\mathbf{r}_2 - \mathbf{r}_1) \times (\mathbf{r}_2 - \mathbf{r}_8)|}{|\mathbf{r}_2 - \mathbf{r}_8|}$$

Now,

$$\mathbf{r}_1 = \frac{a}{2}\,\mathbf{i} + \frac{a}{2}\,\mathbf{j} + a\,\mathbf{k}$$

$$\mathbf{r}_2 = -\frac{a}{2}\,\mathbf{i} + \frac{a}{2}\,\mathbf{j} + a\,\mathbf{k}$$

$$\mathbf{r}_8 = \frac{a}{2}\,\mathbf{i} - \frac{a}{2}\,\mathbf{j}$$

$$\mathbf{r}_2 - \mathbf{r}_8 = -a\,\mathbf{i} + a\,\mathbf{j} + a\,\mathbf{k}$$

$$|\mathbf{r}_2 - \mathbf{r}_8| = a\sqrt{3}$$

$$\mathbf{r}_2 - \mathbf{r}_1 = -a\,\mathbf{i}$$

$$(\mathbf{r}_2 - \mathbf{r}_1) \times (\mathbf{r}_2 - \mathbf{r}_8) = a^2\,\mathbf{j} - a^2\,\mathbf{k}$$

$$|(\mathbf{r}_2 - \mathbf{r}_1) \times (\mathbf{r}_2 - \mathbf{r}_8)| = a^2\sqrt{2}$$

Thus, $d = a\sqrt{\frac{2}{3}}$

By symmetry, the other 5 distances are the same.

Thus, $I = 6md^2 = 6ma^2\left(\frac{2}{3}\right) = 4ma^2$

19. The force exerted by the cylinder on the astronaut is $m\omega^2 R$. To make this equal to her earth weight,

$$mg = m\omega^2 R$$

$$\omega^2 = \frac{g}{R}$$

$$\omega = \sqrt{\frac{g}{R}} = \sqrt{\frac{9.8\text{ m/s}^2}{100\text{ m}}} = 0.31\text{ rad/s}$$

(or nearly 3 rpm).

CHAPTER 13 ROTATION II

Questions:

1. The angular momentum of a particle is the cross product of its position vector and its momentum. Thus it is at right angles to these and to the velocity of the particle (which is parallel to its momentum).

3. The angular momentum is in a plane perpendicular to that containing the point P and the straight line that the particle is describing. Thus the direction of the angular momentum is fixed, but its magnitude is changing.

5. If the applied torque is parallel to the angular momentum of a particle, then the direction of the angular momentum is fixed, but the magnitude changes.

7. Conservation of linear momentum requires that the forces between the particles in the isolated system be equal and opposite, but conservation of angular momentum requires that, in addition, the forces be along the line joining the particles. A system that violates this more stringent statement of Newton's third law would have a changing angular momentum without any external torque.

   In the system envisioned in this problem, the particles would tend to rotate about the center of mass but, lacking a force towards each other to hold them in a circular orbit, they would spiral outward, with ever increasing angular momenta. This is clearly not a physical system.

9. "Choking up" on a bat shortens the moment arm between the point of impact of the ball and the batter's hands, so the torque felt by the batter is decreased. By decreasing the moment of inertia, it also allows the angular acceleration of the batter to be larger for the same torque.

11. Many small animals use their tails to make adjustments to their altitude in mid-flight. By rotating its tail in one direction, the cat's body would rotate in the other, allowing it to land on its feet. A bobtailed cat would have to extend its paws and rotate them and then bring them in in order to accomplish this maneuver. This would have to be a learned behavior, acquired at the expense of some unpleasant falls.

13. When a skater extends his arms, his potential energy increases because his center of gravity is raised. His angular momentum is unchanged because there is no net external torque, and his mechanical energy is conserved. The increase in potential energy is accompanied by a decrease in rotational kinetic energy (and in rotational speed).

15. Since a particle moving with a constant velocity is not being subjected to net external forces, its angular momentum is conserved. The law of areas is a geometric statement of the conservation of angular momentum, so the line joining O to the particle sweeps equal areas in equal times. Geometrically, the area of the triangle defined by the particle, point O, and the point on the trajectory closest to O, is increasing at a uniform rate since its base is constant and its height is increasing at a constant rate. Thus the increase in area per unit time is constant.

17. If the directions of $\mathbf{p}_i$ and $\mathbf{p}_j$ are reversed, then $\boldsymbol{\ell}_i$ and $\boldsymbol{\ell}_j$ are reversed also, so the only change in the figure is that the arrows are reversed. The radial components of angular momentum cancel.

19. When a basketball is not spinning and is being balanced on one finger, it is in unstable equilibrium because a small displacement will result in a torque which will increase that displacement. When the ball is spinning, the torque is at right angles to its angular momentum and the ball will precess rather than roll off.

21. The angular speed for precession is given by

$$\omega_p = \frac{MgD}{I\omega_s}$$

The mass of the wheel affects both the torque because of the weight of the wheel (Mg) and the magnitude of its moment of inertia. If the mass of the wheel is increased without changing its dimensions, then both terms are proportional to M and ω_p does not change. If a larger wheel is used, then I increases both from the change in M and from the increased dimensions, so ω_p decreases.

23. The porter is in for a surprise. Because of the gyroscopic effect, the suitcase will precess about an axis perpendicular to the axis of the spinner and the applied torque.

Exercises:

1. (a) Taking the origin at the position of the observer,

$$\mathbf{r}_c = 125\,\mathbf{j} + (36.4\,t)\,\mathbf{i}\ \text{m}$$

(b) $\mathbf{L} = \mathbf{r}_c + m\mathbf{v}_c$

$$= \left[(125\ \text{m})\,\mathbf{j} + (325\ \text{m})\,\mathbf{i}\right] \times (1340\ \text{kg})(36.4\ \text{m/s})\,\mathbf{i}$$

$$= -(6.1 \times 10^6\ \text{kg}\cdot\text{m}^2/\text{s})\,\mathbf{k}$$

3. $$\mathbf{L} = \mathbf{r} \times m\mathbf{v} = 4.1\text{ kg}\begin{vmatrix} \mathbf{i} & \mathbf{j} & \mathbf{k} \\ -3.5\text{ m} & 1.4\text{ m} & 0\text{ m} \\ -2.0\text{ m/s} & -6.3\text{ m/s} & 0\text{ m/s} \end{vmatrix}$$

$$= \mathbf{k}\left((3.5\text{ m})(6.3\text{ m/s}) + (2.0\text{ m/s})(1.4\text{ m})\right) 4.1\text{ kg}$$

$$= (100\text{ kg}\cdot\text{m}^2/\text{s})\,\mathbf{k}$$

5. Using conservation of energy,

$$mgR\sin\theta = \frac{1}{2}mv^2$$

$$v = \sqrt{2gR\sin\theta}$$

$$= \left[(2)(9.8\text{ m/s}^2)(0.93\text{ m})(\sin 0.87)\right]^{1/2}$$

$$= 3.73\text{ m/s}$$

$$\mathbf{L} = \mathbf{r}\times\mathbf{p} = (0.93\text{ m }\mathbf{i})\times(0.072\text{ kg})(-3.73\text{ m/s }\mathbf{j})$$

$$= -(0.25\text{ kg}\cdot\text{m}^2/\text{s})\,\mathbf{k}$$

7. $$\ell_{\text{TOTAL}} = \ell_a + \ell_b + \ell_c$$

$$= \left((2.4\,\mathbf{i}) - (6.1\,\mathbf{k}) + (-4.8\,\mathbf{i} + 1.6\,\mathbf{j})\right)\text{ kg}\cdot\text{m}^2/\text{s}$$

$$= (-2.4\,\mathbf{i} + 1.6\,\mathbf{j} - 6.1\,\mathbf{k})\text{ kg}\cdot\text{m}^2/\text{s}$$

9. Let the origin be located at a distance x from one line and x + D from the other. Then the angular momentum components are $m(x + D)v$ and $-mxv$, so the total is mvD.

11. $$L = I\omega = \left(\frac{1}{2}mR^2\right)\omega$$

$$= \frac{1}{2}(0.085\text{ kg})(0.150\text{ m})^2\left(\frac{78\text{ rev/min}\times 2\pi\text{ rad/rev}}{60\text{ s/min}}\right)$$

$$= 0.0078\text{ kg}\cdot\text{m}^2/\text{s}$$

Looking from above, the rotation is clockwise, so the angular momentum is directed downward.

13. $\tau = 18\text{ N}\cdot\text{m}, \quad I = 0.15\text{ kg}\cdot\text{m}^2$

$\tau = I\alpha$

$$\therefore \quad \alpha = \frac{18\text{ N}\cdot\text{m}}{0.15\text{ kg}\cdot\text{m}^2} = 120\text{ rad/s}^2$$

$$\omega = \omega_0 + \alpha t = (120\text{ rad/s}^2)t$$

$$\theta = \theta_0 + \omega_0 t + \tfrac{1}{2}\alpha t^2 = \theta_0 + (60\text{ rad/s}^2)t^2$$

15. (a) $\tau = TR = I\alpha$

$$\therefore \quad \alpha = \frac{TR}{I} = \frac{32\text{ N }(0.170\text{ m})}{0.085\text{ kg}\cdot\text{m}^2} = 64\text{ rad/s}^2$$

(b) In this case the tension is less because the mass is accelerating

$$mg - T = ma \quad \text{or} \quad T = mg - ma$$

The acceleration of the mass is connected to the angular acceleration of the pulley by

$$a = R\alpha$$

so $$mg - T = mR\alpha; \quad T = \frac{I\alpha}{R}$$

$$mg = \alpha\left(\frac{I}{R} + mR\right)$$

$$\alpha = \frac{mgR}{I + mR^2} = \frac{32\text{ N }(0.170\text{ m})}{0.085\text{ kg}\cdot\text{m}^2/\text{s} + \frac{32\text{ N}}{9.8\text{ m/s}^2}(0.170\text{ m})^2}$$

$$= 30\text{ rad/s}^2$$

17. $\tau = TR - \tau_f$

Following the development of Exercise 15,

$$I\alpha = (mg - mR\alpha)R - \tau_f$$

so $$I = \frac{mgR - \tau_f}{\alpha} - mR^2$$

Here $m = 7.3\text{ kg}, \quad R = 0.120\text{ m},$

and α is found from

$$x = \tfrac{1}{2}at^2,$$

which gives $a = \frac{2x}{t^2} = \frac{2(0.450\text{ m})}{(0.33\text{ s})^2}$

so $a = 8.26\text{ m/s}^2; \quad \alpha = \frac{a}{R} = 68.9\text{ rad/s}^2$

Hence, $I = \frac{(7.3\text{ kg})(9.8\text{ m/s}^2)(.120\text{ m}) - (.23\text{ N}\cdot\text{m})}{68.9\text{ rad/s}^2} - (7.3\text{ kg})(.120\text{ m})^2$

$= 0.016\text{ kg}\cdot\text{m}^2$

This value for I is smaller than the result for Exercise 16 because in this problem some of the slowing down of the motion is due to friction rather than to the inertia of the mass and pulley alone.

19. The free-body diagrams for this problem are:

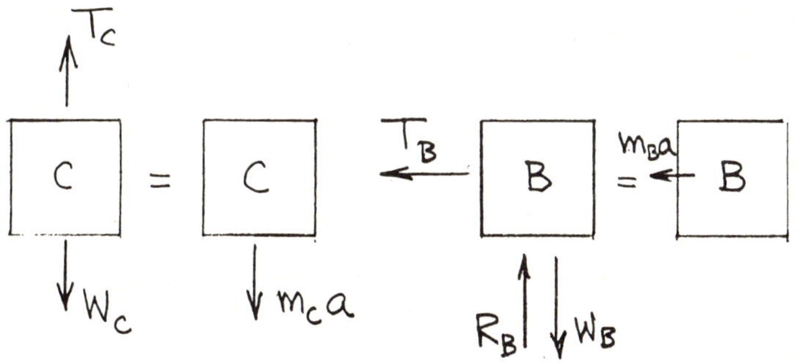

and

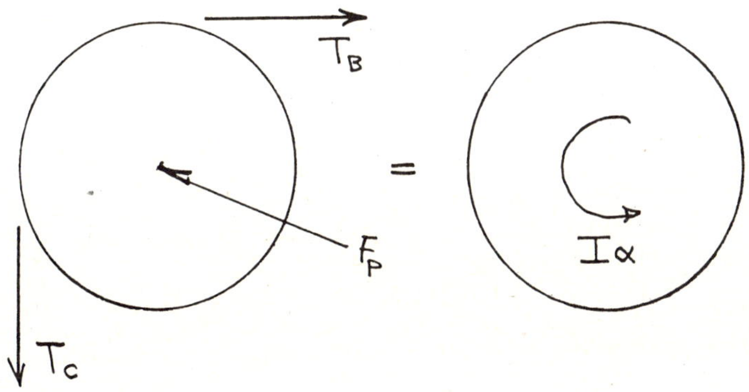

where $\alpha = \frac{R}{a}$

Thus, $m_C a = m_C g - T_C$

$m_B a = T_B$

and $$I\left(\frac{a}{R}\right) = T_C R - T_B R$$

Thus, $$(m_B + m_C)a = m_C g + T_B - T_C$$

$$= m_C g - \frac{Ia}{R^2}$$

so $$a = \frac{m_C g}{m_B + m_C + \frac{I}{R^2}}$$

$$T_B = \frac{m_B m_C g}{m_B + m_C + \frac{I}{R^2}}$$

$$T_C = \frac{m_C g\left(m_B + \frac{I}{R^2}\right)}{m_B + m_C + \frac{I}{R^2}}$$

21. From Fig. 13–10b,

$$F_B = M_w g + F_T$$

$$= (94\text{ kg})(9.8\text{ m/s}^2) + 200\text{ N}$$

$$= 1100\text{ N}$$

23. For a rigid body rotating about a fixed axis,

$$K = \frac{1}{2} I\omega^2$$

$$L = I\omega$$

so $\omega = \frac{L}{I}$, and $K = \frac{1}{2} I\left(\frac{L}{I}\right)^2 = \frac{1}{2} \cdot \frac{L^2}{I}$

25. The kinetic energy of the system is

$$K = \frac{1}{2} m_B v^2 + \frac{1}{2} m_C v^2 + \frac{1}{2} I\omega^2$$

and the potential energy is

$$U = m_C g y - m_B g y,$$

where y is the distance C has risen (and B dropped).

The initial energy is zero since the blocks start from rest.

Now, $\omega = \frac{v}{R}$ and $I = \frac{1}{2} m_p R^2$,

so $$\frac{1}{2}\left(m_B + m_C + \frac{I}{R^2}\right)v^2 + (m_C - m_B)gy = 0$$

$$v^2 = \frac{2\,(0.83\text{ kg} - 0.57\text{ kg})(9.8\text{ m/s}^2)(0.012\text{ m})}{0.83\text{ kg} + 0.57\text{ kg} + \frac{1}{2}\,(0.74\text{ kg})}$$

$$v = 0.186\text{ m/s}$$

27. (a) Angular momentum is conserved,

so $$I_1\omega_1 = I_2\omega_2$$

$$\therefore \quad \omega_2 = \frac{I_1\omega_1}{I_2} = \frac{(1.72\text{ kg}\cdot\text{m}^2)(15\text{ rad/s})}{0.61\text{ kg}\cdot\text{m}^2} = 42.3\text{ rad/s}$$

(b) The change in kinetic energy is

$$\Delta K = \frac{1}{2}I_2\omega_2^2 - \frac{1}{2}I_1\omega_1^2$$

$$= \frac{1}{2}\,(0.61\text{ kg}\cdot\text{m}^2)(42.3\text{ rad/s})^2 - \frac{1}{2}\,(1.72\text{ kg}\cdot\text{m}^2)(15\text{ rad/s})^2$$

$$= 352\text{ J}$$

(c) The increase in energy is accounted for by the work done when the ice skater pulls her arms in.

29. (a) Since the hinges exert a force, the linear momentum is not conserved.

(b) Since the hinges exert negligible torque, angular momentum about the hinge is conserved.

(c) Using conservation of angular momentum,

$$mv_iD = I\omega - mv_fD$$

Here $$I = \frac{1}{3}Mw^2 = \frac{1}{3}\,(35\text{ kg})(0.73\text{ m})^2 = 6.22\text{ kg}\cdot\text{m}^2$$

Hence $$\omega = \frac{m\,(v_i + v_f)D}{I}$$

(d) $$\omega = \frac{(1.1\text{ kg})(27\text{ m/s} + 16\text{ m/s})\,(0.62\text{ m})}{6.22\text{ kg}\cdot\text{m}^2}$$

$$= 4.72\text{ rad/s}$$

31. (a) Since there is no external torque, angular momentum is conserved.

$$m v_i R_i = m v_f R_f, \quad \text{or} \quad \frac{v_f}{v_i} = \frac{R_i}{R_f}$$

Now $$T = \frac{m v^2}{R},$$

so $$\frac{T_f}{T_i} = \frac{v_f^2}{v_i^2} \cdot \frac{R_i}{R_f} = \left(\frac{R_i}{R_f}\right)^3$$

(b) If $T_i = 3.4\ \text{N}$ and $R_f = \frac{R_i}{2}$

then $T_f = 3.4\ \text{N}(2)^3 = 27\ \text{N}$

33. (a) Assuming that the forces exerted by the ice can be neglected, then angular momentum is conserved

$$m v D = I\omega = 2m\left(\frac{D}{2}\right)^2 \omega$$

so $$\omega = \frac{2v}{D}$$

(i.e., the skaters are still moving with speed v in a circle of radius $\frac{D}{2}$)

(b) Since the speed of the skaters has not changed, there is no change in their kinetic energy.

35. (a) $\mathbf{F}_1 = (12\ \text{N})(-\mathbf{j})$ $\quad \mathbf{r}_1 = -(0.38\ \text{m})\,\mathbf{i}$

$\mathbf{F}_2 = (24\ \text{N})(-\mathbf{i})$ $\quad \mathbf{r}_2 = (0.19\ \text{m})\,\mathbf{j}$

$\mathbf{F}_3 = (18\ \text{N})(\mathbf{i})$ $\quad \mathbf{r}_3 = (0.38\ \text{m})\,\mathbf{j}$

$$\vec{\tau} = \mathbf{r}_1 \times \mathbf{F}_1 + \mathbf{r}_2 \times \mathbf{F}_2 + \mathbf{r}_3 \times \mathbf{F}_3$$

$$= (12\ \text{N})(0.38\ \text{m})\,\mathbf{k} + (24\ \text{N})(0.19\ \text{m})\,\mathbf{k} - (18\ \text{N})(0.38\ \text{m})\,\mathbf{k}$$

$$= (2.3\ \text{N}\cdot\text{m})\,\mathbf{k}$$

(b) The moment of inertia of the disk is

$$I = \frac{1}{2} mR^2 = \frac{1}{2}(2.6\ \text{kg})(0.38\ \text{m})^2 = 0.19\ \text{kg}\cdot\text{m}^2$$

and the angular acceleration is

$$\alpha = \frac{\tau}{I} = 12\ \text{rad/s}^2$$

37. Neglecting any frictional torque at the hinges, the angular momentum about the hinges is conserved. Therefore

$$m v_i = I\omega, \quad \text{where} \quad I = \tfrac{1}{3} M w^2 + m d^2$$

Here $\quad m = 20\text{ kg}, \quad v_i = 4.8\text{ m/s}, \quad M = 30\text{ kg}, \quad w = d = 1.5\text{ m},$

so
$$\omega = \frac{(20\text{ kg})(4.8\text{ m/s})}{\left[\frac{1}{3}(30\text{ kg}) + 20\text{ kg}\right](1.5\text{ m})^2} = 1.4\text{ rad/s}$$

39. (a) Assume that the star is initially a uniform sphere with a radius of 1×10^7 m and after the collapse it is a sphere with a radius of 5×10^3 m. Since there are no external forces involved in the collapse, its angular momentum remains constant.

$$\therefore \quad I_i\omega_i = I_f\omega_f, \quad I = \tfrac{2}{5} m R^2$$

$$\omega_i = \frac{I_f\omega_f}{I_i} = \frac{R_f^2\omega_f}{R_i^2} = \frac{(5 \times 10^3\text{ m})^2}{(1 \times 10^7\text{ m})^2}(1000\text{ rad/s})$$

$$= 2.5 \times 10^{-4}\text{ rad/s}$$

(b) The sun's rotational period is 25 days,

so
$$\omega_{sun} = \frac{2\pi\text{ rad/rev}}{25\text{ days/rev} \times 24\text{ hr/day} \times 3600\text{ s/hr}} = 2.9 \times 10^{-6}\text{ rad/s}$$

This is 100 times slower than the initial angular speed of the neutron star.

41. (a) The kinetic energy of the yo-yo is

$$K = \tfrac{1}{2} M v^2 + \tfrac{1}{2} I\omega^2$$

where $v = \omega R_1$ is the forward speed of the yo-yo. In this problem,

$$I = \tfrac{1}{2} M R_2^2,$$

so
$$K = \tfrac{1}{2} M\left(R_1^2 + \tfrac{1}{2} R_2^2\right)\omega^2 = \tfrac{1}{2} M\left(1 + \frac{R_2^2}{2R_1^2}\right)v^2$$

The potential energy is $+Mgy$. Conservation of energy then gives

$$\tfrac{1}{2} M\left(1 + \frac{R_2^2}{2R_1^2}\right)v^2 = Mgh$$

or $$v^2 = \frac{2gh}{1 + \frac{R_2^2}{2R_1^2}}$$

For motion with constant acceleration starting from rest, $v^2 = 2ax$.

Thus, $$a = \frac{g}{1 + \frac{R_2^2}{2R_1^2}}$$

(b) If $\frac{R_2}{R_1} = 4$, $a = \frac{g}{1+8} = 1.1\text{ m/s}^2$

43. $$\ell_x = m(yv_z - zv_y)$$
$$= 2.7\text{ kg}((1.1\text{ m})(-3.3\text{ m/s}) - (2.8\text{ m})(0\text{ m/s}))$$
$$= -9.8\text{ kg}\cdot\text{m}^2/\text{s}$$

$$\ell_y = m(zv_x - xv_z)$$
$$= 2.7\text{ kg}((2.8\text{ m})(1.4\text{ m/s}) - (-3.5\text{ m})(-3.3\text{ m/s}))$$
$$= -20.6\text{ kg}\cdot\text{m}^2/\text{s}$$

$$\ell_z = m(xv_y - yv_x)$$
$$= 2.7\text{ kg}((-3.5\text{ m})(0\text{ m/s}) - (1.1\text{ m})(1.4\text{ m/s}))$$
$$= -4.2\text{ kg}\cdot\text{m}^2/\text{s}$$

$$\boldsymbol{\ell} = -(9.8\text{ kg}\cdot\text{m}^2/\text{s})\,\mathbf{i} - (20.6\text{ kg}\cdot\text{m}^2/\text{s})\,\mathbf{j} - (4.2\text{ kg}\cdot\text{m}^2/\text{s})\,\mathbf{k}$$

45. The external forces on the system are: the gravitational force on the rod; the gravitational force on the putty; and the reaction at the axle. The gravitational force on the rod acts at the center, so there is no unbalanced moment due to it. The only unbalanced moment about the axle is from the gravitational force on the putty. The net torque is $m_P g \frac{\ell}{2}$.

Using $\tau = \frac{d\ell}{dt}$ or $\Delta\ell = \int \tau\, dt$

we see that this torque will not produce an impulsive change in angular momentum. The angular momentum can be regarded as constant over the short time needed for the collision.

Thus, $$m_P v_P \frac{\ell}{2} = \left[I_{rod} + m_P\left(\frac{\ell}{2}\right)^2\right]\omega$$

Here $$I_{rod} = \frac{1}{12} M\ell^2 = \frac{1}{12}(0.12\ \text{kg})(1.0\ \text{m})^2 = 0.010\ \text{kg}\cdot\text{m}^2$$

$$v_P = \sqrt{2gh} = \sqrt{2(9.8\ \text{m/s}^2)(0.50\ \text{m})} = 3.13\ \text{m/s}$$

and $$\ell = 1.0\ \text{m}, \quad m_P = 0.040\ \text{kg},$$

so $$\omega = \frac{(0.040\ \text{kg})(3.13\ \text{m/s})(0.50\ \text{m})}{0.010\ \text{kg}\cdot\text{m}^2 + 0.040\ \text{kg}\,(0.50\ \text{m})^2} = 3.1\ \text{rad/s}$$

47. (a) In this problem, there are no external torques about the vertical axis through the center of the disk, so angular momentum about that axis is conserved:

$$I_i\omega_i = I_f\omega_f$$

Here, $$I_i = \frac{1}{2}MR_o^2, \quad I_f = I_i + m\left(\frac{R_o}{2}\right)^2$$

so $$\frac{\omega_f}{\omega_i} = \frac{I_i}{I_f} = \frac{\frac{1}{2}MR_o^2}{\frac{1}{2}MR_o^2 + \frac{1}{4}mR_o^2} = \frac{1}{1+\frac{m}{2M}}$$

and $$\omega_f = \frac{\omega_i}{1+\frac{m}{2M}}$$

(b) When $M = 2.0$ kg, $m = 1.0$ kg, and $\omega_i = 5.0$ rad/s,

$$\omega_f = \frac{5.0\ \text{rad/s}}{1+\frac{1.0\ \text{kg}}{2(2.0\ \text{kg})}} = 4.0\ \text{rad/s}$$

49. The torque on the pulley is

$$\tau = Fr = (7.7\ \text{N})(0.19\ \text{m}) = 1.46\ \text{N}\cdot\text{m}$$

The power delivered is

$$\begin{aligned} p &= \tau\omega \\ &= (1.46\ \text{N}\cdot\text{m})(2.6\ \text{rad/s}) \\ &= 3.8\ \text{W} \end{aligned}$$

Problems:

1. Choose a coordinate system so the circle lies on the x–y plane and the z–axis is parallel to $\vec{\omega}$ and passes through the center of the circle. The vector **r** from the origin to the particle is in the x–y plane. The vector $\vec{\omega} \times \mathbf{r}$ has a magnitude ωr and it is at right angles to $\vec{\omega}$ and **r**, so it is in the x–y plane and perpendicular to **r**. Its direction, using the right-hand rule, is in the direction of motion of the particle. Thus $\vec{\omega} \times \mathbf{r}$ is the velocity of the particle.

It is not necessary that the origin be located at the center of the circle. Any point along the axis will give the same result.

3. The free-body diagram in this case is

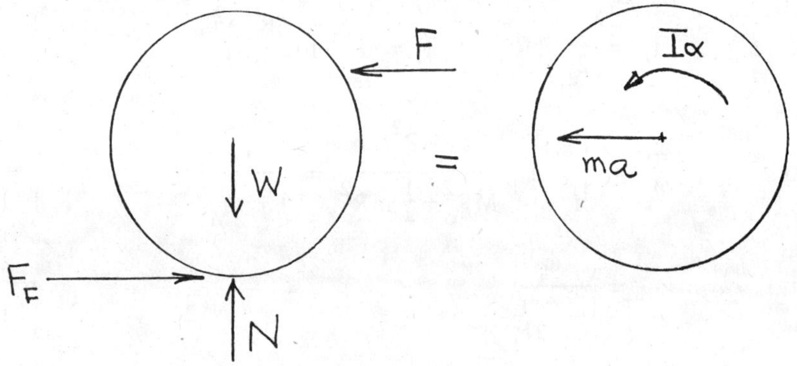

If the ball is not slipping, $a = \alpha r_o$.

Thus, $F - F_F = ma$

$$Fh - F_F r_o = I\alpha$$

so $a = \dfrac{F - F_F}{m} = r_o\alpha = \dfrac{r_o(Fh - F_F r_o)}{I}$

Now, $I = \frac{2}{5} m r_o^2$

so $\frac{2}{5} r_o^2 (F - F_F) = r_o(Fh - F_F r_o)$

$$\left(r_o h - \frac{2}{5} r_o^2\right) F = \frac{3}{5} r_o^2 F_F$$

If $F_F = 0$, then $h = \frac{2}{5} r_o$.

5. While the ball is sliding, the torque on the ball is $\mu m g r_0$,

so $\quad I\alpha = \mu m g r_0, \quad$ with $\quad I = \frac{2}{5} m r_0^2.$

Thus,
$$\alpha = \frac{\mu m g r_0}{\frac{2}{5} m r_0^2}$$
$$= \frac{5}{2} \cdot \frac{\mu g}{r_0}$$

and
$$\omega = \frac{5}{2} \cdot \frac{\mu g t}{r_0}$$

The force on the ball is $-\mu m g$,

so
$$m a = -\mu m g$$
$$a = -\mu g$$
$$v = v_0 - \mu g t$$

(b) The ball stops sliding when $v = \omega r_0$; that is, when

$$v_0 - \mu g t = \frac{5}{2} \mu g t$$

or
$$t = \frac{v_0}{\frac{7}{2} \mu g} = \frac{2}{7}\left(\frac{v_0}{\mu g}\right)$$

(a) At that instant,

$$v = v_0 - \mu g \left(\frac{2}{7} \cdot \frac{v_0}{\mu g}\right) = \frac{5}{7} v_0$$

and the distance travelled is

(c) $x = v_0 t + \frac{1}{2} a t^2$

$$= v_0 \left(\frac{2}{7} \cdot \frac{v_0}{\mu g}\right) - \frac{1}{2} \mu g \left(\frac{2}{7} \cdot \frac{v_0}{\mu g}\right)^2$$
$$= \left(\frac{2}{7} - \frac{2}{49}\right) \frac{v_0^2}{\mu g} = \frac{12}{49} \cdot \frac{v_0^2}{\mu g}$$

7. In order for the ball to remain in contact with the track, at the top of the loop

$$mg = \frac{mv^2}{R_o}$$

so $\quad v^2 = gR_o$

Using conservation of energy,

$$mg(h + r_o) = \frac{1}{2}mv^2 + \frac{1}{2}I\omega^2 + mg(2R_o - r_o)$$

Here $\quad I = \frac{2}{5}mr_o^2 \quad$ and $\quad \omega = \frac{v}{r_o},$

so $\quad mg(h - 2R_o + 2r_o) = \frac{1}{2}(m + \frac{2}{5}m)v^2$

Thus, $\quad g(h - 2R_o + 2r_o) = \frac{7}{10}v^2 = \frac{7}{10}gR_o$

or $\quad h = \frac{27}{10}R_o,$

neglecting terms in r_o.

(Note that the relation $\omega = \frac{v}{r_o}$ neglects terms involving $\frac{r_o}{R_o}$.)

9. Using conservation of energy,

$$\frac{1}{2}I\omega^2 = Mg \cdot \frac{h}{2}(1 - \cos\theta)$$

where I is the moment of inertia about the hinge, $\frac{1}{3}Mh^2$.

Thus, $\quad \omega^2 = \frac{Mgh(1 - \cos\theta)}{\frac{1}{3}Mh^2} = \frac{3g(1 - \cos\theta)}{h}$

and $\quad \omega = \sqrt{\frac{3g(1 - \cos\theta)}{h}}$

11. (a) In Exercise 31, it was shown that

$$T = T_i\left(\frac{R_i}{R}\right)^3 = \frac{mv_i^2R_i^2}{R^3}$$

(b) The work done on the block is

$$W = -\int_{R_i}^{R_f} T\,dR = m v_i^2 R_i^2 \left(\frac{1}{2R^2}\right)\Bigg|_{R_i}^{R_f}$$

$$= \frac{m v_i^2 R_i^2}{2}\left(\frac{1}{R_f^2} - \frac{1}{R_i^2}\right)$$

$$= \frac{1}{2} m v_i^2 \left(\frac{R_i^2}{R_f^2} - 1\right)$$

(c) The change in kinetic energy is

$$\Delta K = \frac{1}{2} m v_f^2 - \frac{1}{2} m v_i^2$$

$$= \frac{1}{2} m v_i^2 \left[\left(\frac{v_f}{v_i}\right)^2 - 1\right]$$

$$= \frac{1}{2} m v_i^2 \left(\frac{R_i^2}{R_f^2} - 1\right),$$

which is equal to the work done on the block.

13. With the gyroscope at an angle θ above the x–z plane,

$$\vec{\tau} = F_e (D\cos\theta)\,\mathbf{i} = MgD\cos\theta\,\mathbf{i}$$

$$\mathbf{L_s} = I_s\omega_s(\cos\theta\,\mathbf{k} + \sin\theta\,\mathbf{j})$$

The precessional angular momentum is

$$\mathbf{L_p} = I_p\omega_p\mathbf{j}$$

and the total angular momentum

$$\mathbf{L} = \mathbf{L_s} + \mathbf{L_p}$$

is still perpendicular to $d\mathbf{L} = \vec{\tau}\,dt$

The infinitesimal precession angle is

$$d\phi_z = \frac{dL}{L_z} = \frac{\tau\, dt}{L_z} \simeq \frac{\tau\, dt}{L_{sz}}$$

$$= \frac{M g D \cos\theta\, dt}{I_s \omega_s \cos\theta}$$

$$= \left(\frac{M g D}{I_s \omega_s}\right) dt$$

So $$\omega_z = \frac{M g D}{I_s \omega_s}$$

15. At the instant shown in Fig. 13–18, the angular momentum of the gyroscope is

$$\mathbf{L} = \mathbf{L}_s + \mathbf{L}_p = L_s\,\mathbf{k} + L_p\,\mathbf{j}$$

and $$\frac{d\mathbf{L}}{dt} = \vec{\tau} = M g D\,\mathbf{i}$$

(a) The center of mass of the gyroscope moves in a horizontal circle of radius D with an angular velocity

$$\omega_p = \frac{M g D}{I_s \omega_s}$$

The centripetal force is F_{pz},

so $$F_{pz} = -M\omega_p^2 D = -MD\left(\frac{M g D}{I_s \omega_s}\right)^2$$

$$= -\frac{M^3 g^2 D^3}{I_s^2 \omega_s^2} = -Mg\left(\frac{M^2 g D^3}{I_s^2 \omega_s^2}\right)$$

and there is no force along the x–direction.

(b) Along the vertical, $F_{py} = Mg$,

so $$F_p = \sqrt{F_{py}^2 + F_{pz}^2} = Mg\sqrt{1 + \left(\frac{M^2 g D^3}{I_s^2 \omega_s^2}\right)^2}, \quad \text{and}$$

(c) $$\tan\theta = \frac{F_{pz}}{F_{py}} = -\frac{M^2 g D^3}{I_s^2 \omega_s^2}$$

17. (a) At the instant shown, the velocities of the spheres are

$$\mathbf{v}_{upper} = \omega r \sin\theta\, \mathbf{i}$$

$$\mathbf{v}_{lower} = -\omega r \sin\theta\, \mathbf{i},$$

so the angular momentum about the origin is

$$\begin{aligned}\mathbf{L} &= (r\cos\theta\,\mathbf{k} - r\sin\theta\,\mathbf{j}) \times M\omega r\sin\theta\,\mathbf{i} \\ &\quad + (-r\cos\theta\,\mathbf{k} + r\sin\theta\,\mathbf{j}) \times (-M\omega r\sin\theta)\,\mathbf{i} \\ &= 2M\omega r^2 \sin\theta\,(\mathbf{j}\cos\theta + \mathbf{k}\sin\theta)\end{aligned}$$

(b) The accelerations of the spheres are

$$\mathbf{a}_{upper} = \omega^2 r\sin\theta\,\mathbf{j}$$

$$\mathbf{a}_{lower} = -\omega^2 r\sin\theta\,\mathbf{j},$$

so the rate of change of angular momentum is

$$\begin{aligned}\frac{d\mathbf{L}}{dt} &= 2M(r\cos\theta\,\mathbf{k} - r\sin\theta\,\mathbf{j}) \times \omega^2 r\sin\theta\,\mathbf{j} \\ &= -2M\omega^2 r^2\,(\mathbf{i}\cos\theta)\sin\theta\end{aligned}$$

Thus, $$\vec{\tau}_b = -2M\omega^2 r^2 \sin\theta\cos\theta\,\mathbf{i}$$

(c) This torque is provided by equal and opposite forces at the bearings,

$$\tau_b = 2DF_b,$$

so $$F_b = \frac{M\omega^2 r^2}{D}\sin\theta\cos\theta$$

19. (a) If energy is conserved, then

$$\frac{1}{2}mv_i^2 = \frac{1}{2}I\omega^2 + \frac{1}{2}mv_f^2$$

In Exercise 29, it was found that

$$\omega = \frac{m(v_i + v_f)D}{I},$$

so $$\frac{1}{2}mv_i^2 = \frac{1}{2}\cdot\frac{m^2(v_i+v_f)^2D^2}{I} + \frac{1}{2}mv_f^2$$

or $$v_i^2 - v_f^2 = \frac{mD^2}{I}(v_i+v_f)^2$$

Dividing by $v_i + v_f$,

$$v_i - v_f = \frac{mD^2}{I}(v_i+v_f)$$

so $$v_i\left(1-\frac{mD^2}{I}\right) = v_f\left(1+\frac{mD^2}{I}\right)$$

$$v_f = \frac{v_i\left(1-\frac{mD^2}{I}\right)}{1+\frac{mD^2}{I}} = \frac{v_i\,(Mw^2-3mD^2)}{Mw^2+3mD^2},$$

since $$I = \frac{1}{3}Mw^2$$

Also, $$\omega = \frac{mD}{I}v_i\left(1+\frac{1-\frac{mD^2}{I}}{1+\frac{mD^2}{I}}\right)$$

or $$\omega = \frac{v_i\left(\frac{2mD}{I}\right)}{1+\frac{mD^2}{I}} = \frac{v_i\,(6mD)}{Mw^2+3mD^2}$$

(b) With $m = 1.1\text{ kg}$, $M = 35\text{ kg}$, $w = 0.73\text{ m}$, $D = 0.62\text{ m}$, $v_i = 27\text{ m/s}$,

$$v_f = \frac{27\text{ m/s}\,(35\text{ kg}\,(0.73\text{ m})^2 - 3(1.1\text{ kg})(0.62\text{ m})^2)}{(35\text{ kg}\,(0.73\text{ m})^2 + 3(1.1\text{ kg})(0.62\text{ m})^2)}$$

$$= 23.6\text{ m/s}$$

$$\omega = \frac{27\text{ m/s}\,(6)(1.1\text{ kg})(0.62\text{ m})}{35\text{ kg}\,(0.73\text{ m})^2 + 3(1.1\text{ kg})(0.62\text{ m})^2} = 5.55\text{ rad/s}$$

In Exercise 29, $v_f = 16\text{ m/s}$ and $\omega = 4.72\text{ rad/s}$,

so the collision was not elastic.

21. The free-body diagram is as shown.

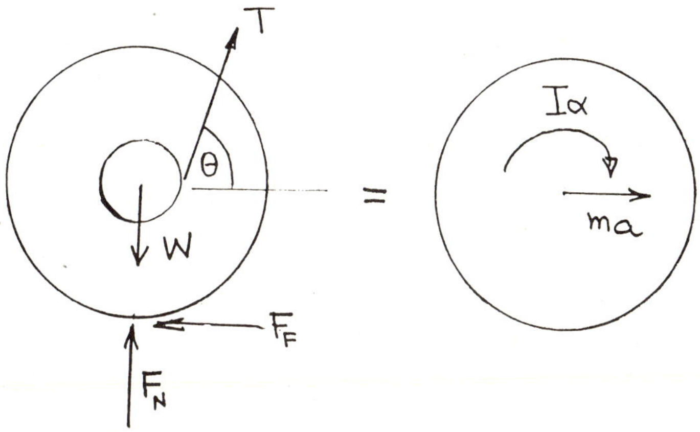

The net force in the forward direction is

$$T\cos\theta - F_F = ma_x$$

The net torque about the center is

$$F_F R_o - TR_1 = I_C\alpha$$

Since the yo-yo does not slip, $a_x = R_o\alpha$

Eliminating F_F,

$$TR_o\cos\theta - mR_o^2\alpha - TR_1 = I_C\alpha$$

or $$T(R_o\cos\theta - R_1) = (I_C + mR_o^2)\alpha$$

Thus, $$a_x = \frac{T\left(\cos\theta - \frac{R_1}{R_o}\right)}{m + \frac{I_C}{R_o^2}}$$

The only way that the string can keep from winding or unwinding is if

$$a_x = 0 \quad \text{or} \quad \cos\theta = \frac{R_1}{R_o}$$

If $\cos\theta > \frac{R_1}{R_o}$, then $a_x > 0$, $\alpha > 0$, and the string winds onto the yo-yo; if $\cos\theta < \frac{R_1}{R_o}$, then $a_x < 0$, $\alpha < 0$, and the string unwinds.

CHAPTER 14 OSCILLATIONS

Questions:

1. Examples of oscillating systems are:

(a) The motion of the flywheel in a wristwatch. This is SHM since the motion is governed by the torsion in the spring and the moment of inertia of the flywheel.

(b) A ball rolling back and forth on a circular bowl. These oscillations are nonlinear, so the system is not an SH oscillator.

(c) A ball bouncing back and forth between two walls. This is not SHM.

3. (a) $\nu = \frac{\omega}{2\pi}$, so ν is doubled.

(b) $T = \frac{2\pi}{\omega}$, so T is halved.

(c) The amplitude is independent of the frequency, so it would not be affected directly.

(d) The phase is also independent of the frequency and would not be affected directly.

If the oscillator suddenly underwent a change in restoring force with no change in position and momentum, then both A and ϕ would change, but not in a simple way.

5. The velocity and displacement are both positive for a quarter of every period and both negative for another quarter period. The same holds for the velocity and acceleration. The displacement and the acceleration are never in the same direction.

7. The mass of the spring would add inertia to the system, but not as much as if its mass were added directly to m because not all parts of the spring would be moving with the same velocity. In any case, the frequency would be decreased.

9. When the springs are side by side the restoring force is $2kx$, so the angular frequency is

$$\omega = \sqrt{\frac{2k}{m}}$$

When the springs are end to end, the displacement is

$$\frac{F}{k} + \frac{F}{k}, \quad \text{so} \quad F = \frac{kx}{2}$$

Thus, $$\omega = \sqrt{\frac{k}{2m}}$$

11. Assume that the car has a mass of 500 kg and that it is supported by four springs and that the period of oscillation is 1 second.

Then $$4k = m\omega^2 = \frac{4\pi^2 m}{T^2}$$

or $$k \simeq 5000 \text{ N/m}$$

13. A long pendulum tends to lose energy to friction more slowly than a short one, so a clock with a long pendulum will need winding less frequently. On the other hand, a 1 meter pendulum will only fit in a grandfather clock, while a 25 cm one will fit in a table clock.

15. The pendulum would expand at higher temperatures, thus increasing its period. The percent change in its length would translate into half that percent change in period. This is the same effect that a percentage decrease in gravity would have.

17. If the soldiers are marching so their steps are at the same frequency as the natural frequency of oscillation of the bridge, the resonant forced oscillations could damage the structure.

It is very unlikely that a cat could cause a bridge to collapse, but it is conceivable. The cat would have to excite a mode of oscillation of the bridge and maintain that excitation until the oscillations build to very large amplitudes. The energy for these oscillations would have to come from the cat, however.

19. If the conical pendulum were generating a right circular cone, both the periods and the amplitudes of the x and y oscillations are the same. Otherwise, for small angles, the two periods are the same but the amplitudes are independent of each other. By changing one amplitude with respect to the other, the bob can be made to trace circles, ellipses, or lines. The phase would determine the orientation of the ellipses.

Exercises:

1. Given $\omega = 5.8 \text{ rad/s}$

$$T = \frac{2\pi}{\omega} = \frac{2\pi \text{ rad}}{5.8 \text{ rad/s}} = 1.08 \text{ s}$$

$$\nu = \frac{\omega}{2\pi} = 0.92 \text{ Hz}$$

3. (a)

$$x = (0.063\ \text{m}) \cos ((4.1\ \text{rad/s})\,t)$$

$$v_x = (-0.26\ \text{m/s}) \sin ((4.1\ \text{rad/s})\,t)$$

$$a_x = (-1.06\ \text{m/s}) \cos ((4.1\ \text{rad/s})\,t)$$

(b) When $t = 1.7\ \text{s}$,

$$x = 0.049\ \text{m}$$

$$v_x = -0.16\ \text{m/s}$$

$$a_x = -0.82\ \text{m/s}^2$$

5.

$$v_x = (1.8\ \text{m/s}) \sin ((7.1\ \text{rad/s})\,t)$$

$$= -(1.8\ \text{m/s}) \cos ((7.1\ \text{rad/s})\,t + \pi)$$

(a)

$$\omega = 7.1\ \text{rad/s}$$

$$\omega A = 1.8\ \text{m/s}, \quad \text{so} \quad A = \frac{1.8\ \text{m/s}}{7.1\ \text{rad/s}} = 0.25\ \text{m}$$

$$\nu = \frac{\omega}{2\pi} = 1.13\ \text{Hz}$$

$$T = \frac{2\pi}{\omega} = 0.88\ \text{s}$$

$$\phi = \pi\ \text{rad}$$

(b)

$$x = (-0.25\ \text{m}) \cos ((7.1\ \text{rad/s})\,t)$$

$$a_x = (13\ \text{m/s}^2) \cos ((7.1\ \text{rad/s})\,t)$$

(c) When $t = 0.25\ \text{s}$,

$$x = (-0.25\ \text{m}) \cos (1.775\ \text{rad}) = 0.051\ \text{m}$$

$$v_x = (1.8\ \text{m/s}) \sin (1.78\ \text{rad}) = 1.8\ \text{m}$$

$$a_x = (13\ \text{m/s}^2) \cos (1.78\ \text{rad}) = -2.6\ \text{m/s}^2$$

7. (a)

$$a_{max} = 13\ \text{m/s}^2$$

$$T = 0.94\ \text{s}, \quad \text{so} \quad \omega = \frac{2\pi\ \text{rad}}{0.94\ \text{s}} = 6.7\ \text{rad/s}$$

$$a_{max} = A\omega^2, \quad \text{so} \quad A = \frac{13\ \text{m/s}^2}{(6.7\ \text{rad/s})^2} = 0.29\ \text{m}$$

$$\phi = \frac{\pi}{2},$$

$$\text{so} \quad x = 0.29\ \text{m} \cos\left((6.7\ \text{rad/s})\,t + \frac{\pi}{2}\right)$$

$$v_x = -1.9\ \text{m/s} \sin\left((6.7\ \text{rad/s})\,t + \frac{\pi}{2}\right)$$

$$a_x = -13\ \text{m/s}^2 \cos\left((6.7\ \text{rad/s})\,t + \frac{\pi}{2}\right)$$

(b) When $t = 0.54\ \text{s}, \quad (6.7\ \text{rad/s})\,t + \frac{\pi}{2}\ \text{rad} = 5.2\ \text{rad}$

$$x = 0.29\ \text{m} \cos(5.2\ \text{rad}) = 0.13\ \text{m}$$

$$v_x = -1.9\ \text{m/s} \sin(5.2\ \text{rad}) = 1.7\ \text{m/s}$$

$$a_x = -13\ \text{m/s}^2 \cos(5.2\ \text{rad}) = -5.9\ \text{m/s}^2$$

9.

$$A\cos\phi = x_o$$

$$-A\omega\sin\phi = v_{x_o}$$

$$\tan\phi = -\frac{v_{x_o}}{\omega x_o} = -\frac{1.3\ \text{m/s}}{(2\pi\ \text{rad})(0.021\ \text{m})(0.42\ \text{Hz})}$$

$$= -23$$

$$\phi = -1.5\ \text{rad}$$

$$A = \frac{x_o}{\cos\phi} = \frac{0.021\ \text{m}}{\cos(-1.5\ \text{rad})}$$

$$= 0.49\ \text{m}$$

$$v_{max} = A\omega = (0.49\ \text{m})(2\pi)(0.42\ \text{rad/s})$$

$$= 1.3\ \text{m/s}$$

$$a_{max} = A\omega^2 = 3.4\ \text{m/s}^2$$

11. (a) $\omega = \sqrt{\frac{k}{m}} = \sqrt{\frac{45\text{ N/m}}{0.88\text{ kg}}} = 7.2\text{ rad/s}$

(b) $\nu = \frac{\omega}{2\pi} = 1.14\text{ Hz}$

(c) $T = \frac{1}{\nu} = 0.88\text{ s}$

13. $k = m\omega^2 = \frac{4\pi^2 m}{T^2} = \frac{4\pi^2(0.62\text{ kg})}{(0.87\text{ s})^2} = 32\text{ N/m}$

15. $m = \frac{kT^2}{4\pi^2} = \frac{k}{(2\pi\nu)^2} = \frac{26\text{ N/kg}}{(2\pi(1.4\text{ Hz}))^2} = 0.34\text{ kg}$

17. $E = \frac{1}{2}kA^2 = \frac{1}{2}(22\text{ N/m})(0.087\text{ m})^2 = 0.083\text{ J}$

19. $\omega = \sqrt{\frac{k}{m}} = \sqrt{\frac{31\text{ N/m}}{0.74\text{ kg}}} = 6.5\text{ rad/s}$

$x = (0.039\text{ m})\cos((6.5\text{ rad/s})t)$

$v_x = -(0.25\text{ m/s})\sin((6.5\text{ rad/s})t)$

(a) $U = \frac{1}{2}kx^2 = (0.024\text{ J})\cos^2((6.5\text{ rad/s})t)$

(b) $K = \frac{1}{2}mv^2 = (0.024\text{ J})\sin^2((6.5\text{ rad/s})t)$

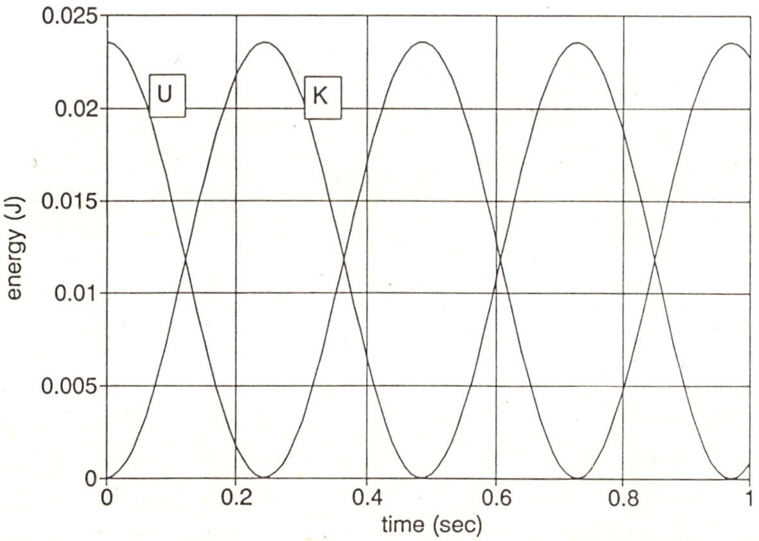

21. $E = 25 \times 10^{-3}\,\text{J} = \frac{1}{2}kA^2$

(a) $\therefore \quad A = \sqrt{\frac{2E}{k}} = \sqrt{\frac{2(25 \times 10^{-3}\,\text{J})}{23\,\text{N/m}}}$

$= 0.047\,\text{m}$

(b) $v_{max} = \sqrt{\frac{2E}{m}} = \sqrt{\frac{2(25 \times 10^{-3}\,\text{J})}{0.47\,\text{kg}}}$

$= 0.33\,\text{m/s}$

(c) $E = \frac{1}{2}mv^2 + \frac{1}{2}kx^2$

$v = \sqrt{\frac{2E}{m} - \frac{kx^2}{m}} = \left(\frac{50 \times 10^{-3}\,\text{J} - 23\,\text{N/m}\,(0.011\,\text{m})^2}{0.47\,\text{kg}}\right)^{1/2}$

$= 0.317\,\text{m/s}$

(d) $x = \sqrt{\frac{2E}{k} - \frac{mv^2}{k}} = \left(\frac{50 \times 10^{-3}\,\text{J} - 0.47\,\text{kg}(.25\,\text{m/s})^2}{23\,\text{N/m}}\right)^{1/2}$

$= 0.030\,\text{m}$

23. (a) $T = 2\pi\sqrt{\frac{m}{k}}, \quad \frac{T_1}{T_2} = \sqrt{\frac{m_1}{m_2}}, \quad \text{or} \quad \frac{m_1}{m_2} = \left(\frac{T_1}{T_2}\right)^2$

so $\frac{m}{1\,\text{kg}} = \left(\frac{T_1}{T_2}\right)^2 = \left(\frac{1.85\,\text{s}}{1.43\,\text{s}}\right)^2, \quad m = 1.67\,\text{kg}$

(b) $k = \frac{4\pi^2 m}{T^2} = \frac{4\pi^2(1\,\text{kg})}{(1.43\,\text{s})^2} = 19.3\,\text{N/m}$

25. (a) At $y = 0, \quad E_o = \frac{1}{2}ky_o^2,$

where $y_o = \frac{mg}{k}$

So $E_o = \frac{1}{2} \cdot \frac{m^2 g^2}{k} = 1.0\,\text{J}$

(b) $E - E_o = \frac{1}{2} kA^2 = \frac{1}{2}(1200\ \text{N/m})(0.025\ \text{m})^2$

$$= 0.38\ \text{J}$$

$$E = 1.0\ \text{J} + 0.38\ \text{J} = 1.4\ \text{J}$$

(c) The block will have its maximum speed at $y = 0$.

$$v_{max} = \sqrt{\frac{2(E - E_o)}{m}} = \sqrt{\frac{2(0.38\ \text{J})}{5.0\ \text{kg}}}$$

$$= 0.39\ \text{m/s}$$

(d) The maximum gravitational energy is

$$mgA = (5.0\ \text{kg})(9.8\ \text{m/s}^2)(0.025\ \text{m})$$

$$= 1.23\ \text{J}$$

27. (a) $\omega = \sqrt{\frac{g}{\ell}} = \sqrt{\frac{9.8\ \text{m/s}^2}{3.0\ \text{m}}} = 1.8\ \text{rad/s}$

$$\nu = \frac{\omega}{2\pi} = 0.29\ \text{Hz}$$

$$T = \frac{1}{\nu} = 3.47\ \text{s}$$

(b) $v_{max} = \omega \ell \theta_{max} = (1.8\ \text{rad/s})(3.0\ \text{m})(0.07\ \text{rad})$

$$= 0.38\ \text{m/s}$$

(The mass of the child does not affect these results.)

29. The moment of inertia of a stick of length ℓ about its center is $\frac{1}{12} M\ell^2$. Shifting the axis to one end makes this

$$I_{end} = \frac{1}{12} M\ell^2 + M\left(\frac{\ell}{2}\right)^2 = \frac{1}{3} M\ell^2$$

and shifting it to $\frac{\ell}{4}$ away from the center makes it

$$I_{\ell/4} = \frac{1}{12} M\ell^2 + M\left(\frac{\ell}{4}\right)^2 = \frac{7}{48} M\ell^2$$

The period of a physical pendulum is given by

$$T = 2\pi\sqrt{\frac{I}{MgL}},$$

where L is the distance between the point of suspension and the center of gravity.

(a) For a stick suspended at one end,

$$L = \frac{\ell}{2},$$

so $$T = 2\pi\sqrt{\frac{\frac{1}{3}M\ell^2}{Mg\frac{\ell}{2}}} = 2\pi\sqrt{\frac{2}{3}\cdot\frac{\ell}{g}}$$

With $\ell = 1$ m, $T = 1.64$ s

(b) With $L = \frac{\ell}{4}$,

$$T = 2\pi\sqrt{\frac{\frac{7}{48}M\ell^2}{Mg\frac{\ell}{4}}} = 2\pi\sqrt{\frac{7}{12}\cdot\frac{\ell}{g}}$$

$$= 1.53 \text{ s}$$

(See Problem 19 for further discussion.)

31. (a) For a torsional pendulum,

$$T = 2\pi\sqrt{\frac{I}{\kappa}}$$

or $$\kappa = \frac{4\pi^2 I}{T^2}$$

$$= \frac{4\pi^2(4.20\times 10^{-8}\text{ kg}\cdot\text{m}^2)}{(0.250\text{ s})^2}$$

$$= 2.65\times 10^{-5}\text{ N}\cdot\text{m/rad}$$

(b) $$\alpha_{max} = \theta_{max}\omega^2 = \theta_{max}\left(\frac{2\pi}{T}\right)^2$$

$$= (0.45\text{ rad})\left(\frac{2\pi}{0.250\text{ s}}\right)^2 = 280\text{ rad/s}^2$$

33. (a) At equilibrium,

$$\tau = \kappa\theta$$

Thus, $$(0.075\ \text{m})(2.2 \times 10^{-3}\ \text{N}) = \kappa\left(\frac{\pi}{4\ \text{rad}}\right)$$

so $$\kappa = 0.00021\ \text{N} \cdot \text{m/rad}$$

Now, $$T = 2\pi\sqrt{\frac{I}{\kappa}}$$

and $$I = \frac{1}{12}M\ell^2 = \frac{1}{12}(0.025\ \text{kg})(0.075\ \text{m})^2$$

$$= 1.17 \times 10^{-5}\ \text{kg} \cdot \text{m}^2$$

so $$T = 1.48\ \text{s}$$

(b) The maximum rotational kinetic energy after the pencil is released is

$$K_{max} = \frac{1}{2}I\omega_{max}^2 = \frac{1}{2}I\left(\frac{2\pi\theta_{max}}{T}\right)^2$$

$$= 6.47 \times 10^{-5}\ \text{J}$$

35. The shadow of Q has an x–coordinate given by

$$x = R\cos\theta, \quad \text{where} \quad \theta = \omega t$$

Here $$R = 0.150\ \text{m} \quad \text{and} \quad \omega = \frac{(33\frac{1}{3}\ \text{rev/min})\,2\pi\ \text{rad/rev}}{60\ \text{s/min}} = 3.49\ \text{rad/s}$$

Thus, $$x = (0.150\ \text{m})\cos((3.49\ \text{rad/s})t),$$

$$v_x = \frac{dx}{dt} = (-0.524\ \text{m/s})\sin((3.49\ \text{rad/s})t),$$

and $$a_x = \frac{dv_x}{dt} = (-1.83\ \text{m/s}^2)\cos((3.49\ \text{rad/s})t)$$

37. (a) In the conical pendulum, the bob moves in a circle of radius $R = L\sin\theta$. The centripetal force is provided by the horizontal component of the tension

$$T_x = m\omega^2 R = m\omega^2 L\sin\theta,$$

while the vertical component supports the weight

$$T_y = mg$$

But $\frac{T_x}{T_y} = \tan\theta$,

since the tension is along the string.

Thus, $m\omega^2 L \sin\theta = mg\tan\theta$

or $\omega^2 = \frac{g}{L\cos\theta}$,

where $\omega = \frac{d\phi}{dt}$

and ϕ is the angular location of the bob in the x–y plane.

Now,
$$x = R\cos\phi$$
$$= L\sin\theta\cos\omega t$$
$$v_x = -\omega L\sin\theta\sin\omega t$$
$$a_x = -\omega^2 L\sin\theta\cos\omega t$$

This is SHM.

(b) The amplitude of the SHM is L sin θ and the angular frequency is

$$\omega = \sqrt{\frac{g}{L\cos\theta}}$$

39. (a) Since the amplitude is halved every 2.4 minutes, the amplitude will be 30 mm after 4.8 minutes.

(b) $A = A_o e^{-\gamma t}$

$$60\text{ mm} = 120\text{ mm}\, e^{-\gamma(2.4\text{ min})(60\text{ s/min})}$$

Taking natural logarithms,

$$-\gamma(2.4\text{ min})(60\text{ s/min}) = \ln\left(\frac{60\text{ mm}}{120\text{ mm}}\right)$$

$$\therefore \quad \gamma = 0.0048\text{ s}^{-1}$$

41. (a) $A = A_o e^{-\gamma t}$

$$95\text{ mm} = 135\text{ mm}\, e^{-\gamma(5.0\text{ min})(60\text{ s/min})}$$

$$\gamma = -\frac{1}{300\text{ s}}\ln\left(\frac{95\text{ mm}}{135\text{ mm}}\right) = 0.00117\text{ s}^{-1}$$

(b) The time required for A to reach 50 mm is found from

$$50\text{ mm} = 135\text{ mm}\, e^{-(0.00117\text{ s}^{-1})t}$$

or
$$t = -\frac{1}{.00117\text{ s}^{-1}} \ln\left(\frac{50\text{ mm}}{135\text{ mm}}\right) = 849\text{ s}$$

which is 14 minutes after the weight reached the bottom, namely at 9:31.

43. From equation 14–28,

$$y = e^{-\gamma t} A \cos(\omega_D t + \phi)$$

$$v_y = \frac{dy}{dt} = e^{-\gamma t} A\left[-\gamma\cos(\omega_D t + \phi) - \omega_D \sin(\omega_D t + \phi)\right]$$

$$a_y = \frac{dv_y}{dt} = e^{-\gamma t} A\left[(\gamma^2 - \omega_D^2)\cos(\omega_D t + \phi) + 2\omega_D \gamma \sin(\omega_D t + \phi)\right]$$

Using the first two equations,

$$e^{-\gamma t} A \sin(\omega_D t + \phi) = -\frac{v_y + \gamma y}{\omega_D}$$

and thus

$$a_y = (\gamma^2 - \omega_D^2)y - 2\omega_D \gamma \frac{v_y + \gamma y}{\omega_D}$$

$$= (\gamma^2 - \omega_D^2 - 2\gamma^2)y - 2\gamma v_y$$

$$= -(\omega^2 + \gamma^2)y - 2\gamma v_y$$

$$= -\left(\frac{k}{m} - \frac{b^2}{4m^2} + \frac{b^2}{4m^2}\right)y - \frac{2b}{2m} v_y$$

$$= -\frac{k}{m} y - \frac{b}{m} v_y,$$

which is equation 14–27.

45. If $y = e^{-\omega t}(A + Bt)$,

then $v_y = \frac{dy}{dt} = e^{-\omega t}(-\omega A - \omega B t + B)$

and $a_y = \frac{dv_y}{dt} = e^{-\omega t}\left[-\omega(-\omega A + B - \omega B t) - \omega B\right]$

Simplifying,

$$v_y = (B - \omega A)\,e^{-\omega t} - \omega B\,t e^{-\omega t}$$

$$a_y = (\omega^2 A - 2\omega B)\,e^{-\omega t} + \omega^2 B\,t e^{-\omega t}$$

The goal is to express a_y in terms of y and v_y. We do this by solving for $e^{-\omega t}$ and $te^{-\omega t}$ in terms of y and v_y:

$$v_y + \omega y = B e^{-\omega t}$$

and

$$\begin{aligned} B\,t e^{-\omega t} &= y - A e^{-\omega t} \\ &= y - \left(\tfrac{A}{B}\right)(v_y + \omega y) \\ &= \left(1 - \tfrac{\omega A}{B}\right) y - \tfrac{A}{B}\,v_y \end{aligned}$$

Thus,

$$\begin{aligned} a_y &= \frac{(\omega^2 A - 2\omega B)(v_y + \omega y)}{B} + \omega^2\left[\left(1 - \tfrac{\omega A}{B}\right) y - \tfrac{A}{B}\,v_y\right] \\ &= \left(\frac{\omega^3 A}{B} - 2\omega^2 + \omega^2 - \frac{\omega^3 A}{B}\right) y + \left(\frac{\omega^2 A}{B} - 2\omega - \frac{\omega^2 A}{B}\right) v_y \\ &= -\omega^2 y - 2\omega v_y \end{aligned}$$

This is identical to equation 14–27, provided that

$$\frac{k}{m} = \omega^2 \quad \text{and} \quad \frac{b}{m} = 2\omega,$$

which are the conditions for critical damping.

47. From Exercise 40, $\gamma = \frac{b}{2m} = 10\ s^{-1}$

$$\omega = \sqrt{\frac{k}{m}} = \sqrt{\frac{1250\ N/m}{2.5\ kg}} = 22.4\ rad/s$$

Using equation 14–31,

$$\text{(a)}\quad A_o = \frac{\frac{F_o}{m}}{\sqrt{(\omega_E^2 - \omega^2)^2 + 4\gamma^2\omega_E^2}}$$

$$= \frac{\frac{12\ N}{2.5\ kg}}{\sqrt{(625\ rad^2/s^2 - 500\ rad^2/s^2)^2 + 4(10\ s^{-1})^2(25\ rad/s)^2}} = 0.00931\ m$$

and

(b) $$\tan\phi_E = \frac{2\gamma\omega_E}{\omega^2 - \omega_E^2} = \frac{2\,(10\ \text{s}^{-1})(25\ \text{rad/s})}{500\ \text{rad}^2/\text{s}^2 - 625\ \text{rad}^2/\text{s}^2} = -4.0$$

$$\phi_E = -1.33\ \text{radians}$$

(c) If the system is driven at resonance,

$$A_o = \frac{\frac{F_o}{m}}{2\gamma\omega} = \frac{\frac{12\ \text{N}}{2.5\ \text{kg}}}{2\,(10\ \text{s}^{-1})(22.4\ \text{rad/s})}$$

$$= 0.0107\ \text{m}$$

49. If $x = A\sin(\omega t + \delta)$, then

(a) $v_x = \frac{dx}{dt} = A\omega\cos(\omega t + \delta)$, and

(b) $a_x = \frac{dv_x}{dt} = -A\omega^2\sin(\omega t + \delta)$

(c) Evaluating these expressions at $t = 0$,

$$x_o = A\sin\delta$$

$$v_o = A\omega\cos\delta$$

Therefore $\frac{\omega x_o}{v_o} = \tan\delta$,

so $\delta = \tan^{-1}\left(\frac{\omega x_o}{v_o}\right)$, and

(d) $A^2 = x_o^2 + \frac{v_o^2}{\omega^2}$

51. Given $x = (0.27\ \text{m})\cos((9.3\ \text{rad/s})t - 2.4\ \text{rad})$,

(a) the amplitude of the oscillations is 0.27 m

(b) the angular frequency is 9.3 rad/s

(c) $\nu = \frac{\omega}{2\pi} = 1.48\ \text{Hz}$

(d) $T = \frac{1}{\nu} = 0.676\ \text{s}$

(e) At $t = 0.33$ s, the phase is

$$(9.3 \text{ rad/s})(0.33 \text{ s}) - 2.4 \text{ rad} = 0.67 \text{ radians}$$

and

(f) $\mathbf{x} = \mathbf{i}\,(0.27 \text{ m}) \cos(0.67 \text{ rad})$

$= (0.21 \text{ m})\,\mathbf{i}$

53. From the graph, when $x = 15$ mm, $U = 22.5$ mJ

When $x = 0$ mm, $U = 0$ mJ

Thus,

(a) $K = \frac{1}{2} m v^2 = 22.5 \times 10^{-3}$ J

(b) $v = \sqrt{\dfrac{2 \times 22.5 \times 10^{-3} \text{ J}}{0.87 \text{ kg}}} = 0.227$ m/s

When $x = 10$ mm, $U = 10$ mJ

Thus,

(c) $K = 22.5 \text{ mJ} - 10 \text{ mJ} = 12.5$ mJ,

and

(d) $v = \sqrt{\dfrac{2 \times 12.5 \times 10^{-3} \text{ J}}{0.87 \text{ kg}}} = 0.170$ m/s

55. For a disc, $I_C = \frac{1}{2} m R^2$. Using the parallel axis theorem, the moment of inertia about an axis through its edge is

$$I = \frac{1}{2} m R^2 + m R^2 = \frac{3}{2} m R^2$$

From equation 14–23,

$$\omega = \sqrt{\frac{m g L}{I}} = \sqrt{\frac{m g R}{\frac{3}{2} m R^2}} = \sqrt{\frac{2 g}{3 R}}$$

57. $I = \frac{1}{2} M\ell^2, \quad \omega^2 = \frac{\kappa}{I},$

so
$$\kappa = I\omega^2 = \frac{1}{2} M\ell^2\omega^2$$
$$= \frac{1}{2}(34\text{ kg})(0.92\text{ m})^2\left(\frac{2\pi\text{ rad}}{1.4\text{ s}}\right)^2$$
$$= 290\text{ N/m}$$

59. Drawing a straight line through the origin and passing as close as possible to the plotted points in Fig. 14–29, the value of T^2 corresponding to m = 1.00 kg is found to be 0.40 s^2.

Thus, $$T^2 = (0.40\text{ s}^2/\text{kg})m$$

Now, $$\omega^2 = \frac{k}{m} = \frac{4\pi^2}{T^2},$$

so $$k = 4\pi^2\left(\frac{m}{T^2}\right) = \frac{4\pi^2}{0.40\text{ s}^2/\text{kg}} = 99\text{ N/m}$$

61. Estimate a person's height with arms outstretched at 2.5 m and the mass at 70 kg with the center of gravity at approximately 1.25 m. Also estimate the central moment of inertia at $\frac{1}{20} M\ell^2$, somewhat less than for a uniform bar, since the trunk will have a larger mass than the limbs. Then

$$I_{end} = M\left(\frac{\ell}{2}\right)^2 + \frac{1}{20}M\ell^2 = \frac{3}{10}M\ell^2$$

and $$T = 2\pi\sqrt{\frac{I}{Mg\left(\frac{\ell}{2}\right)}} = 2\pi\sqrt{\frac{3}{5}\cdot\frac{\ell}{g}}$$
$$\simeq 2\text{ s}$$

63. In this case,

$$I = M\left(\frac{D}{2}\right)^2 + M(D)^2$$
$$= \frac{5}{4}MD^2$$

and $$L = \frac{3D}{4}$$

Thus, $$T = 2\pi\sqrt{\frac{I}{MgL}} = 2\pi\sqrt{\frac{5}{3}\cdot\frac{D}{g}}$$

65. Let θ be the angle subtended at the sun by the radii at t = 0 and t. Then

$$\theta = \omega t, \qquad \omega = \frac{2\pi}{1\text{ year}} = 0.20 \times 10^{-6}\text{ rad/s}$$

Thus,

(a) $$x = (1.5 \times 10^{11}\text{ m}) \cos\left((0.20 \times 10^{-6}\text{ rad/s})t\right)$$

$$y = (1.5 \times 10^{11}\text{ m}) \sin\left((0.20 \times 10^{-6}\text{ rad/s})t\right)$$

(b) $$v_x = \frac{dx}{dt} = (-3.0 \times 10^{5}\text{ m/s}) \sin\left((0.20 \times 10^{-6}\text{ rad/s})t\right)$$

$$v_y = \frac{dy}{dt} = (3.0 \times 10^{5}\text{ m/s}) \cos\left((0.20 \times 10^{-6}\text{ rad/s})t\right)$$

(c) $$a_x = \frac{dv_x}{dt} = (-0.0060\text{ m/s}^2) \cos\left((0.20 \times 10^{-6}\text{ rad/s})t\right)$$

and

$$a_y = \frac{dv_y}{dt} = (-0.0060\text{ m/s}^2) \sin\left((0.20 \times 10^{-6}\text{ rad/s})t\right)$$

Problems:

1. $x = A \cos \dfrac{2\pi t}{T}$

$x = A$ when $t = 0$

$x = \dfrac{A}{2}$ when $\cos \dfrac{2\pi t}{T} = \dfrac{1}{2}$

that is, when $\dfrac{2\pi t}{T} = \dfrac{\pi}{3}$

Thus it takes $t = \dfrac{T}{6}$ for the oscillator to travel from $x = A$ to $x = \dfrac{A}{2}$.

3. When the elevator has an upward acceleration a, the pendulum experiences an effective value of gravity of g + a.

Let $T_o = 2\pi\sqrt{\dfrac{\ell}{g}}$ (the period when unaccelerated)

(a) $$T_{up} = 2\pi\sqrt{\frac{\ell}{g+a}} = T_o\sqrt{\frac{g}{\frac{3}{2}g}} = T_o\sqrt{\frac{2}{3}}$$

$$= 0.090\text{ s}$$

(b) When the elevator is moving at constant velocity, $a = 0$

so $T = T_o = 0.11$ s

(c) When $a = -\frac{g}{2}$

$$T_{down} = 2\pi\sqrt{\frac{\ell}{g - \frac{g}{2}}} = T_o\sqrt{2} = 0.16 \text{ s}$$

(d) The elevator clock runs fast when it is accelerating upward and slow when it is accelerating downward.

Assuming that the upward and downward accelerations are $\pm\frac{g}{2}$ and have equal durations T_A, then the elevator clock will have ticked

$$\frac{T_A}{T_{up}} + \frac{T_A}{T_{down}} = \sqrt{\frac{3}{2}}\frac{T_A}{T_o} + \sqrt{\frac{1}{2}}\frac{T_A}{T_o}$$

$$= 1.932\frac{T_A}{T_o} \text{ ticks}$$

in the time that the lobby clock has ticked

$$\frac{2T_A}{T_o} \text{ ticks}$$

Thus, the elevator clock will be slow because the lobby clock averages more ticks per unit time.

5. (a) $K_s = \frac{1}{2}\int_0^{\ell} \mu v_s^2 \, d\xi = \frac{1}{2}\cdot\frac{m_s}{\ell} v^2 \int_0^{\ell} \left(\frac{\xi}{\ell}\right)^2 d\xi$

$$= \frac{1}{6} m_s v^2$$

(b) $E = \frac{1}{2}mv^2 + \frac{1}{6}m_s v^2 + \frac{1}{2}kx^2$

$$= \frac{1}{2}\left(m + \frac{1}{3}m_s\right)v^2 + \frac{1}{2}kx^2$$

Thus, the system behaves like a block of mass $m + \frac{1}{3}m_s$ connected to a linear spring with spring constant k and

$$\omega = \sqrt{\frac{k}{m + \frac{m_s}{3}}}$$

7. The system is the same as a pendulum of length = 140 mm executing small oscillations. (The normal force exerted by the bowl replaces the tension in the string.)

Thus, $$T = 2\pi\sqrt{\frac{\ell}{g}} = 2\pi\sqrt{\frac{0.140\text{ m}}{9.8\text{ m/s}^2}}$$

$$= 0.75\text{ s}$$

9. (a) From equations 14–29 and 14–31,

$$F_{Ex} = F_o \cos \omega_E t$$

$$x = A_o \cos(\omega_E t - \phi_E)$$

$$v_x = \frac{dx}{dt} = -A_o \omega_E \sin(\omega_E t - \phi_E)$$

Thus, $$P = F_{Ex} v_x = -A_o F_o \omega_E \cos \omega_E t \sin(\omega_E t - \phi_E)$$

$$= -A_o F_o \omega_E \cos \omega_E t\,[\sin \omega_E t \cos \phi_E - \cos \omega_E t \sin \phi_E]$$

(b) The average value of $\sin \omega_E t \cos \omega_E t$ is zero and that of $\cos^2 \omega_E t$ is $\frac{1}{2}$,

so $$P_{av} = \frac{1}{2} A_o F_o \omega_E \sin \phi_E$$

Now, $$\tan \phi_E = \frac{2\gamma\omega_E}{\omega^2 - \omega_E^2}$$

so $$\sin \phi_E = \frac{2\gamma\omega_E}{\sqrt{(\omega^2 - \omega_E^2)^2 + 4\gamma^2\omega_E^2}}$$

and using $$A_o = \frac{\frac{F_{Ex}}{m}}{\sqrt{(\omega_E^2 - \omega^2)^2 + 4\gamma^2\omega_E^2}},$$

$$P_{av} = \frac{\frac{\gamma\omega_E F_{Ex}^2}{m}}{(\omega_E^2 - \omega^2)^2 + 4\gamma^2\omega_E^2}$$

The maximum power occurs when the denominator is a minimum, that is, when $\omega = \omega_E$ (the denominator is the sum of two positive terms, one of which is constant so the smallest value occurs when the first term is zero).

$$(P_{av})_{max} = \frac{1}{4} \cdot \frac{F_{Ex}^2}{\gamma\omega_E m}$$

11. Using the spreadsheet from Chapter 4, the resulting graphs are shown below:

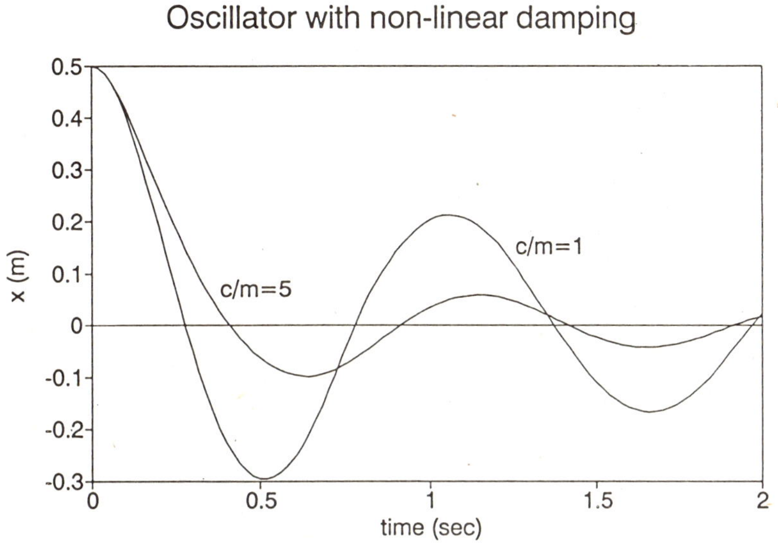

13. The central moments of inertia for the figures shown are:

$$I_{ac} = \tfrac{1}{2}MR^2,$$

$$I_{bc} = MR^2, \quad \text{and}$$

$$I_{cc} = \tfrac{1}{2}M_1R^2 - \tfrac{1}{2}M_2\left(\tfrac{R}{2}\right)^2,$$

where $M_1 = \rho\pi R^2, \quad M_2 = \rho\pi\left(\tfrac{R}{2}\right)^2,$

and $M_1 - M_2 = M$

Thus, $\tfrac{3}{4}\rho\pi R^2 = M, \quad I_{cc} = \tfrac{15}{32}\rho\pi R^4 = \tfrac{5}{8}MR^2$

Using the parallel axis theorem, the moments about point O become

$$I_a = \tfrac{3}{2}MR^2, \quad I_b = 2MR^2, \quad I_c = \tfrac{13}{8}MR^2$$

In all three cases $L = R$, so

(a) $T_a = 2\pi\sqrt{\dfrac{I_a}{MgL}} = 2\pi\sqrt{\dfrac{3R}{2g}}$

(b) $T_b = 2\pi\sqrt{\frac{2R}{g}}$

(c) $T_c = 2\pi\sqrt{\frac{13R}{8g}}$

15. The resulting map is shown below.

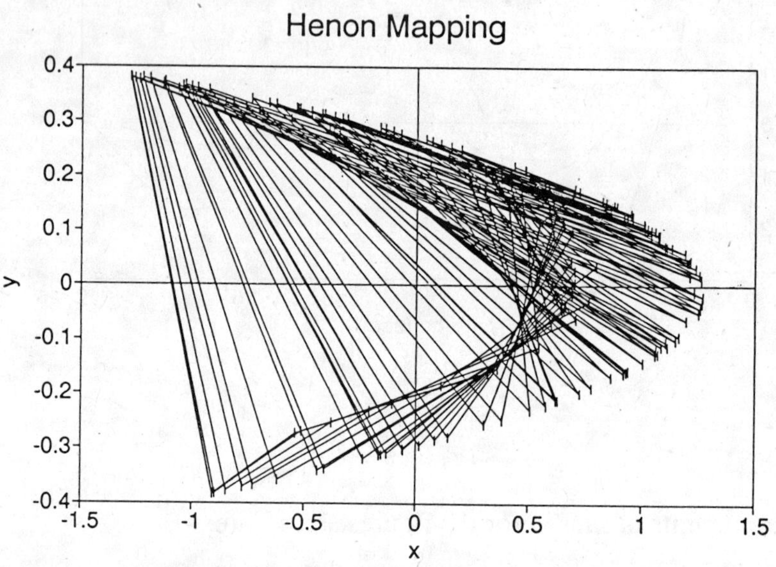

17. (a) For the real pendulum,

$$T = 2\pi\sqrt{\frac{I}{mgL}},$$

while for the simple pendulum

$$T = 2\pi\sqrt{\frac{d}{g}}$$

To make these expressions equivalent,

$$\frac{I}{mgL} = \frac{d}{g}$$

or $$d = \frac{I}{mL}$$

(b) For a uniform rod suspended from one end,

$$I = \frac{1}{3}mD^2, \quad L = \frac{D}{2},$$

so $$d = \frac{\frac{1}{3} mD^2}{\frac{mD}{2}} = \frac{2}{3} D$$

(c) This is the same result as that obtained in example 14–6.

19. The relevant equations for this problem were derived in Exercise 29.

$$I = \frac{1}{12} MD^2 + Mx^2, \qquad L = x$$

(a) $$T = 2\pi\sqrt{\frac{I}{MgL}} = 2\pi\sqrt{\frac{x + \frac{1}{12}\cdot\frac{D^2}{x}}{g}}$$

(b) T will be a minimum when

$$\frac{d}{dx}\left(x + \frac{1}{12}\cdot\frac{D^2}{x}\right) = 0, \qquad \text{which is when} \qquad x^2 = \frac{D^2}{12}$$

or $$x = \frac{D}{\sqrt{12}}$$

(c) $$T_{min} = 2\pi\sqrt{\frac{D}{g\sqrt{3}}}$$

21. Measuring the rotation of the sphere with respect to the line joining its center to the center of the bowl by ϕ, the sphere advances a distance $r_1\phi$ along the bowl, which subtends an angle $\theta = \frac{r_1\phi}{r_2}$ at the center of the bowl. The angle that the sphere rotates about a fixed reference frame is $\phi - \theta$.

Thus, $$\omega = \frac{d}{dt}(\phi - \theta) = \frac{r_2 - r_1}{r_1}\cdot\frac{d\theta}{dt}$$

and the speed of the sphere is ωr_1.

Thus, $$K = \frac{1}{2} mv^2 + \frac{1}{2}\left(\frac{2}{5} mr_1^2\right)\omega^2$$

$$= \frac{1}{2}\left(\frac{7}{5} mr_1^2\right)\omega^2 = \frac{1}{2}\left(\frac{7}{5} m(r_2 - r_1)^2\right)\left(\frac{d\theta}{dt}\right)^2$$

$$U = mg(r_2 - r_1)(1 - \cos\theta)$$

$$\simeq \frac{1}{2} mg(r_2 - r_1)\theta^2$$

This corresponds to a real pendulum with

$$I = \frac{7}{5}\, m(r_2 - r_1)^2$$

and $$L = (r_2 - r_1),$$

so $$T = 2\pi \sqrt{\frac{\frac{7}{5}(r_2 - r_1)}{g}}$$

which is the period of a simple pendulum of length $\frac{7}{5}(r_2 - r_1)$.

CHAPTER 15 SOLIDS AND FLUIDS

Questions:

1. Ice, a glacier, and gelatin are solids. Water vapor is a gas. Fog and clouds are water droplets in air, but could be considered as a gas for most purposes. Taffy, bread dough, and honey all flow, so they are liquids. Sugar is a solid.

3. Breaking up the stresses into components along the joint and normal to the joint, it can be seen that the stresses are both tensile and shear stresses and they are of equal magnitude, that is, their ratio is equal to one.

5. The material is a combination of a high Young's modulus material — the epoxy — which, however, is brittle (low yield point), and a material with the opposite set of characteristics. Thus, the composite has better structural properties.

7. The least dense gas under standard conditions is hydrogen. The densest substance is iridium.

9. The diver is more compressible than its surrounding liquid. When the pressure is increased the density of the diver becomes greater and it sinks. When the pressure is reduced the density of the diver decreases and it rises to the surface. When the diver is stationary in the fluid, it has the same density as the fluid. The density and compressibility of the fluid (and the diver) are important in the operation of this apparatus.

11. The tankers load with water so that they will be floating at the same water line as in the loaded condition. If they did not, they would be less stable and could capsize.

13. As the water falls away from the faucet it speeds up. Since the volume of water flowing per unit time is the same at every section, the area of the stream has to decrease as the velocity increases.

15. Water appears to be flowing uphill in an open channel (at constant pressure).

17. The air flow past the car decreases the pressure outside. Thus the inside pressure is larger and the convertible top bulges out.

19. No; in laminar flow there is internal shear. Turbulent flow comes closer to being irrotational except near the walls of the pipe.

21. The air flow draws you toward the truck. This is because the higher air velocity translates into a decrease in pressure. People changing a tire by the side of the road have been killed by being sucked under a passing truck.

23. The statement that there is a buoyant force on the rope is incorrect. The rope is not completely immersed in water, thus the net force on the rope is a purely compressive, horizontal force squeezing the rope. The key components of the hydrostatic force, in the vertical direction, are missing.

25. Once flow is established, all that matters for a siphon to work is that the flow be from the higher reservoir to the lower. The siphon will stop functioning if the absolute pressure drops below the vapor pressure of water. Then a cavity (a vapor bubble) develops and the siphon no longer works. This occurs when

$$\rho g h = p_{atm} - p_{vap}$$

Since $p_{vap} \ll p_{atm}$ this occurs when

$$h = \frac{p_{atm}}{\rho g} = \frac{101\text{ k Pa}}{(1000\text{ kg/m}^3)(9.8\text{ m/s}^2)}$$

$$= 10.3\text{ m}$$

Exercises:

1. The compressive stress is

$$\sigma_c = \frac{F}{A} = \frac{215\text{ N}}{(0.0127\text{ m})^2} = 1.33\text{ MPa}$$

3. (a) The aluminum is under a tensile stress

$$\sigma_t = \frac{F}{A} = \frac{(22.4\text{ kg})(9.8\text{ m/s}^2)}{\left(\frac{\pi}{4}(.091\text{ m})^2\right)}$$

$$\sigma_t = 34\text{ k Pa}$$

(b) The brass cube is under a shear stress

$$\sigma_s = \frac{F}{A} = \frac{(22.4\text{ kg})(9.8\text{ m/s}^2)}{(0.0850\text{ m})^2} = 30\text{ k Pa}$$

5. (a) $\varepsilon = \dfrac{\sigma}{Y_{copper}} = \dfrac{1.33 \times 10^6 \text{ Pa}}{110 \times 10^9 \text{ Pa}} = 1.21 \times 10^{-5}$

(b) $\Delta x = \varepsilon \ell = (1.21 \times 10^{-5})(0.0127 \text{ m})$

$= 1.54 \times 10^{-7} \text{ m}$

7. (a) The strain on the brass cube is

$$\varepsilon = \frac{\sigma}{S_{brass}} = \frac{30 \text{ k Pa}}{36 \times 10^9 \text{ Pa}} = 8.3 \times 10^{-7}$$

(b) $\Delta\theta = \varepsilon = 8.3 \times 10^{-7}$ radians

(c) The block moves down as a result of the shear deformation of the brass cube and the tensile elongation of the aluminum wire:

$$\Delta x = \ell \Delta\theta + \frac{\ell_{Al}\sigma_t}{Y_{Al}}$$

$$= (0.0850 \text{ m})(8.3 \times 10^{-7}) + \frac{(0.750 \text{ m})(33.7 \times 10^3 \text{ Pa})}{70 \times 10^9 \text{ Pa}}$$

$$= 7.1 \times 10^{-8} \text{ m} + 3.6 \times 10^{-7} \text{ m}$$

$$= 4.3 \times 10^{-7} \text{ m}$$

9. $\rho = \frac{m}{V}$; $V = \frac{4}{3}\pi r^3$

$\therefore$ $\rho = \dfrac{32.3 \text{ kg}}{\frac{4}{3}\pi(0.0730 \text{ m})^3} = 1.98 \times 10^4 \text{ kg/m}^3$

11. The volume of 1.00 kg of water at sea level is $1.00 \times 10^{-3} \text{ m}^3$. At a pressure of 30 MPa, the volume is decreased by

$$\Delta V = \frac{Vp}{B}$$

$$= \frac{(1.00 \times 10^{-3} \text{ m}^3)(30 \times 10^6 \text{ Pa})}{(2.2 \times 10^9 \text{ Pa})}$$

$$= 1.36 \times 10^{-5} \text{ m}^3$$

The density at that pressure is

$$\rho = \frac{m}{V - \Delta V} = \frac{\frac{m}{V}}{1 - \frac{\Delta V}{V}} \simeq \frac{m}{V}\left(1 + \frac{\Delta V}{V}\right)$$

$$= 1000\ \text{kg/m}^3 (1 + 1.36 \times 10^{-2})$$

$$= 1014\ \text{kg/m}^3$$

Repeating the calculation for lead,

$$\rho \simeq \rho_o\left(1 + \frac{p}{B}\right)$$

$$= (11.3 \times 10^3\ \text{kg/m}^3)\left(1 + \frac{30 \times 10^6\ \text{Pa}}{7.7 \times 10^9\ \text{Pa}}\right)$$

$$= 11.3 \times 10^3\ \text{kg/m}^3$$

13. Neglecting the additional effect due to the compressibility of the water,

$$p = \rho g h + p_{atm} \qquad (p_{atm} \text{ is negligible})$$

$$= (1.03 \times 10^3\ \text{kg/m}^3)(9.8\ \text{m/s}^2)(10 \times 10^3\ \text{m})$$

$$= 1.0 \times 10^8\ \text{Pa}$$

15. For an isothermal atmosphere,

$$p = p_o e^{-\rho_o g h/p_o}$$

Here $$\frac{\rho_o g h}{p_o} = \frac{(1.29\ \text{kg/m}^3)(9.8\ \text{m/s}^2)(8 \times 10^3\ \text{m})}{1.01 \times 10^5\ \text{Pa}} = 1.0$$

$\therefore$ $$p = p_o e^{-1.0}$$

$$= (1.01 \times 10^5\ \text{Pa})(.368)$$

$$= 0.37 \times 10^5\ \text{Pa} = 37\ \text{k Pa}$$

17. The pressures at the bottom of the pistons are equal because they are at the same elevation.

Thus $$\frac{F_A}{A_A} = \frac{F_B}{A_B} \quad \text{or} \quad F_A = F_B \frac{A_A}{A_B} = F_B \left(\frac{d_A}{d_B}\right)^2$$

$$F_A = (0.85 \times 10^3 \text{ kg})(9.8 \text{ m/s}^2)\left(\frac{17 \text{ mm}}{300 \text{ mm}}\right)^2$$

$$= 27 \text{ N}$$

19. The weight of the iceberg is

$$W = \rho_{ice}\, g\, V_{iceberg}$$

The buoyant force is

$$B = \rho_{seawater}\, g\, V_{disp}$$

Now, $$V_{iceberg} = V_{disp} + 100 \text{ m}^3,$$

$$\rho_{ice} = 0.92 \times 10^3 \text{ kg/m}^3,$$

and $$\rho_{seawater} = 1.03 \times 10^3 \text{ kg/m}^3$$

At equilibrium $B = W$,

$$\therefore \quad \frac{V_{iceberg}}{V_{disp}} = \frac{\rho_{seawater}}{\rho_{ice}} = \frac{1.03 \times 10^3 \text{ kg/m}^3}{0.92 \times 10^3 \text{ kg/m}^3}$$

$$1 + \frac{100 \text{ m}^3}{V_{disp}} = 1.12$$

$$V_{disp} = 100 \text{ m}^3\left(\frac{1}{1.12 - 1}\right) = 840 \text{ m}^3$$

21. Since a person can float or sink simply by controlling the amount of air in the lungs, it follows that the density of a human body is approximately equal to the density of water. For a person having a mass of 70 kg, the volume would be

$$V = \frac{m}{\rho} = \frac{70 \text{ kg}}{1000 \text{ kg/m}^3}$$

$$= 0.07 \text{ m}^3$$

(or 70 liters)

23. $F_e = \rho_{water} g V$

$F'_e = F_e - \rho_{air} g V$

$$f = \frac{F_e}{F'_e} = \frac{\rho_{water}}{\rho_{water} - \rho_{air}} = \frac{1.000 \times 10^3 \text{ kg/m}^3}{1.000 \times 10^3 \text{ kg/m}^3 - 1.29 \text{ kg/m}^3}$$

$= 1.001$

25. (a) $Q_A = V_A A_A = (1.7 \text{ m/s}) \left(\frac{\pi}{4} (0.050 \text{ m})^2\right)$

$= 0.0033 \text{ m}^3/\text{s}$

$Q_B = \frac{1}{2} Q_A = 0.0017 \text{ m}^3/\text{s}$

(b) $V_B = \frac{Q_B}{A_B} = \frac{0.0017 \text{ m}^3/\text{s}}{\frac{\pi}{4} (0.025 \text{ m})^2} = 3.4 \text{ m/s}$

(c) Using Bernoulli's equation,

$$p_A + \frac{1}{2} \rho V_A^2 = p_B + \frac{1}{2} \rho V_B^2$$

$$p_B = p_A + \frac{1}{2} \rho (V_A^2 - V_B^2)$$

$$= 75 \text{ kPa} + \frac{1}{2} (1000 \text{ kg/m}^3)((1.7 \text{ m/s})^2 - (3.4 \text{ m/s})^2)$$

$$= 75 \text{ kPa} - 4.3 \text{ kPa}$$

$$= 71 \text{ kPa}$$

27. Bernoulli's equation states that

$$p_1 + \rho g y_1 + \frac{1}{2} \rho v_1^2 = p_2 + \rho g y_2 + \frac{1}{2} \rho v_2^2$$

If $v_1 = v_2$, this reduces to

$$p_1 + \rho g y_1 = p_2 + \rho g y_2$$

or $p_2 - p_1 = \rho g (y_1 - y_2) = \rho g h$

29. The volume of air flowing through the engine is

$$Q = \left(\frac{1.6\text{ L}}{2\text{ revs}}\right) \times \left(\frac{3500\text{ revs}}{\text{min}}\right) \times \left(\frac{1\text{ m}^3}{1000\text{ L}}\right) \times \left(\frac{1\text{ min}}{60\text{ s}}\right)$$

$$= 47 \times 10^{-3}\text{ m}^3/\text{s}$$

The speed of the air through the venturi tube is

$$v = \frac{Q}{A} = \frac{47 \times 10^{-3}\text{ m}^3/\text{s}}{\pi (0.0091\text{ m})^2} = 180\text{ m/s}$$

Using the Bernoulli equation,

$$p_1 + \frac{1}{2}\rho v_1^2 = p_2 + \frac{1}{2}\rho v_2^2$$

Here $p_1 = 0$ (atmospheric), and $v_1 \simeq 0$,

so $p_2 = -\frac{1}{2}(1.29\text{ kg/m}^3)(180\text{ m/s})^2 = -21\text{ k Pa}$

This is approximately $\frac{1}{5}$ of an atmosphere below the ambient pressure and would provide the suction to insert fuel into the engine.

31. (a) Applying the Bernoulli equation between a point at the surface of the reservoir (where $v = 0$, $p = 0$) and C,

$$\rho g h = \frac{1}{2}\rho v_C^2$$

$$\therefore \quad v_C = \sqrt{2gh} = \sqrt{2(9.8\text{ m/s}^2)(3.0\text{ m})}$$

$$= 7.7\text{ m/s}$$

(b) $Q = A_C v_C = (0.025\text{ m}^2)(7.7\text{ m/s})$

$$= 0.19\text{ m}^3/\text{s}$$

(c) $$v_A = \frac{Q}{A_A} = \frac{0.19\text{ m}^3/\text{s}}{0.055\text{ m}^2} = 3.5\text{ m/s}$$

$$v_B = \frac{Q}{A_B} = \frac{0.19\text{ m}^3/\text{s}}{0.040\text{ m}^2} = 4.8\text{ m/s}$$

Applying the Bernoulli equation between C and A and between C and B,

$$p_A = p_C + \frac{1}{2}\rho(v_C^2 - v_A^2)$$

$$= \frac{1}{2}(1000\ \text{kg/m}^3)((7.7\ \text{m/s})^2 - (3.5\ \text{m/s})^2) = 23\ \text{k Pa}$$

$$p_B = \frac{1}{2}(1000\ \text{kg/m}^3)((7.7\ \text{m/s})^2 - (4.8\ \text{m/s})^2) = 18\ \text{k Pa}$$

33. Assuming laminar flow,

$$F = \frac{\eta A v}{\ell} = \frac{(1.0 \times 10^{-3}\ \text{N} \cdot \text{s/m}^2)(30\ \text{m}^2)(1.5\ \text{m/s})}{0.140\ \text{m}}$$

$$= 0.32\ \text{N}$$

35. The flow rate is given by

$$Q = \frac{\pi(\Delta p)R^4}{8\eta L}$$

$$= \frac{\pi(0.305 \times 10^{-3}\ \text{Pa})(0.422 \times 10^{-3}\ \text{m})^4}{8(2.5 \times 10^{-3}\ \text{Pa s})(0.010\ \text{m})}$$

$$= 1.51 \times 10^{-13}\ \text{m}^3/\text{s}$$

$$= 1.51 \times 10^{-13}\ \text{m}^3/\text{s} \cdot \frac{3600\ \text{s}}{\text{h}}\left(\frac{1000\ \text{mm}}{\text{m}}\right)^3$$

$$= 0.544\ \text{mm}^3/\text{h}$$

37. Let V = volume of block, x = fraction of volume made of wood

Then $m = \rho_{wood} x V + \rho_{copper} V(1 - x)$

For the block to be neutrally buoyant,

$$\frac{m}{V} = \rho_{water}$$

Thus, $\rho_{water} = x\rho_{wood} + (1 - x)\rho_{copper}$

or $\rho_{copper} - \rho_{water} = x(\rho_{copper} - \rho_{wood})$

$$\therefore \quad x = \frac{\rho_{copper} - \rho_{water}}{\rho_{copper} - \rho_{wood}}$$

$$= \frac{8.9 \times 10^3\ kg/m^3 - 1.00 \times 10^3\ kg/m^3}{8.9 \times 10^3\ kg/m^3 - 0.710 \times 10^3\ kg/m^3}$$

$$= 0.96$$

So, the block should be 96% wood by volume.

39. The weight of the displaced fluid is

$$W_{disp} = mg = \left(\rho_{oil}\frac{V}{2} + \rho_{water}\frac{V}{2}\right) g$$

By Archimedes' principle, this is equal to the buoyant force.

Thus, $\rho_{wood} V g = W_{wood} = B$

so $$\rho_{wood} = \frac{1}{2}(\rho_{oil} + \rho_{water})$$

$$= \frac{1}{2}(840\ kg/m^3 + 1000\ kg/m^3)$$

$$= 920\ kg/m^3$$

41. For a person 1.7 m tall, using

$$\rho_{blood} = 1050\ kg/m^3,$$

the difference in pressure would be

$$\Delta p = \rho_{blood} \Delta h g$$

$$= (1050\ kg/m^3)(1.7\ m)(9.8\ m/s^2)$$

$$= 17\ k\,Pa$$

43. At lift-off, the buoyant force is equal to the weight of the balloon.

$$B = \rho_{air}\, g V$$

$$W = \rho_{hot\ air}\, g V + mg$$

$$\therefore \quad \rho_{hot\ air} = \rho_{air} - \frac{m}{V}$$

$$= 1.29\ kg/m^3 - \frac{200\ kg}{3000\ m^3}$$

$$= 1.22\ kg/m^3$$

45. Assuming that the dead air inside the building can be considered to be part of the same irrotational flow as the wind outside, Bernoulli's equation gives

$$p_{in} = p_{out} + \frac{1}{2}\rho v^2$$

$$\therefore \quad p_{in} - p_{out} = \Delta p = \frac{1}{2}(1.29\ \text{kg/m}^3)(5\ \text{m/s})^2$$

$$= 16\ \text{N/m}^2$$

The outward force on the window is

$$F = (\Delta p)A = (16\ \text{N/m}^2)(2\ \text{m})(3\ \text{m}) = 96\ \text{N}$$

47. For laminar pipe flow,

$$Q = \frac{\pi(\Delta p)R^4}{8\eta\ell}$$

or

$$\eta = \frac{\pi\left(\frac{\Delta p}{\ell}\right)R^4}{8Q}$$

$$= \frac{\pi(0.3\ \text{Pa/m})(0.2\ \text{m})^4}{8(0.25\ \text{m}^3/\text{s})}$$

$$= 0.00075\ \text{Pa s}$$

Problems:

1. This problem is best answered by showing the calculations in a table:

Symbol	ℓ	A	$V = \ell A$	ρ	$M = \rho V$
Bar #	Length (m)	Area (m^2)	Volume (m^3)	Density (kg/m^3)	Mass (kg)
1	0.55	420×10^{-6}	2.3×10^{-4}	8.9×10^3	2.1 kg
2	0.75	390×10^{-6}	2.9×10^{-4}	7.8×10^3	2.3 kg
3	0.45	405×10^{-6}	1.8×10^{-4}	2.7×10^3	0.5 kg
			7.0×10^{-4}		4.9 kg

(a) Totals $M_{TOT} = 4.9$ kg

(b) $\rho_{av} = \dfrac{4.9\text{ kg}}{7.0 \times 10^{-4}\text{ m}^3} = 6.8 \times 10^3\text{ kg/m}^3$

(c)

Bar #	ℓ Length (m)	A Area (m^2)	Y (N/m^2)	$\Delta\ell = \frac{\ell F}{AY}$ (m)
1	0.55	420×10^{-6}	110×10^9	1.2×10^{-4}
2	0.75	390×10^{-6}	100×10^9	1.9×10^{-4}
3	0.45	405×10^{-6}	70×10^9	1.6×10^{-4}
Totals				4.7×10^{-4}

The total elongation is 4.7×10^{-4} m. The elongations can be added because the three sections of the bar are under the same tension.

3. Let V = volume of the cork, x = fraction of the volume under water

The buoyant force is equal to the weight of the fluid displaced by the cork.

$$B = \rho_{H_2O}\, g\, x\, V + \rho_o\, g\, (1 - x)\, V$$

The weight of the cork is

$$W = \rho_C\, g\, V$$

At equilibrium, $W = B$

$$\therefore \quad \rho_C = \rho_{H_2O}\, x + \rho_o (1 - x)$$

Solving for x,

$$x = \frac{\rho_C - \rho_o}{\rho_{H_2O} - \rho_o}$$

5. If $\left(\frac{\Delta\ell}{\ell}\right)_{max} = .001$, then $\left(\frac{F}{AY}\right)_{max} = .01$

The force near the bottom of the building is $\rho\, g\, A\, h$

so $\rho\, g\, A\, h < (.001) A\, Y$

$$h < \frac{.001\, Y}{\rho\, g}$$

For concrete, $h_{max} = \dfrac{(.001)(25 \times 10^9 \text{ Pa})}{(2.4 \times 10^3 \text{ kg/m}^3)(9.8 \text{ m/s}^2)} = 1.1 \text{ km}$

For steel, $h_{max} = \dfrac{(.001)(200 \times 10^9 \text{ Pa})}{(7.8 \times 10^3 \text{ kg/m}^3)(9.8 \text{ m/s}^2)} = 2.6 \text{ km}$

7. Applying the Bernoulli equation between the surface of the tank (where the gauge pressure $p = 0$, $v = 0$, and $y = h_1$) and the outlet (where $p = 0$, $y = 0$)

$$\rho g h_1 = \tfrac{1}{2}\rho v^2, \qquad v = \sqrt{2gh_1}$$

Thus, $Q = Av = A\sqrt{2gh_1}$

At the venturi constriction,

$$Q = A_v v_v$$

Now, $A_v = \tfrac{1}{9}A$, so $v_v = 9v$

Applying the Bernoulli equation between the venturi and the outlet,

$$\tfrac{1}{2}\rho v_v^2 + p_v = \tfrac{1}{2}\rho v^2$$

so
$$p_v = -\tfrac{1}{2}\rho(v_v^2 - v^2) = -40\rho v^2$$
$$= -80\rho g h_1$$

The hydrostatic pressure in the fluid column increases the pressure from this value back to zero gauge pressure,

so
$$\rho g h_2 = 80\rho g h_1$$
$$h_2 = 80 h_1$$

9. The flow is governed by the viscous flow in the tube, which is driven by the hydrostatic pressure at the bottom of the tank. Thus if the depth of liquid is h,

$$\Delta p = \rho g h$$

and
$$Q = \frac{\pi(\Delta p)R^4}{8\eta\ell}$$

Now, $Q = A\dfrac{dh}{dt}$, where A = area of tank,

so $$\frac{dh}{dt} = \frac{Q}{A} = \frac{\pi(\Delta p)R^4}{8\eta \ell A} = \frac{\pi \rho g h R^4}{8\eta \ell A}$$

$$\frac{1}{h} \cdot \frac{dh}{dt} = \frac{\pi(1000\ \text{kg/m}^3)(9.8\ \text{m/s}^2)(.0019\ \text{m})^4}{8(1.8 \times 10^{-3}\ \text{N} \cdot \text{s/m}^2)(1.0\ \text{m})(\pi(.250\ \text{m})^2)}$$

$$= .00014\ \text{s}^{-1}$$

Thus, $$h = h_o e^{-(.00014\ \text{s}^{-1})t}$$

and $$h = \frac{h_o}{2} \quad \text{when}$$

$$-(.00014\ \text{s}^{-1})t = \ln\left(\frac{1}{2}\right)$$

$$t = 4900\ \text{s}$$

$$= 1.4\ \text{hr}$$

11. (a) The free-body diagram for this problem is shown below.

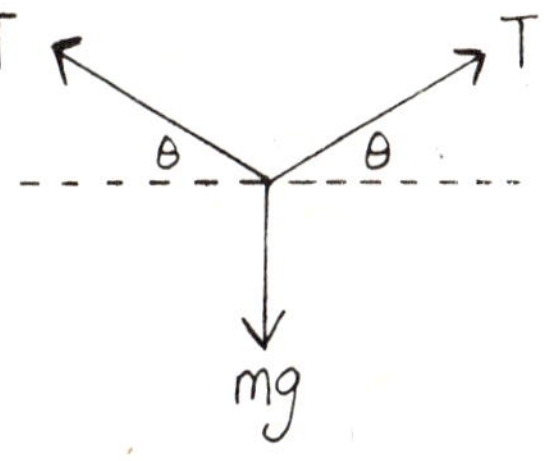

Setting $\sum F_y = 0$ gives

$$2T \sin\theta = mg$$

or $$T = \frac{mg}{2\sin\theta}$$

If the original length of the wire is $2\ell_o$ and the new length is

$$2\ell = 2(\ell_o + \Delta\ell)$$

then $$\cos\theta = \frac{\ell_o}{\ell}$$

Now, $\ell = \ell_o\left(1 + \frac{T}{YA}\right) = \ell_o + \Delta\ell$

so $\frac{\ell}{\ell_o} = \frac{1}{\cos\theta} = 1 + \frac{mg}{2\sin\theta\, YA}$

or $2\sin\theta(\sec\theta - 1) = \frac{mg}{YA}$

(b) If θ is very small, $\sin\theta \simeq \theta$ and $\cos\theta \simeq 1 - \frac{1}{2}\theta^2$,

so $(\sec\theta - 1) \simeq \frac{1}{2}\theta^2$

Thus, for small θ,

$$\theta^3 = \frac{mg}{YA} \quad \text{or} \quad \theta = \left(\frac{mg}{YA}\right)^{1/3}$$

If $\frac{mg}{YA} = .01$, then $\theta = (.01)^{1/3} = .21 \text{ rad} = 12°$

13. From equation 15–11,

$$F = \frac{\eta A v}{\ell} \quad \text{or} \quad \frac{v}{\ell} = \frac{F}{\eta A}$$

(a) Applying this to the current situation,

$$A \to 2\pi R L, \quad \frac{v}{\ell} \to \frac{dv}{dR},$$

so $F = \eta(2\pi R L)\frac{dv}{dR}$

(b) Now, the viscous force on the fluid must be balanced by the force due to the pressure change,

so $F = -(\Delta p)(\pi R^2)$

so $2\pi R L\eta \frac{dv}{dR} = -(\Delta p)\pi R^2$

$$\frac{dv}{dR} = -\frac{(\Delta p)R}{2\eta L}$$

(c) Integrating, $\displaystyle\int_{R_{pipe}}^{R} \frac{dv}{dR} = -\frac{(\Delta p)}{2\eta L}\int_{R_{pipe}}^{R} R\, dR$

$$v(R) = -\frac{\Delta p}{2\eta L}\cdot\frac{1}{2}(R^2 - R_{pipe}^2)$$

$$= \frac{\Delta p}{4\eta L}(R_{pipe}^2 - R^2)$$

15. Applying the Bernoulli equation,

$$\frac{1}{2}\rho v_{main}^2 + p_{main} = \frac{1}{2}\rho v_{user}^2 + p_{user} + \rho g y_{user}$$

$$v_{main} = \frac{Q}{A_{main}} = \frac{0.20\ m^3/s}{\frac{\pi}{4}(.25\ m)^2} = 4.1\ m/s$$

$$v_{user} = \frac{Q}{A_{user}} = \frac{0.20\ m^3/s}{\frac{\pi}{4}(.10\ m)^2} = 25\ m/s$$

Using gauge pressures, $p_{user} = 0$

$$\therefore \quad p_{main} = \frac{1}{2}\rho(v_{user}^2 - v_{main}^2) + \rho g y_{user}$$

$$= \frac{1}{2}(1000\ kg/m^3)((25\ m/s)^2 - (4.1\ m/s)^2)$$

$$+ (1000\ kg/m^3)(9.8\ m/s^2)(5.0\ m)$$

$$= 316\ kPa + 49\ kPa$$

$$= 3.7\times 10^5\ Pa$$

17. $\rho = \rho_o\left(1 - \frac{y}{h}\right)$

$$\frac{dp}{dy} = -\rho g = -\rho_o g\left(1 - \frac{y}{h}\right)$$

Integrating, $p = p_o - \rho_o g\left(y - \frac{y^2}{2h}\right)$

$$p(h) = 0, \quad \text{so} \quad p_o = \rho_o g\left(h - \frac{h}{2}\right) = \frac{\rho_o g h}{2}$$

$$h = \frac{2p_o}{\rho_o g} = \frac{2(1.01\times 10^5\ Pa)}{(1.29\ kg/m^3)(9.8\ m/s^2)}$$

$$= 16\ km$$

19. Let A be the cross-sectional area of the tube.

$$V_1 = \left(h_1 + \frac{1}{3}h_1\right)A; \qquad V_2 = h_2 A$$

But $\quad V_1 = V_2, \quad$ so $\quad h_2 = \frac{4}{3}h_1$

The pressure at the interface between the two liquids is the same on both sides,

so $\quad \rho_1 g h_1 = \rho_2 g h_2$

$$\rho_1 = \frac{\rho_2 h_2}{h_1} = \frac{4}{3}\rho_2$$

$$= \frac{4}{3}(1200\ \text{kg/m}^3)$$

$$= 1600\ \text{kg/m}^3$$

CHAPTER 16 TEMPERATURE AND HEAT TRANSFER

Questions:

1. (a) An adiabatic wall would be desired between the refrigerator and the kitchen and between the oven and the kitchen.

(b) A diathermic wall would be desired between the stove top and the pots placed on the burners and between the cooling coils in the back of the refrigerator and the air in the room.

3. Pressure is always a state variable, but its importance decreases as the substance becomes less compressible. It is less important in liquids than in gases, and it can be ignored for most purposes in the case of solids.

5. In a constant-pressure gas thermometer, the volume increases with increasing temperature.

7. At very low temperatures the gas can become a liquid or solid; at very high temperatures the container may melt or it may not be able to withstand the pressure and explode.

9. Yes, if two systems are at the same temperature they are in thermal equilibrium with each other. If two systems are at different temperatures they are not in thermal equilibrium with each other.

11. The inside diameter increases also. Suppose that a solid sphere is actually made up of a hollow sphere with a smaller solid sphere fitting snugly inside it. When this combination is heated, it should expand in exactly the same way as a solid sphere, without developing internal stresses. The small sphere expands, and so the cavity in the hollow sphere must expand the same amount.

13. Heat is not a state variable. Heat is energy in transit and is not a property that can be attributed to a substance or system.

15. The metal rod feels cooler because it conducts heat away from your hand.

17. The SI unit of R-value is $\mathrm{K\,m^2\,s/J}$.

19. The insulating value of fiberglass batts or rolls comes from the air trapped in the space between the fibers. Air is a good insulator, but an air gap alone would not work. Convective currents would be set up because of the buoyant force on the warmer air. The circulating air carries heat well. To prevent convection it is necessary to put an obstacle in the way. This is the role of the fiberglass.

21. The walls are silvered to prevent the absorption of radiant heat.

23.

Symbol	Represents	Type	SI unit
p	pressure	scalar	Pa
V	volume	scalar	m^3
n	number of moles	scalar	mol
T	temperature	scalar	K
t_C	Celsius temperature	scalar	°C
α	coefficient of linear expansion	scalar	K^{-1}
β	volume coefficient of expansion	scalar	K^{-1}
Q	heat	scalar	J
H	heat current	scalar	W
k	thermal conductivity	scalar	$W\,m^{-1}\,K^{-1}$
R	universal gas constant	scalar	$J\,K^{-1}\,mol^{-1}$
σ	Stefan's constant	scalar	$W\,m^{-2}\,K^{-4}$

Exercises:

1. $$t = (1 \times 10^{-9}\ \text{s/molecule}) \times (10^{23}\ \text{molecules}) \times \left(\frac{1\ \text{h}}{3600\ \text{s}}\right) \times \left(\frac{1\ \text{day}}{24\ \text{h}}\right) \times \left(\frac{1\ \text{yr}}{365\ \text{day}}\right)$$

$$= 3 \times 10^{6}\ \text{years}$$

3. (a) $22.4\text{ L} = 22.4\text{ L} \times \frac{1\text{ m}^3}{1000\text{ L}} = 0.0224\text{ m}^3$

$$x^3 = 0.0224\text{ m}^3$$

$$x = (0.0224\text{ m}^3)^{1/3} = 0.282\text{ m}$$

The length of the cube edge is 28.2 cm.

(b) $F = pA = (101\text{ kPa})(0.282\text{ m})^2$

$$= 8.03\text{ kN}$$

5. For a constant volume gas thermometer,

$$\frac{T_B}{T_A} = \frac{p_B}{p_A}$$

Here $T_A = 273.16\text{ K}$, $p_A = 1439\text{ Pa}$

(a) When $T_B = 693\text{ K}$, then

$$p_B = \frac{T_B}{T_A} p_A = \frac{693\text{ K}}{273.16\text{ K}}(1439\text{ Pa})$$

$$= 3650\text{ Pa}$$

(b) When $p_B = 406\text{ Pa}$, then

$$T_B = \frac{p_B}{p_A} T_A = \frac{406\text{ Pa}}{1439\text{ Pa}}(273.16\text{ K})$$

$$= 77.1\text{ K}$$

7. $\frac{T_B}{T_A} = \frac{p_B}{p_A}$

Here $T_A = 373.15\text{ K}$, $p_A = 24.5\text{ mmHg}$

(a) When $T_B = 273.16\text{ K}$,

$$p_B = p_A \frac{T_B}{T_A} = (24.5\text{ mmHg})\left(\frac{273.16\text{ K}}{373.15\text{ K}}\right)$$

$$= 17.9\text{ mmHg}$$

(b) When $T_B = 273.15\text{ K}$, $p_B = 17.9\text{ mmHg}$

(c) When $t_B = 37°C, \quad T_B = 273.15\text{ K} + 37\text{ K} = 310\text{ K}$

$$p_B = (24.5\text{ mmHg})\left(\frac{310\text{ K}}{373\text{ K}}\right) = 20.4\text{ mmHg}$$

9. Assuming that the length of the column of mercury is a linear function of the temperature,

$$\frac{t_C - 37°C}{39°C - 37°C} = \frac{x - 43\text{ mm}}{67\text{ mm} - 43\text{ mm}}$$

Thus when $x = 16\text{ mm}$,

$$t_C = 37°C + 2°C\left(\frac{16\text{ mm} - 43\text{ mm}}{24\text{ mm}}\right)$$

$$= 35°C$$

11. (a) $t_C = \frac{5}{9}(t_F - 32°F)$

$t_C = t_F$ when $9t = 5t - 160°$

or $t = -40°F = -40°C$

(b) $T = 273.15\text{ K} + \frac{5}{9}(t_F - 32°F)$

$T = t_F$ when $\frac{4}{9}T = 273.15\text{ K} - \frac{5}{9}(32°F)$

$T = 575\text{ K}, \quad t_F = 575°F$

(c) $T_R = \frac{9}{5}T$

$T_R = T$ when $T_R = 0°\text{ R} \quad T = 0\text{ K}$

(d) T and t_C cannot have the same numerical value at any temperature.

13. When the steel rule is cooled from 22°C to –5°C, it shrinks according to the expression

$$\Delta L = L_o \alpha \Delta T$$

(a) Taking $L_o = 10.00\text{ mm}, \quad \alpha = 1.1 \times 10^{-5}\,°C^{-1}$, and $\Delta T = -27°C$,

$$\Delta L = (10.00\text{ mm})(1.1 \times 10^{-5}\,°C^{-1})(-27°C) = -0.003\text{ mm}$$

This difference is smaller than the significant figures indicate for the precision of the rule; however what was exactly 1 cm at 22°C becomes 9.997 mm at –5°C.

(b) For a nominal 1 m rule,

$$\frac{\Delta L}{L} = -0.0003$$

so the percent error is 0.03%

(c) The absolute error in 100 m is (measured value – actual value)

$$= 0.03 \text{ m} \quad \text{or} \quad 3 \text{ cm}$$

15. (a) All the linear dimensions of the shaft change according to the equation

$$\Delta L = L_o \alpha \Delta T$$

Here $\Delta L = -0.01$ mm, $L_o = 42.51$ mm, $\alpha = 1.1 \times 10^{-5}\ {}^\circ C^{-1}$,

so $$\Delta T = \frac{\Delta L}{\alpha L_o} = \frac{-0.01 \text{ mm}}{(1.1 \times 10^{-5}\ {}^\circ C^{-1})(42.51 \text{ mm})}$$

$$= -20^\circ C$$

So the shaft must be cooled 20°C to fit into the hole.

(b) If the entire structure is then cooled to –5°C, the shaft will be tight because both shrink at the same rate, but the hole would shrink more because it is being cooled through a larger temperature difference. It is unable to shrink as much as it should because of the presence of the shaft, so it squeezes tight around the shaft, and the shaft cannot come loose.

17. (a) The length of the pendulum changes by

$$\Delta L = L_o \alpha \Delta T$$

$$= (0.2482 \text{ m})(1.1 \times 10^{-5}\ {}^\circ C^{-1})(-5^\circ C)$$

$$= -0.000014 \text{ m}$$

The period of the pendulum is given by

$$P = 2\pi \sqrt{\frac{L}{g}} = 1.00 \text{ s}$$

Thus, $$\frac{\Delta P}{P} = \frac{1}{2} \cdot \frac{\Delta L}{L} = \frac{\alpha \Delta T}{2} = -0.000028$$

or $$\Delta P = -0.000028 \text{ s}$$

one day, the number of swings of the pendulum is

$$n = \frac{(24\text{ h/day})(3600\text{ s/h})}{P} = \frac{86400\text{ s}}{P\text{ day}}$$

So at 27°, the clock ticks 86400 ticks per day.

$$\text{Now} \quad \frac{\Delta n}{n} = -\frac{\Delta P}{P} = 0.000028$$

so it ticks an additional 2.4 ticks per day; that is, it gains 2.4 seconds per day.

19. The volume of the mercury increases by

$$\Delta V_{Hg} = \beta_{Hg} V \Delta T$$

$$= (18 \times 10^{-5}\,{}^\circ C^{-1})(176.2\text{ mL})(50.0^\circ C)$$

$$= 1.59\text{ mL}$$

The volume of the inside of the container increases by

$$\Delta V_{cont} = (2.2 \times 10^{-5}\,{}^\circ C^{-1})(176.2\text{ mL})(50.0^\circ C)$$

$$= 0.19\text{ mL}$$

Thus 1.4 mL overflows into the glass tube.

The length of the column of mercury in the tube is

$$L = \frac{\Delta V_{Hg} - \Delta V_{cont}}{A_{tube}}$$

$$= \frac{1.4\text{ cm}^3}{\frac{\pi}{4}(0.25\text{ cm})^2} = 28\text{ cm}$$

The expansion of the glass tube amounts to $\frac{\beta \Delta T}{3} = 0.0004$ in each linear dimension, or 0.08% in the calculation of the area, so it can be neglected.

21. The change in the radius of the sphere is

$$\Delta R = \alpha R \Delta T$$

$$= (2.4 \times 10^{-5}\,{}^\circ C^{-1})(3.000\text{ mm})(-100^\circ C)$$

$$= -0.0072\text{ mm},$$

so $$R = 2.993\text{ mm at } 0^\circ C$$

The volume of 0°C is

$$V = \frac{4}{3}\pi R^3 = \frac{4}{3}\pi(2.993\ \text{mm})^3$$

$$= 112.3\ \text{mm}^3,$$

which agrees with the answer found in example 16–5.

23. The heat current is given by

$$H = \frac{kA(T_2 - T_1)}{L}$$

$$= \frac{(237\ \text{W/m K})(1.5\times 10^4\ \text{mm}^2)(2\ \text{K})}{12\ \text{mm}\times 1000\ \text{mm/m}}$$

$$= 590\ \text{W}$$

25. (a) The temperature gradient is

$$\frac{\Delta T}{\Delta x} = \frac{25^\circ\text{C}}{15\ \text{mm}} = 1.7^\circ\text{C/mm}$$

(b) $$k = \frac{H\Delta x}{A\Delta T} = \frac{(14.3\ \text{W})(15\ \text{mm})}{(350\ \text{mm})^2(25\ \text{K})}\times\frac{1000\ \text{mm}}{\text{m}}$$

$$= 0.070\ \text{W/mK}$$

(c) This material is a fairly good insulator.

27. (a) The heat current is

$$H = \frac{Q}{\Delta t} = 1100\ \text{W}\times\frac{1\ \text{Btu}}{1055\ \text{J}}\times\frac{3600\ \text{s}}{1\ \text{h}}$$

$$= 3800\ \text{Btu/h}$$

(b) $$R = \frac{A\Delta T}{H} = \frac{(96\ \text{ft}^2)(30^\circ\text{F})}{(3800\ \text{Btu/h})}$$

$$= 0.77^\circ\text{F ft}^2\ \text{h/Btu}$$

(c) $$k = \frac{H\Delta x}{A\Delta T} = \frac{(1100\ \text{W})(0.75\ \text{in.})(3.28\ \text{ft/m})}{(96\ \text{ft}^2)(30^\circ\text{F})\left(\frac{5}{9}\cdot\frac{^\circ\text{C}}{^\circ\text{F}}\right)(12\ \text{in./ft})}$$

$$= 0.14\ \text{W m}^{-1}\,{}^\circ\text{C}^{-1}$$

29. $H_1 = \frac{A_1 \Delta T}{R_1}$, $H_2 = \frac{A_2 \Delta T}{R_2}$

$$H = H_1 + H_2 = \frac{(A_1 + A_2)\Delta T}{R_{eff}}$$

$$\therefore \quad \frac{A_1 + A_2}{R_{eff}} = \frac{A_1}{R_1} + \frac{A_2}{R_2}$$

If $A_1 = A_2$, then

$$\frac{2}{R_{eff}} = \frac{1}{R_1} + \frac{1}{R_2}$$

31. (a) For a parallel arrangement,

$$H_1 = \frac{k_1 A_1 \Delta T}{\Delta x}, \quad H_2 = \frac{k_2 A_2 \Delta T}{\Delta x}$$

$$H = H_1 + H_2 = \frac{(k_1 A_1 + k_2 A_2)\Delta T}{\Delta x}$$

$$k_1 A_1 = (0.19\ \mathrm{Wm^{-1}K^{-1}})(\pi(0.020\ \mathrm{m})^2 - \pi(0.010\ \mathrm{m})^2)$$

$$= 0.00018\ \mathrm{W\,m\,K^{-1}}$$

$$k_2 A_2 = (237\ \mathrm{Wm^{-1}K^{-1}})\pi(0.010\ \mathrm{m})^2$$

$$= 0.074\ \mathrm{W\,m\,K^{-1}}$$

$$H = (0.074\ \mathrm{W\,m/K})\left(\frac{150\ \mathrm{K}}{0.250\ \mathrm{m}}\right) = 45\ \mathrm{W}$$

(b) The energy transferred in one hour is

$$(45\ \mathrm{W})(3600\ \mathrm{s}) = 160\ \mathrm{kJ}$$

33. $\frac{P}{A} = e\sigma T^4$, $e = 1$, $\sigma = 5.67 \times 10^{-8}\ \mathrm{Wm^{-2}K^{-4}}$

(a) $$\left(\frac{P}{A}\right)_{sun} = (5.67 \times 10^{-8}\ \mathrm{W/m^2K^4})(6000\ \mathrm{K})^4$$

$$= 7.34 \times 10^7\ \mathrm{W/m^2}$$

(b) $$\left(\frac{P}{A}\right)_{earth} = (5.67 \times 10^{-8}\ \mathrm{W/m^2K^4})(300\ \mathrm{K})^4$$

$$= 0.46\ \mathrm{kW/m^2}$$

(c) $$\left(\frac{P}{A}\right)_{moon} = (5.67 \times 10^{-8}\ W/m^2K^4)(200\ K)^4$$

$$= 91\ W/m^2$$

(d) $$\left(\frac{P}{A}\right)_{star} = (5.67 \times 10^{-8}\ W/m^2K^4)(3\ K)^4$$

$$= 4.6 \times 10^{-6}\ W/m^2$$

35. (a) $m = m_{He} + m_{Ne}$

$$= (1.2\ mol)(4.00\ g/mol) + (0.8\ mol)(20.2\ g/mol)$$

$$= 21\ g$$

(b) There are 2.0 mols of gas altogether,

so $$N = (2.0\ mol)(6.02 \times 10^{23}\ molecules/mol)$$

$$= 12 \times 10^{23}\ molecules$$

37. The change in length of the 40 m bridge in going from one extreme of temperature to the other is

$$\Delta L = \alpha L_o \Delta T$$

$$= (1.1 \times 10^{-5}\ {}^\circ C^{-1})(40\ m)(60^\circ C)$$

$$= 0.026\ m$$

39. (a) Let t be the temperature at which the rods have the same length L

$$L = 1.0041\ m(1 + 1.1 \times 10^{-5}\ {}^\circ C^{-1} t)$$

$$= 1.0038\ m(1 + 2.4 \times 10^{-5}\ {}^\circ C^{-1} t)$$

$$\therefore \quad 1.0041\ m - 1.0038\ m = 1.30 \times 10^{-5}\ m^\circ C^{-1} t$$

$$t = \frac{0.0003\ m}{1.3 \times 10^{-5}\ m^\circ C^{-1}} = 23^\circ C$$

(b) $L = 1.0041\ m(1 + 1.1 \times 10^{-5}\ {}^\circ C^{-1}(23^\circ C))$

$$= 1.0044\ m$$

41. $L_o = 364\text{ mm}, \quad W_o = 448\text{ mm} \quad \text{at} \quad 0°C$

(a) $A_o = L_o W_o = (0.364\text{ m})(0.448\text{ m}) = 0.163\text{ m}^2$

(b) At 150°C,

$$L = L_o(1 + \alpha\Delta T)$$

$$= (0.364\text{ m})(1 + 2.4 \times 10^{-5}\ °C^{-1}(150°C))$$

$$= (0.364\text{ m})(1.0036) = 0.365\text{ m}$$

$$W = W_o(1 + \alpha\Delta T)$$

$$= (0.448\text{ m})(1.0036) = 0.450\text{ m}$$

(c) $A = LW = 0.164\text{ m}^2$

43. The area of a sphere is $4\pi R^2$. From Exercise 33,

$$\left(\frac{P}{A}\right)_{sun} = 7.34 \times 10^7\ \text{W/m}^2$$

$$\therefore \quad P = (7.34 \times 10^7\ \text{W/m}^2)(4\pi(7 \times 10^8\ \text{m})^2)$$

$$= 4.5 \times 10^{26}\ \text{W}$$

45. $H_{glass} = H_{air} = H, \quad \Delta T_{glass} + \Delta T_{air} = 20°C$

$$H = \frac{k_{glass} A \left(\frac{dT}{dx}\right)_{glass}}{L_{glass}} = \frac{k_{air} A \left(\frac{dT}{dx}\right)_{air}}{L_{air}}$$

$$\frac{\left(\frac{dT}{dx}\right)_{air}}{\left(\frac{dT}{dx}\right)_{glass}} = \frac{L_{air}}{L_{glass}} \bullet \frac{k_{glass}}{k_{air}}$$

$$= \left(\frac{12\text{ mm}}{6\text{ mm}}\right)\left(\frac{1\text{ W m}^{-1}\text{K}^{-1}}{0.02\text{ Wm}^{-1}\text{K}^{-1}}\right) = 100$$

$$\Delta T = \left(\frac{dT}{dx}\right)_{air} L_{air} + \left(\frac{dT}{dx}\right)_{glass} L_{glass}$$

$$20°C = \left(\frac{dT}{dx}\right)_{glass} (100\, L_{air} + L_{glass})$$

$$\left(\frac{dT}{dx}\right)_{glass} = \frac{20^\circ C}{100 \times 12\ mm + 6\ mm} = 0.017^\circ C/mm$$

$$\left(\frac{dT}{dx}\right)_{air} = 1.7^\circ C/mm$$

Problems:

1. $$A_0 + \Delta A = (L_0 + \Delta L)(W_0 + \Delta W)$$

$$= L_0 W_0 + W_0 \Delta L + L_0 \Delta W + \Delta L \Delta W$$

neglecting the product $\Delta L \Delta W$, and taking into account that $A_0 = L_0 W_0$,

$$\Delta A = W_0 \Delta L + L_0 \Delta W$$

$$= W_0(\alpha L_0 \Delta T) + L_0(\alpha W_0 \Delta T)$$

$$= A_0(2\alpha \Delta T)$$

3. $$H = -kA\frac{dT}{dr}, \quad A = 2\pi rL$$

(a) Since energy is conserved, $H(r_1) = H(r_2)$,

so $$H = -k(2\pi rL)\frac{dT}{dr}$$

or

(b) $$\frac{dT}{dr} = -\frac{H}{2\pi rLk}$$

(c) $$\int dT = -\frac{H}{2\pi Lk}\int \frac{dr}{r}$$

$$T(r) = -\frac{H}{2\pi Lk}\ln r + \text{constant}$$

(d) $$\int_{T_a}^{T_b} dT = -\frac{H}{2\pi Lk}\int_a^b \frac{dr}{r}$$

$$T_b - T_a = -\frac{H}{2\pi Lk}\ln\left(\frac{b}{a}\right)$$

$$\therefore \qquad H = \frac{2\pi k L (T_a - T_b)}{\ln\left(\frac{b}{a}\right)}$$

Also, $$T - T_a = -\frac{H}{2\pi k L} \ln\left(\frac{r}{a}\right)$$

so $$T(r) = T_a + \frac{(T_b - T_a)\ln\left(\frac{r}{a}\right)}{\ln\left(\frac{b}{a}\right)}$$

5. (a) Using the results of Problem 3,

$$\frac{H}{L} = \frac{2\pi k (T_a - T_b)}{\ln\left(\frac{b}{a}\right)}$$

$$= \frac{2\pi (0.11 \text{ W/mK})(140°\text{C} - 35°\text{C})}{\ln\left(\frac{28 \text{ mm}}{12 \text{ mm}}\right)}$$

$$= 86 \text{ W/m}$$

(b) In the insulating medium,

$$T(r) = 140°\text{C} - 105°\text{C}\,\frac{\ln\left(\frac{r}{12 \text{ mm}}\right)}{\ln\left(\frac{28 \text{ mm}}{12 \text{ mm}}\right)}$$

$$= 140°\text{C} - 124°\text{C} \ln\left(\frac{r}{12 \text{ mm}}\right)$$

(c)

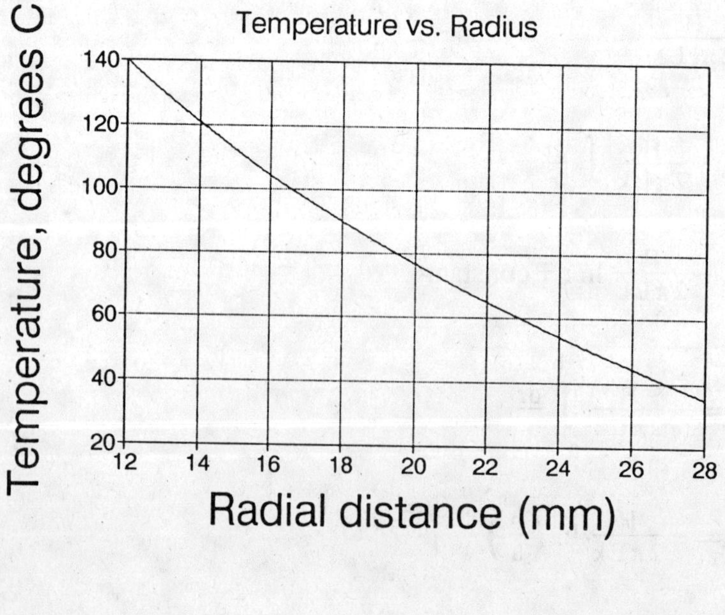

(d) When $r = 20$ mm,

$$\frac{dT}{dr} = \frac{-\left(\frac{H}{L}\right)}{2\pi k r} = \frac{-86\ \text{W/m}}{2\pi(0.11\ \text{W/mK})(0.020\ \text{m})}$$

$$= -6200°\text{C/m} \quad \text{or} \quad -6.2°\text{C/mm}$$

7. The lengths of the strips when heated are

$$L_1 = L_o(1 + \alpha_1 \Delta T)$$

$$L_2 = L_o(1 + \alpha_2 \Delta T)$$

Since both subtend the same angle,

$$\theta = \frac{L_1}{R_1} = \frac{L_2}{R_2}$$

$$\frac{L_1}{L_2} = \frac{R_1}{R_2} = \frac{R - \frac{1}{2}d}{R + \frac{1}{2}d}$$

$$\therefore \quad \left(R - \tfrac{1}{2}d\right)(1 + \alpha_2 \Delta T) = \left(R + \tfrac{1}{2}d\right)(1 + \alpha_1 \Delta T)$$

$$R - \tfrac{1}{2}d + R\alpha_2 \Delta T - \tfrac{1}{2}\alpha_2 d \Delta T = R + \tfrac{1}{2}d + R\alpha_1 \Delta T + \tfrac{1}{2}\alpha_1 d \Delta T$$

Collecting terms,

$$d = R(\alpha_2 - \alpha_1)\Delta T - \tfrac{1}{2}(\alpha_2 - \alpha_1) d \Delta T$$

The last term can be neglected since

$$\left|(\alpha_2 - \alpha_1)\Delta T\right| << 1$$

so

$$R = \frac{d}{(\alpha_2 - \alpha_1)\Delta T}$$

9. Given: $\frac{d^2T}{dx^2} = 0$, with $T(x = 0) = T_2$ and $T(x = L) = T_1$

Integrating,

$$\frac{dT}{dx} = a \quad (a = \text{constant})$$

and

$$T = ax + b \quad (b = \text{constant})$$

Setting $x = 0,$

$$T_2 = b, \quad \text{so} \quad T = T_2 + ax$$

and setting $x = L,$

$$T_1 = T_2 + aL, \quad \text{so} \quad a = \frac{T_1 - T_2}{L}$$

Thus, $$T = T_2 + (T_1 - T_2)\frac{x}{L}$$

11. In Problem 4,

$$\frac{dT}{dr} = -\frac{H}{4\pi k r^2} = -\frac{ab(T_a - T_b)}{(b-a)r^2}$$

Thus, $$\frac{d}{dr}\left(r^2 \frac{dT}{dr}\right) = \frac{d}{dr}\left(-\frac{H}{4\pi k}\right) = 0,$$

so the temperature distribution does satisfy Laplace's equation.

13. At equilibrium, the energy absorbed must be equal to the energy radiated. The area exposed to the sun is πR^2 while energy is radiated by the entire surface, $4\pi R^2$.

$$1400\ \text{W/m}^2 (0.64)\pi R^2 = (0.64)(5.67 \times 10^{-8}\ \text{W/m}^2\text{K}^4) 4\pi R^2 T^4$$

$$T = \left(\frac{1400\ \text{W/m}^2}{4\,(5.67 \times 10^{-8}\ \text{W/m}^2\text{K}^4)}\right)^{1/4}$$

$$= 280\ \text{K},$$

which is quite close to the mean surface temperature of 285 K.

15. Using 15 cells by 15 cells, the result is the spreadsheet shown below.

	100.00	100.00	100.00	100.00	100.00	100.00	100.00	100.00	100.00	100.00	100.00	100.00	100.00	
100.00	98.88	97.75	96.62	95.48	94.29	93.01	91.55	89.80	87.52	84.27	79.11	69.78	50.00	0.00
100.00	97.75	95.51	93.27	91.00	88.67	86.18	83.41	80.14	76.01	70.44	62.39	50.00	30.22	0.00
100.00	96.62	93.27	89.93	86.59	83.20	79.65	75.77	71.32	65.94	59.09	50.00	37.61	20.89	0.00
100.00	95.48	91.00	86.59	82.24	77.89	73.43	68.69	63.44	57.34	50.00	40.90	29.56	15.73	0.00
100.00	94.29	88.67	83.20	77.89	72.70	67.50	62.14	56.39	50.00	42.65	34.05	23.99	12.48	0.00
100.00	93.01	86.18	79.65	73.43	67.50	61.73	55.97	50.00	43.60	36.56	28.68	19.86	10.20	0.00
100.00	91.55	83.41	75.77	68.69	62.14	55.97	50.00	44.03	37.86	31.30	24.23	16.59	8.44	0.00
100.00	89.80	80.14	71.32	63.44	56.39	50.00	44.03	38.26	32.49	26.56	20.35	13.81	6.99	0.00
100.00	87.52	76.01	65.94	57.34	50.00	43.60	37.86	32.49	27.30	22.10	16.80	11.33	5.71	0.00
100.00	84.27	70.44	59.09	50.00	42.65	36.56	31.30	26.56	22.10	17.75	13.40	9.00	4.52	0.00
100.00	79.11	62.39	50.00	40.90	34.05	28.68	24.23	20.35	16.80	13.40	10.07	6.73	3.37	0.00
100.00	69.78	50.00	37.61	29.56	23.99	19.86	16.59	13.81	11.33	9.00	6.73	4.49	2.25	0.00
100.00	50.00	30.22	20.89	15.73	12.48	10.20	8.44	6.99	5.71	4.52	3.37	2.25	1.12	0.00
	0.00	0.00	0.00	0.00	0.00	0.00	0.00	0.00	0.00	0.00	0.00	0.00	0.00	

17. Assuming that e = 1 for a light bulb,

$$P = \sigma A T^4$$

or

$$T = \left(\frac{P}{\sigma A}\right)^{1/4}$$

$$= \left(\frac{100\ \text{W}}{(5.67 \times 10^{-8}\ \text{W/m}^2\text{K}^4)(1 \times 10^{-5}\ \text{m}^2)}\right)^{1/4}$$

$$= 3600\ \text{K}$$

(Only one significant figure is justified in this problem but the answer is given to two figures to make a comparison possible with the melting point of tungsten, 3650 K. Thus the filament is very close to the melting point.)

19. From Problem 4,

$$\frac{dT}{dr} = -\frac{H}{4\pi r^2 k}$$

Integrating,

$$T = -\int \frac{H\,dr}{4\pi r^2 k} = \frac{H}{4\pi k r} + \text{constant}$$

or

$$T - T_s = \frac{H}{4\pi k}\left(\frac{1}{r} - \frac{1}{r_E}\right)$$

$$T = 300\ \text{K} + \frac{(20 \times 10^{-3}\ \text{W/m}^2)(4\pi r_E^2)}{4\pi(2\ \text{W m}^{-1}\text{K}^{-1})}\left(\frac{r_E - r}{r r_E}\right)$$

$$= 300\ \text{K} + 1 \times 10^{-2}\ \text{K/m}\left(\frac{r_E^2}{r}\right)\left(\frac{r_E - r}{r_E}\right)$$

(a) With $r_E = 6371$ km, this becomes

$$T = 300\ \text{K} + 1 \times 10^{-2}\ \text{K/m}\ (r_E - r),$$

provided $|r_E - r| << r_E$

Let $d = r_E - r$ (the depth)

Thus, $T \simeq 300\text{ K} + 10\text{ K/km } d$

$$T(1\text{ km}) = 310\text{ K}$$

$$T(10\text{ km}) = 400\text{ K}$$

$$T(100\text{ km}) = 1300\text{ K}$$

(b) $d = \dfrac{T - 300\text{ K}}{10\text{ K/km}} = \dfrac{1300\text{ K}}{10\text{ K/km}}$

$= 130\text{ km}$

CHAPTER 17 THE FIRST LAW OF THERMODYNAMICS

Questions:

1. (a) The equation $pV = nRT$ gives an accurate description of the behavior of He gas provided the temperature is not too low and the pressure is not too high.

   (b) The equation does not describe the liquid phase because the predicted volumes would be too high and the compressibility would also be too high.

3. The law of partial pressures states that the pressure of a mixture of gases is the sum of the pressures that the individual gases exert in the same volume and at the same temperature. Thus,

$$p = p_1 + p_2 + \dots,$$

where $p_1 = n_1 \frac{RT}{V}, \; p_2 = n_2 \frac{RT}{V} \dots,$

so $$p = (n_1 + n_2 + \dots)\frac{RT}{V}$$

Thus n represents the total number of moles of each gas present.

Also, $$n = \frac{m_1}{M_1} + \frac{m_2}{M_2} + \dots$$

$$= m\left(\frac{f_1}{M_1} + \frac{f_2}{M_2} + \dots\right) = \frac{m}{M}$$

where m_1 denotes the mass of gas 1 and M_1 its molecular mass. m denotes the total mass of the mixture and f_1 the fraction $\frac{m_1}{m}$. M is thus an equivalent molecular mass for a mixture of that composition.

5. A process which is not quasi-static cannot be shown in a p–V diagram because there is not single pressure that describes the gas during the process, and the gas may not actually occupy the entire container, so the volume is not known either.

7. Work cannot be performed by a system without a change in volume because mechanical work requires a displacement and there is no net displacement if the volume does not change.

9. For an ideal gas, the internal energy is proportional to the temperature, so its internal energy would have to increase if the temperature increases. For a real substance, the internal energy is a function of both temperature and pressure, and it is possible to vary both in such a way as to keep the internal energy constant.

11. The work is not performed directly on the gas. It is being converted into heat by the rubbing action and thus the container is being heated. Therefore, the energy added to the gas is in the form of heat.

13. A bowl of soup can be heated by performing work on the soup. The temperature of a system does not have to increse when heat is added. An example is adding heat to water at 100° C. The result is that some of the water changes into steam but the temperature remains the same.

15. The process is adiabatic since it takes place in a well-insulated container. The energy required to melt the ice came from the internal energy of the water.

17. Phase transitions are well illustrated in a p–T diagram.

19. No. The information given is not sufficient to determine the heat added or the work done. To determine the work done, it is necessary to know the entire path followed in the p–V diagram. Only the difference between the heat gained and the work done can be found from a knowledge of the initial and final states or from the change in internal energy.

21. In a truly cyclical process the system returns to its initial state at the end of each cycle, thus there is no way to determine the number of cycles performed previously. Even a cycle counter would have to return to its original reading at the end of the cycle.

23. (a) For a process to be isobaric, a unique pressure must be definable for the entire system. The process would thus have to be quasi-static.

(b) Similarly, in an isothermal process, the temperature of every part of the system would have to remain constant, so the change would have to be quasi-static.

(c) In an isochoric process, only the total volume has to remain constant. Parts of the system may be at one temperature, parts at another, and different pressures could occur at different places in the volume. Therefore, the change in the system would not have to be quasi-static.

25. The analogy is quite good. The equation

$$q - w = \Delta u$$

would be true, and both q and w represent money in transit: income is money flowing in and expenditures are money flowing out. There are no "expenditures on hand" just as a system does not possess work. The analogy does not extend to the government, which can increase the amount of cash on hand without increasing its income or decreasing its expenditures, by manufacturing more money.

27.

Symbol	Represents	Type	SI Unit
R	Ideal gas constant	scalar	$J\ mol^{-1}\ K^{-1}$
C_p	Specific heat at constant pressure	scalar	$J\ kg^{-1}\ K^{-1}$
C_v	Specific heat at constant volume	scalar	$J\ kg^{-1}\ K^{-1}$
$\mathcal{C}_v$	Molar heat capacity at constant volume	scalar	$J\ mol^{-1}1\ K^{-1}$
L	Latent heat	scalar	$J\ kg^{-1}$
W	Work	scalar	J
ΔU	Change in internal energy	scalar	J
U	Internal energy	scalar	J

Exercises:

1. $\frac{T_2}{T_1} = 2, \quad \frac{p_2}{p_1} = 3$

(a) $\frac{p_2V_2}{T_2} = \frac{p_1V_1}{T_1} \quad \therefore \quad \frac{V_2}{V_1} = \left(\frac{p_1}{p_2}\right)\left(\frac{T_2}{T_1}\right) = \frac{2}{3}$

(b)

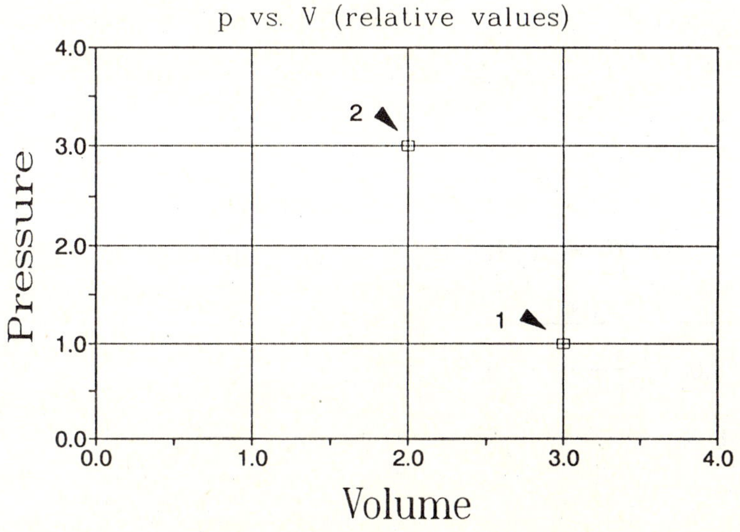

3. $p_1 = 0.73$ kPa, $V_1 = 12$ L, $T_1 = 320$ K

(a) $n = \dfrac{pV}{RT} = \dfrac{(0.73 \times 10^3 \text{ Pa})(12 \times 10^{-3} \text{m}^3)}{(8.31 \text{ J mol}^{-1} \text{ K}^{-1})(320 \text{ K})}$

$= 0.0033$ mol

(b) $V_2 = 18$ L, $T_2 = T_1$

$$p_2 = \frac{p_1 V_1}{V_2} = 0.73 \text{ kPa} \left(\frac{12 \text{ L}}{18 \text{ L}}\right) = 0.49 \text{ kPa}$$

(c)

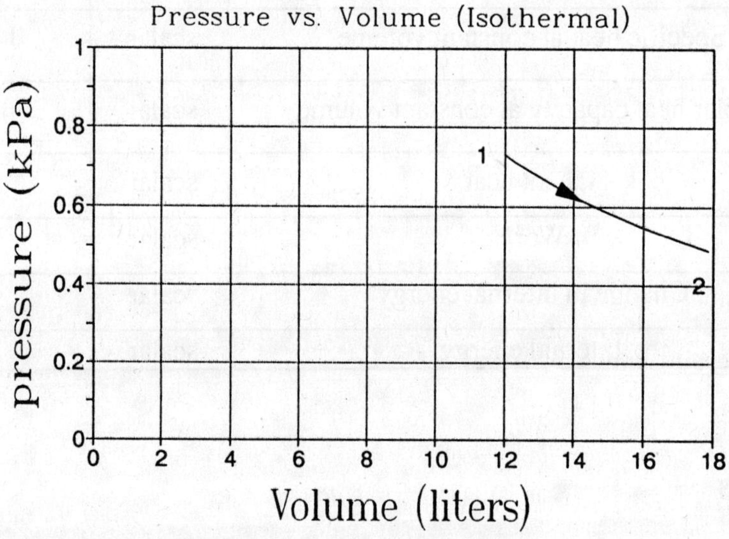

5. $R = \dfrac{8.31 \text{ J}}{\text{mol K}} = \dfrac{8.31 \text{ Pa m}^3}{\text{mol K}} \left(\dfrac{1 \text{ atm}}{101 \times 10^3 \text{ Pa}}\right)\left(\dfrac{10^3 \text{ L}}{\text{m}^3}\right)$

$= 0.0823 \dfrac{\text{L atm}}{\text{K mol}}$

7. From Exercise 5, $R = 0.082 \dfrac{\text{L atm}}{\text{K mol}}$

If $p = 1.0$ atm, $t_c = 0°$ C ($T = 273$ K), and $V = 1 \text{ m}^3 = 1000$ L, then

$$n = \frac{pV}{RT} = \frac{(1.0 \text{ atm})(1000 \text{ L})}{\left(0.082 \dfrac{\text{L atm}}{\text{K mol}}\right)(273 \text{ K})}$$

$= 45$ mol

$$N = nN_o = (45\text{ mol})\left(6.02 \times 10^{23}\ \frac{\text{molecules}}{\text{mol}}\right)$$

$$= 2.7 \times 10^{25}\text{ molecules}$$

9. $Q = mc_v\,\Delta T$

$= (0.003\text{ kg})(447\text{ J kg}^{-1}\text{ K}^{-1})(20\text{ K})$

$= 27\text{ J}$

Alternatively,

$Q = (3\text{ g})(0.107\text{ cal g}^{-1}\,{}^\circ\text{C}^{-1})(20^\circ\text{ C})$

$= 6.4\text{ cal}$

11. (a) Let T = final temperature in °C

Q_w = Heat transferred to water

$= m_w c_w \Delta T_w = (0.75\text{ kg})(4180\text{ J kg}^{-1}\text{ K}^{-1})(T - 20^\circ)$

Q_{Pb} = Heat transferred from lead = $m_{Pb}\, c_{Pb}\, \Delta T_{Pb}$

$= (1.24\text{ kg})(128\text{ J kg}^{-1}\text{ K}^{-1})(95^\circ - T)$

Setting $Q_w = Q_{Pb}$,

$T - 20^\circ = 0.0506\,(95^\circ - T)$

$1.0506\,T = 20^\circ + 4.80^\circ = 24.8^\circ$

$T = 24^\circ\text{ C}$

(b) $Q_w = (0.75\text{ kg})(4180\text{ J kg}^{-1}\text{ K}^{-1})(3.6^\circ)$

$= 11\text{ kJ}$ or $2.7 \times 10^3\text{ cal}$

13. Heat required to melt 0.35 kg of ice at 0° = $(0.35\text{ kg})(0.335\text{ MJ kg}^{-1}) = 0.117\text{ MJ}$
This amount of heat is removed from water at 25° C as it is cooled to 0° C.

$Q = mc_v\,\Delta T$

$$m = \frac{Q}{c_v\,\Delta T} = \frac{0.117\text{ MJ}}{(4180\text{ J kg}^{-1}\text{ K}^{-1})(25\text{ K})}$$

$= 1.1\text{ kg}$

15. (a) Let T be the temperature of the water before the thermometer is inserted. Then the amount of heat lost by the water is

$$Q_w = mc_v\,\Delta T$$
$$= (1.2\ \text{kg})(4180\ \text{J kg}^{-1}\ \text{K}^{-1})(T - 57.9°\ \text{C})$$

and the heat gained by the thermometer is

$$Q_{th} = mc_v\,\Delta T$$
$$= (0.033\ \text{kg})(1070\ \text{J kg}^{-1}\ \text{K}^{-1})(57.9°\ \text{C} - 23.5°\ \text{C})$$
$$= 1.2\ \text{kJ}$$

Setting $Q_w = Q_{th}$,

$$T - 57.9°\ \text{C} = \frac{1.2\ \text{kJ}}{(1.2\ \text{kg})(4180\ \text{J kg}^{-1}\text{K}^{-1})}$$
$$= 0.24°\ \text{C}$$

so, $T = 58.1°\ \text{C}$

(b) If $m_w = 0.012$ kg, the effect is much more pronounced. In fact,

$$T - 57.9°\ \text{C} = 24°\ \text{C},$$

so that the initial temperature of the thermometer becomes very important because the heat capacity of the thermometer is almost as large as that of the water.

17. (a) The work done by the fluid

$$W = p\Delta V$$
$$= (101\ \text{kPa})(0.424\ \text{m}^3 - 0.344\ \text{m}^3)$$
$$= 8.08\ \text{kJ}$$

(b) In the reverse process, the work done by the fluid is –8.08 kJ

19. (a) During the isobaric expansion, the work done by the gas is

$$W = p\Delta V$$
$$= (2 \times 10^6\ \text{Pa})(40 \times 10^{-3}\ \text{m}^3 - 10 \times 10^{-3}\ \text{m}^3)$$
$$= 60\ \text{kJ}$$

(b) During the isothermal compression, the work done by the gas is

$$W = \int_{V_i}^{V_f} p\,dV = nRT \int_{V_i}^{V_f} \frac{dV}{V} = nRT \ln \frac{V_f}{V_i}$$

$$= p_f V_f \ln \frac{V_f}{V_i}$$

$$= (4 \times 10^6 \text{ Pa})(20 \times 10^{-3} \text{ m}^3) \ln \left(\frac{20 \text{ L}}{40 \text{ L}}\right)$$

$$= -55 \text{ kJ}$$

(c) For the entire process,

$$W = 60 \text{ kJ} - 55 \text{ kJ} = 5\text{kJ}$$

(d) The first area is a rectangle, and W = 60 kJ. For the second part, approximating the curve by a straight line,

$$W = \frac{1}{2}(4 \text{ MPa} + 2 \text{ MPa})(20 \text{ L} - 40 \text{ L})$$

$$= -60 \text{ kJ}$$

This is an overestimate, so –55 kJ seems reasonable.

21. (a) During the process, $p = p_i + \frac{p_f - p_i}{V_f - V_i}(V - V_i)$

The work done is

$$W = \int_{V_i}^{V_f} p\,dV = p_i(V_f - V_i) + \left(\frac{p_f - p_i}{V_f - V_i}\right)\frac{(V_f - V_i)^2}{2}$$

$$= \frac{1}{2}(p_f + p_i)(V_f - V_i)$$

(b) $$W = \frac{1}{2}(108 \text{ kPa} + 140 \text{ kPa})(0.096 \text{ m}^3 - 0.064 \text{ m}^3)$$

$$= 4.0 \text{ kJ}$$

(c) $$T_i = \frac{p_i V_i}{nR} = \frac{(140 \text{ kPa})(0.064\text{m}^3)}{(3.0 \text{ mol})(8.31 \text{ J mol}^{-1} \text{ K}^{-1})}$$

$$= 360 \text{ K}$$

$$T_f = \frac{p_f V_f}{nR} = \frac{(108 \text{ kPa})(0.096 \text{ m}^3)}{(3.0 \text{ mol})(8.31 \text{ J mol}^{-1} \text{ K}^{-1})}$$

$$= 420 \text{ K}$$

23. (a) An expression for the work done during a reversible adiabatic process is given following equation 17–11:

$$W_Q = \frac{p_i V_i}{\gamma - 1}\left[1 - \left(\frac{V_i}{V_f}\right)^{\gamma-1}\right]$$

$$= \frac{(202 \text{ kPa})(45 \text{ L})}{1.40 - 1}\left[1 - \left(\frac{45 \text{ L}}{65 \text{ L}}\right)^{.40}\right]$$

$$= 3100 \text{ J}$$

Note: $(1 \text{ kPa})(1\text{L}) = (10^3 \text{ N/m}^2)(10^{-3} \text{ m}^3) = 1\text{J}$

(b) Since the process is reversible,

$$W_{f\to i} = -W_{i\to f}, \text{ so}$$

$$\frac{p_i V_i}{\gamma - 1}\left[1 - \left(\frac{V_i}{V_f}\right)^{\gamma-1}\right] = -\frac{p_f V_f}{\gamma - 1}\left[1 - \left(\frac{V_f}{V_i}\right)^{\gamma-1}\right]$$

$$p_f = p_i\left(\frac{V_i}{V_f}\right)\frac{1 - \left(\frac{V_i}{V_f}\right)^{\gamma-1}}{\left(\frac{V_f}{V_i}\right)^{\gamma-1} - 1}$$

$$= (202 \text{ kPa})\left(\frac{45 \text{ L}}{65 \text{ L}}\right)\frac{1 - \left(\frac{45 \text{ L}}{65 \text{ L}}\right)^{0.40}}{\left(\frac{65 \text{ L}}{45 \text{ L}}\right)^{0.40} - 1}$$

$$= 120 \text{ kPa}$$

25. (a) During process 1→2,

$$W_p = p\Delta V = (100 \text{ kPa})(60 \text{ L} - 40\text{L})$$

$$= 2000 \text{ J}$$

$$U_2 = U_1 + Q_p - W_p$$

$$= 5.0 \text{ kJ} + 10.0 \text{ kJ} - 2.0 \text{ kJ}$$

$$= 13.0 \text{ kJ}$$

(b) During process 2→3,

$$W_v = 0$$

$$U_3 = U_2 + Q_v - W_v$$

$$= 13.0 \text{ kJ} + 11.0 \text{ kJ} - 0.0 \text{ kJ}$$

$$= 24.0 \text{ kJ}$$

(c) For the same initial and final states,

$$\Delta U = U_3 - U_1 = 19.0 \text{ kJ}$$

$$Q_c = W_c + \Delta U$$

$$= 6.6 \text{ kJ} + 19.0 \text{ kJ}$$

$$= 25.6 \text{ kJ}$$

27. First use the ideal gas law to calculate the temperatures at a, i, and f.

$$T_a = \frac{p_a V_a}{nR} = \frac{(100 \text{ kPa})(60 \text{ L})}{(1.0 \text{ mol})(8.31 \text{ J mol}^{-1} \text{ K}^{-1})} = 722 \text{ K}$$

$$T_i = \left(\frac{V_i}{V_a}\right) T_a = \left(\frac{20 \text{ L}}{60 \text{ L}}\right) 722 \text{ K} = 241 \text{ K}$$

$$T_f = \left(\frac{p_f}{p_a}\right) T_a = \left(\frac{200 \text{ kPa}}{100 \text{ kPa}}\right) 722\text{K} = 1440 \text{ K}$$

(a) In the process i→a,

$$Q = nC_p(T_a - T_i)$$

$$= (1.0 \text{ mol})(20.8 \text{ J mol}^{-1} \text{ K}^{-1})(481 \text{ K})$$

$$= 10.0 \text{ kJ}$$

$$W = p\Delta V = (100 \text{ kPa})(40 \text{ L})$$

$$= 4.0 \text{ kJ}$$

$$U_a - U_i = Q - W = 6.0 \text{ kJ}$$

(b) In the process a→f,

$$Q = nC_v(T_f - T_a)$$

$$= (1.0 \text{ mol})(12.5 \text{ J mol}^{-1} \text{ K}^{-1})(722 \text{ K})$$

$$= 9.0 \text{ kJ}$$

$$W = 0$$

$$U_f - U_a = Q - W = 9.0 \text{ kJ}$$

(c) $U_f - U_i = (U_f - U_a) + (U_a - U_i)$

$$= 9.0 \text{ kJ} + 6.0 \text{ kJ}$$

$$= 15.0 \text{ kJ}$$

29. If we heat 1 mol of water from 0° C to 100° C,

$$Q = nC_p \Delta T$$

$$= (1\text{ mol})(75.3\text{ J mol}^{-1}\text{ K}^{-1})(100\text{ K})$$

$$= 7.53\text{ kJ}$$

$$W \approx 0$$

$$\therefore \quad \Delta U = 7.53\text{ kJ}$$

31.

Signs of Q, W, ΔU			
Process Description	**Q**	**W**	**ΔU**
Isochoric pressure drop	–	0	– (given)
Isobaric compression of an ideal gas	–	–	–
Adiabatic expansion	0	+	–
Isothermal expansion of an ideal gas	–	+	0
Isothermal compression of an ideal gas	–	–	0
Free expansion of a gas	0	0	0
Cyclic process	+	+ (given)	0

33. If no heat is lost to the surroundings, the entire initial potential energy is converted into heat.

$$Q = Mgh = M\left(9.8\ \frac{\text{m}}{\text{s}^2}\right)(4\text{m}) = \left(39\ \frac{\text{J}}{\text{kg}}\right)M$$

The corresponding rise in temperature is

$$\Delta T = \frac{Q}{Mc_p} = \frac{\left(39\ \frac{\text{J}}{\text{kg}}\right)}{128\text{ J kg}^{-1}\text{ K}^{-1}} = 0.3\text{ K}$$

It is not necessary to know the mass of the brick because both the initial energy and the heat capacity are proportional to the mass.

35. The amount of heat needed to melt 1.00 kg of ice is

$$Q = mL_f = (1.00\text{ kg})(0.335\text{ MJ kg}^{-1}) = 0.335\text{ MJ}$$

If this amount of heat is obtained from the cooling of water from 50° C to 0° C, the mass of water required is

$$m = \frac{Q}{c_p \Delta T} = \frac{0.335\text{ MJ}}{(4180\text{ J kg}^{-1}\text{K}^{-1})(50\text{K})} = 1.60\text{ kg}$$

37. (a) For an isothermal process, $p = \frac{nRT}{V}$

$$W_T = \int_{V_i}^{V_f} p\,dV = nRT \int_{V_i}^{V_f} \frac{dV}{V}$$

$$= nRT \ln\left(\frac{V_f}{V_i}\right) = nRT \ln\left(\frac{nRT/p_f}{nRT/p_i}\right) = nRT \ln\left(\frac{p_i}{p_f}\right)$$

(b) $W_T = (0.080 \text{ mol})(8.31 \text{ J mol}^{-1} \text{ K}^{-1})(350 \text{ K}) \ln\left(\frac{150 \text{ kPa}}{300 \text{ kPa}}\right)$

$= -160 \text{ J}$

39. $$\frac{\text{Cost}}{\text{day}} = \frac{\text{Cost}}{\text{energy}} \times \frac{\text{heat required to raise temperature}}{\text{mass of water}} \times \frac{\text{mass}}{\text{volume}} \times \frac{\text{volume}}{\text{day}}$$

$$= \frac{\$0.15}{\text{kW h}} \times (45 \text{ K})(4.18 \text{ kJ kg}^{-1} \text{ K}^{-1}) \times \frac{1000 \text{ kg}}{\text{m}^3} \times \frac{1 \text{ m}^3}{\text{day}} \times \frac{1 \text{ h}}{3600 \text{ s}} = \$7.80$$

41. The process described is a mixing process at constant pressure. Let T be the final temperature.

Heat gained by He = $n_{He} C_{p_{He}} (T - 320 \text{ K})$

Heat lost by N_2 = $n_{N_2} C_{p_{N_2}} (380 \text{ K} - T)$

Since the system is insulated these are equal, so

$$T - 320 \text{ K} = \frac{n_{N_2} C_{p_{N_2}}}{n_{He} C_{p_{He}}} (380 \text{ K} - T).$$

$$= \frac{(1.0 \text{ mol})(29.1 \text{ J mol}^{-1} \text{ K}^{-1})}{(2.0 \text{ mol})(20.8 \text{ J mol}^{-1} \text{ K}^{-1})} (380 \text{ K} - T)$$

$$= 266 \text{ K} - 0.70\, T$$

$$T = \frac{320 \text{ K} + 266 \text{ K}}{1 + 0.70} = 345 \text{ K}$$

43. 100 kJ = heat gained by skillet + heat gained by water

$= (2.0 \text{ kg})(447 \text{ J kg}^{-1} \text{ K}^{-1})(\Delta T) + (1.0 \text{ kg})(4180 \text{ J kg}^{-1} \text{ K}^{-1})(\Delta T)$

$= (5070 \text{ JK}^{-1})\Delta T$

$\therefore \quad \Delta T = 20 \text{ K}$

45.

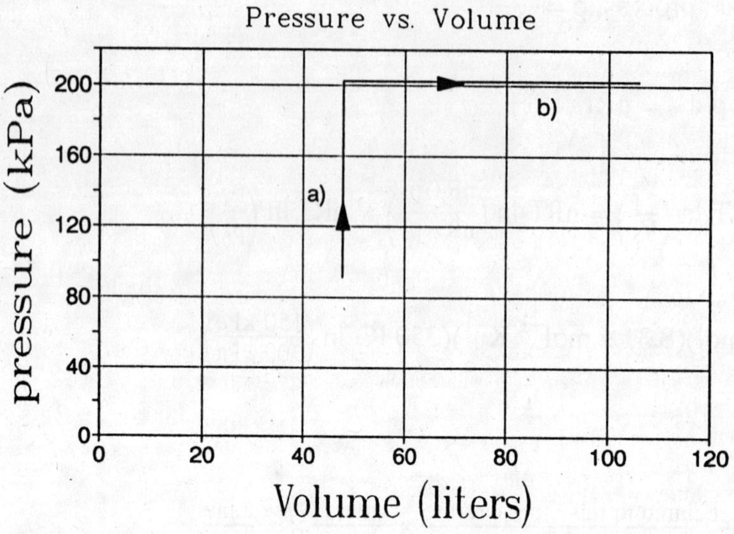

During step (a) no work is done because there is no volume change. In step (b),

$$W = p\Delta V = (202 \text{ kPa})(106 \text{ L} - 48 \text{ L}) = 12 \text{ kJ}$$

The work for the entire process is also 12 kJ.

47. $\Delta U = nC_v\Delta T$

$= (2.50 \text{ mol})(12.5 \text{ J mol}^{-1} \text{ K}^{-1})(10 \text{ K})$

$= 310 \text{ J}$

49. (a) $W_a = p_o\Delta V = 2\,p_oV_o = 2(100 \text{ kPa})(1.00 \text{ L})$

$U_2 = U_1 + Q_a - W_a$

$= 200 \text{ J} + 450 \text{ J} - 200 \text{ J}$

$= 450 \text{ J}$

(b) Since the process is isochoric, $W_b = 0$, and

$U_3 = U_2 + Q_b - W_b = 450 \text{ J} + 200 \text{ J} = 650 \text{ J}$

(c) W_{cyc} = –area inside triangle

$= -\frac{1}{2}(2\,p_o)(2\,V_o) = -2\,p_oV_o = -2(100 \text{ kPa})(1.00 \text{ L}) = -200 \text{ J}$

(d) $Q_{cyc} = W_{cyc} = -200 \text{ J}$

$Q_c = Q_{cyc} - Q_a - Q_b$

$= -200 \text{ J} - 450 \text{ J} - 200 \text{ J}$

$= -850 \text{ J}$

(e) Since $Q_c < 0$, the system rejected heat during process c.

Problems:

1. (a) Since the relationship is linear,

$$\mathcal{C}_p = \mathcal{C}_p(300\text{ K}) + \frac{\mathcal{C}_p(600\text{ K}) - \mathcal{C}_p(300\text{ K})}{600\text{ K} - 300\text{ K}}(\text{T} - 300\text{ K})$$

$$= 24.4\text{ J mol}^{-1}\text{ K}^{-1} + \frac{(28.1 - 24.4)\text{J mol}^{-1}\text{ K}^{-1}}{300\text{ K}}(\text{T} - 300\text{K})$$

$$= 20.7\text{ J mol}^{-1}\text{ K}^{-1} + (0.0123\text{ J mol}^{-1}\text{ K}^{-2})\text{T}$$

$$= \text{A} + \text{BT}$$

Thus, $\text{A} = 20.7\text{ J mol}^{-1}\text{ K}^{-1}$ and $\text{B} = 0.0123\text{ J mol}^{-1}\text{ K}^{-2}$

(b)

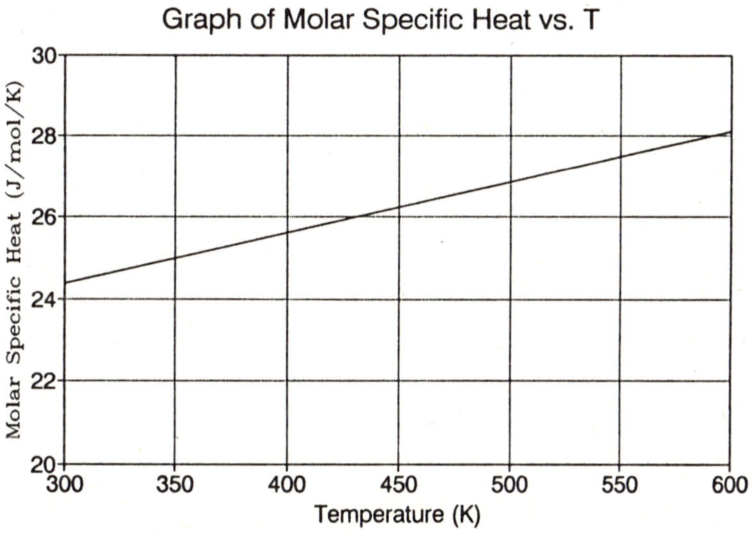

(c) $$\text{Q} = \text{n}\int_{\text{T}_i}^{\text{T}_f} \mathcal{C}_p\,\text{dT} = \text{n}\int_{\text{T}_i}^{\text{T}_f} (\text{A} + \text{BT})\text{dT}$$

$$= \text{nA}(\text{T}_f - \text{T}_i) + \frac{\text{nB}}{2}(\text{T}_f^2 - \text{T}_i^2)$$

$$= (2.50\text{ mol})(20.7\text{ J mol}^{-1}\text{ K}^{-1})(500\text{ K} - 300\text{ K}) + \frac{1}{2}(2.50\text{ mol})(0.0123\text{ J mol}^{-1}\text{K}^{-2})((500\text{ K})^2 - (300\text{ K})^2)$$

$$= 10.3\text{ kJ} + 2.5\text{ kJ} = 12.8\text{ kJ}$$

3. (a) For an ideal gas at constant temperature,

$$V = \frac{nRT}{p}, \quad \frac{dV}{dp} = -\frac{nRT}{p^2}$$

so $$\beta_i = -\frac{1}{V}\frac{dV}{dp} = \frac{nRT}{p^2 V} = \frac{1}{p}\left(\frac{nRT}{pV}\right) = \frac{1}{p}$$

(b) For a van der Waals gas,

$$RT = \left(p + \frac{a}{V^2}\right)(V - b), \text{ so}$$

$$0 = \left(p + \frac{a}{V^2}\right)\frac{dV}{dp} + (V-b)\left(1 - \frac{2a}{V^3}\frac{dV}{dp}\right)$$

so $$\left(p + \frac{a}{V^2} - \frac{2a(V-b)}{V^3}\right)\frac{dV}{dp} + (V-b) = 0$$

$$\therefore \quad \beta_{vdW} = \frac{1}{V}\frac{V-b}{p - \frac{a}{V^2} + \frac{2ab}{V^3}}$$

(c) If a and b → 0, then

$$\beta_{vdW} = \frac{1}{V}\frac{V}{p} = \frac{1}{p} = \beta_i$$

5. Let W_2 = work done on the fluid by the left piston,

W_1 = work done by the fluid on the right piston.

$$W_2 = p_2 V_2, \quad W_1 = p_1 V_1$$

$$W = W_1 - W_2 = p_1 V_1 - p_2 V_2$$

Applying the first law,

$$\Delta U = U_1 - U_2 = Q - W = 0 - p_1 V_1 + p_2 V_2$$

so $$U_1 + p_1 V_1 = U_2 + p_2 V_2$$

7. The note at the end of Exercise 22 indicates that for a quasi-static adiabatic process in an ideal gas,

$$p = p_i\left(\frac{V_i}{V}\right)^{\gamma}$$

(a) Replacing p by $\frac{nRT}{V}$,

$$\frac{nRT}{V} = \frac{p_i V_i^{\gamma}}{V^{\gamma}} \Rightarrow TV^{\gamma - 1} = \frac{p_i V_i^{\gamma}}{nR} = \text{constant}$$

(b) Replacing V by $\frac{nRT}{p}$,

$$p = p_i V_i^\gamma \left(\frac{p}{nRT}\right)^\gamma \Rightarrow \frac{p^{\gamma-1}}{T^\gamma} = \frac{(nR)^\gamma}{p_i V_i^\gamma} = \text{constant}$$

9. For an ideal gas, U = U(T). Thus,

$$U_1 = U_2 = U(T_a) \equiv U_a$$

and $U_3 = U_4 = U(T_b) \equiv U_b$

For process b,

$$Q_b = 0, \text{ so } W_b = -(U_b - U_a)$$

and for process d,

$$Q_d = 0, \text{ so } W_d = -(U_a - U_b) = -W_b$$

11. (a) $$T_1 = \frac{p_1 V_1}{nR} = \frac{(100\text{ kPa})(2.00\text{ L})}{(0.0963\text{ mol})(8.31\text{ J mol}^{-1}\text{ K}^{-1})} = 250\text{ K}$$

$$T_2 = \frac{p_2 V_2}{nR} = \frac{(400\text{ kPa})(2.00\text{ L})}{(0.0963\text{ mol})(8.31\text{ J mol}^{-1}\text{ K}^{-1})} = 1000\text{ K}$$

(b) $$Q_a = nC_v(T_2 - T_1) = (0.0963\text{ mol})(15.0\text{ J mol}^{-1}\text{ K}^{-1})(750\text{ K}) = 1080\text{ J}$$

(c) For process c,

$$W_c = -\tfrac{1}{2}(p_3 + p_1)(V_3 - V_1) = -\tfrac{1}{2}(200\text{ kPa} + 100\text{ kPa})(2.00\text{ L}) = -300\text{ J}$$

$$U_1 - U_3 = nC_v(T_1 - T_3) = -1080\text{ J}$$

$$\therefore \quad Q_c = -1080\text{ J} + (-300\text{ J}) = -1380\text{ J}$$

(d) $$W_b = nRT\ln\left(\frac{V_3}{V_2}\right) = (0.0963\text{ mol})(8.31\text{ J mol}^{-1}\text{ K}^{-1})(1000\text{ K})\ln\left(\frac{4.00\text{ L}}{2.00\text{ L}}\right) = 555\text{ J}$$

$$W_{cyc} = W_b + W_c = 555\text{ J} + (-300\text{ J}) = 255\text{ J}$$

13. The temperatures at 1, 2, and 3 are:

$$T_1 = \frac{p_1 V_1}{nR} = \frac{(100\text{ kPa})(2.08\text{ L})}{(0.0782\text{ mol})(8.31\text{ J mol}^{-1}\text{ K}^{-1})} = 320\text{ K}$$

$$T_2 = \frac{p_2 V_2}{nR} = \frac{p_2}{p_1}T_1 = \left(\frac{200\text{ kPa}}{100\text{ kPa}}\right)320\text{ K} = 640\text{ K}$$

$$T_3 = \frac{p_3 V_3}{nR} = \frac{(300\text{ kPa})(1.00\text{ L})}{(0.0782\text{ mol})(8.31\text{ J mol}^{-1}\text{ K}^{-1})} = 462\text{ K}$$

For process a,

$$W_a = 0 \text{ and } Q_a = n\mathcal{C}_v(T_2 - T_1)$$

$$= (0.0782\ \text{mol})(16.62\ \text{J mol}^{-1}\ \text{K}^{-1})(320\ \text{K}) = 416\ \text{J}$$

$$\Delta U_a = Q_a - W_a = 416\ \text{J}$$

For process b,

$$W_b = \frac{1}{2}(p_2 + p_3)(V_3 - V_2) = \frac{1}{2}(200\ \text{kPa} + 300\ \text{kPa})(1.00\ \text{L} - 2.08\ \text{L}) = -270\ \text{J}$$

Since the gas is ideal, $U = U(T)$, so $\Delta U_b = U(T_3) - U(T_2)$ and this can be calculated from an isochoric process between these temperatures. Thus,

$$\Delta U_b = n\mathcal{C}_v(T_3 - T_2)$$

$$= (0.0782\ \text{mol})(16.62\ \text{J mol}^{-1}\ \text{K}^{-1})(462\ \text{K} - 640\ \text{K}) = -230\ \text{J}$$

$$Q_b = \Delta U_b + W_b = -230\ \text{J} - 270\ \text{J} = -500\ \text{J}$$

For process c, $Q_c = 0$, and

$$W_c = \frac{p_3V_3}{\gamma - 1}\left[1 - \left(\frac{V_3}{V_1}\right)^{\gamma-1}\right],$$

where $$\gamma = \frac{\mathcal{C}_p}{\mathcal{C}_v} = \frac{24.93\ \text{J mol}^{-1}\ \text{K}^{-1}}{16.62\ \text{J mol}^{-1}\ \text{K}^{-1}} = 1.500$$

$$\therefore \quad W_c = \frac{(300\ \text{kPa})(1.00\ \text{L})}{.500}\left[1 - \left(\frac{1.00\ \text{L}}{2.08\ \text{L}}\right)^{.500}\right] = 184\ \text{J}$$

and $$\Delta U_c = Q_c - W_c = -184\ \text{J}$$

Summarizing these findings in a table, we have:

	Energies in J		
Process	**W**	**Q**	**ΔU**
a	0	416	416
b	−270	−500	−230
c	184	0	−184
Cycle	−86	−86	0

15. Given $p = KV^{-\gamma}$,

(a) $$W_Q = \int_{V_i}^{V_f} p\, dV = K \int_{V_i}^{V_f} V^{-\gamma}\, dV = \frac{K}{1-\gamma}(V_f^{1-\gamma} - V_i^{1-\gamma})$$

But $K = p_i V_i^{\gamma}$, so

$$W_Q = \frac{p_i V_i^{\gamma}}{\gamma - 1}(V_i^{1-\gamma} - V_f^{1-\gamma}) = \frac{p_i V_i}{\gamma - 1}\left(1 - \left(\frac{V_i}{V_f}\right)^{\gamma - 1}\right)$$

(b) $$W_Q = \frac{(202\text{ kPa})(45\text{ L})}{0.40}\left(1 - \left(\frac{45\text{ L}}{65\text{ L}}\right)^{0.40}\right) = 3.1\text{ kJ}$$

CHAPTER 18 KINETIC THEORY OF GASES

Questions:

1. That the mean values of the velocity components are zero simply indicates that the molecules are equally likely to be moving in one direction or its opposite. In fact, if the mean velocity components are not zero, the gas as a whole would be moving. The mean speed $\langle v \rangle$ of the molecules can only be zero if none of the molecules are moving.

3. For a mixture of gases in the same volume V, the pressure would be equal to the sum of the partial pressures due to each of the different components. Thus, for two gases,

$$p = p_a + p_b = \frac{1}{3V}(N_a m_a \langle v_a^2 \rangle + N_b m_b \langle v_b^2 \rangle)$$

$$pV = \frac{2}{3}\left\{N_a\left(\frac{1}{2} m_a \langle v_a^2 \rangle\right) + N_b\left(\frac{1}{2} m_b \langle v_b^2 \rangle\right)\right\}$$

$$= NkT$$

$$\therefore \quad \frac{3}{2}kT = \frac{N_a\left(\frac{1}{2} m_a \langle v_a^2 \rangle\right) + N_b\left(\frac{1}{2} m_b \langle v_b^2 \rangle\right)}{N_a + N_b}$$

The right hand side represents the average translational kinetic energy per molecule of the gas mixture, so the relation still holds.

5. While the molecules making up the wall are also in motion, their average velocity component is zero. To simplify matters, suppose that the molecules of the wall and those of the gas are all moving normal to the wall. In a collision with a receding wall molecule, the change in momentum perpendicular to the wall is

$$m[(v - 2v_w) + v] = 2m(v - v_w)$$

while the change in momentum for a collision with an approaching molecule is $2m(v + v_w)$. On the average, the change in momentum is 2mv, just as it is for a stationary wall.

7. Yes. The basketball in the long run will have zero displacement, so the molecules inside it will also.

9. When a gas is heated at constant volume, no work is done, so $Q_V = \Delta U$. When the gas is heated at constant pressure, $Q_p = \Delta U + W$, where W is positive because the gas expands. For a gas, U is strongly dependent on the temperature, so that for corresponding temperature changes there are corresponding changes in U.

Thus, $Q_p > Q_V$ and $C_p > C_V$.

For a solid or liquid these arguments are usually valid. However, it is not necessarily true that a heated substance expands (for example water below 4° C) or that U is basically a function of the temperature only, so these cases would have to be examined more closely.

11. No. If f is time dependent, then the gas is not in equilibrium. Equilibrium implies that things are not changing, at least macroscopically.

13. Looking at a one–dimensional model, a molecule moving with a speed v which collides with a receding wall that has a speed v_w will bounce with a speed $v - 2v_w$. Thus, the mean speed of the molecules of a gas in which the volume is expanding will be losing energy and the temperature of the gas will decrease. If the gas is being compressed the temperature will increase.

15. Temperature is a measure of the average kinetic energy of the molecules. Thus the concept must be applied to a sufficient number such that a statistical average can be applied. The number of molecules needed for the concept to be applicable depends on the precision of the measurements to be made.

17. As the temperature drops the velocity distribution becomes more narrowly peaked and $\langle v \rangle$, v_m, and v_{rms} become nearly equal.

19. In order for the distribution of speeds to be symmetrical about the most probable value, v_m would also have to be the central value. If the highest speed is c, then v_m would have to be $\frac{c}{2}$, regardless of the temperature.

21. For a monatomic gas,

$$C_V = \frac{3}{2}R \text{ and } C_p = \frac{5}{2}R,$$

so $C_V = 12.5\ \text{J mol}^{-1}$ and $C_p = 20.8\ \text{J mol}^{-1}$.

Exercises:

1.

$v(m/s)$	$v^2(m^2/s^2)$
290	84100
47	2209
182	33124
439	192721
330	108900
268	71824
302	91204
372	138384
344	118336
410	168100
2984	1008902

(a) $\langle v\rangle = \dfrac{2984\text{ m/s}}{10} = 298\text{ m/s}$

(b) $\langle v^2\rangle = \dfrac{1008902\text{ m}^2/\text{s}^2}{10} = 1.01 \times 10^5\text{ m}^2/\text{s}^2$

(c) $\langle v\rangle^2 = 0.890 \times 10^5\text{ m}^2/\text{s}^2$

So $\langle v\rangle^2$ is less than $\langle v^2\rangle$.

3. $\vec{v}_{cm} = \dfrac{\sum m_i \vec{v}_i}{\sum m_i} = \dfrac{m\sum \vec{v}}{mN} = \dfrac{1}{N}\sum \vec{v}$

$= \langle \vec{v} \rangle$

That is, $(v_{cm})_x = \langle v_x\rangle$, $(v_{cm})_y = \langle v_y\rangle$, $(v_{cm})_z = \langle v_z\rangle$.

In the description given, $\langle v_x\rangle = \langle v_y\rangle = \langle v_z\rangle = 0$, so the description refers to a reference frame in which the center of mass of the gas is at rest.

5. The expected number of times that the coin will come up heads is 500. One should not expect that there will be precisely 500 heads in 1000 tosses. However, if a million people tossed a coin 1000 times each, then the average among them would be very close to 500 heads.

7. Equation 18–8 is

$$pV = \frac{Nm\langle v^2\rangle}{3},$$

where m = mass of 1 molecule

$$= \frac{\text{mass of 1 mole}}{\text{Avogadro's number}} = \frac{M}{N_A}$$

and N = number of molecules

= (number of moles)(Avogadro's number)

$= nN_A$

$$\therefore \quad pV = (nN_A)\left(\frac{M}{N_A}\right)\frac{\langle v^2\rangle}{3}$$

$$= \frac{nM\langle v^2\rangle}{3}$$

9. $v_{rms} = \sqrt{\dfrac{3kT}{m}}, \quad T = 300\text{ K}$

(a) For N_2, $m = \dfrac{0.0280\text{ kg/mol}}{6.02 \times 10^{23}\text{ molecules/mol}}$

$= 4.65 \times 10^{-26}$ kg

$$v_{rms} = \sqrt{\frac{(3)(1.38 \times 10^{-23}\text{ J/K})(300\text{ K})}{4.65 \times 10^{-26}\text{ kg}}}$$

= 517 m/s

(b) For O_2, $m = \dfrac{0.0320\text{ kg/mol}}{6.02 \times 10^{23}\text{ molecules/mol}}$

$= 5.31 \times 10^{-26}$ kg

$$v_{rms} = \sqrt{\frac{(3)(1.38 \times 10^{-23}\text{ J/K})(300\text{ K})}{5.31 \times 10^{-26}\text{ kg}}}$$

= 483 m/s

(c) For CO_2, $m = \dfrac{0.0440\text{ kg/mol}}{6.02 \times 10^{23}\text{ molecules/mol}}$

$= 7.31 \times 10^{-26}$ kg

v_{rms} = 412 m/s

(d) The average translational kinetic energy is given by $\langle E \rangle = \langle \frac{1}{2} mv^2 \rangle = \frac{1}{2} mv_{rms}^2$

Thus, for N_2,

$$\langle E \rangle = \frac{1}{2}(4.65 \times 10^{-26}\ \text{kg})\left(517\ \frac{\text{m}}{\text{s}}\right)^2$$

$$= 6.21 \times 10^{-21}\ \text{J};$$

For O_2,

$$\langle E \rangle = \frac{1}{2}(5.31 \times 10^{-26}\ \text{kg})\left(483\ \frac{\text{m}}{\text{s}}\right)^2$$

$$= 6.21 \times 10^{-21}\ \text{J};$$

and for CO_2,

$$\langle E \rangle = \frac{1}{2}(7.31 \times 10^{-26}\ \text{kg})\left(412\ \frac{\text{m}}{\text{s}}\right)^2$$

$$= 6.21 \times 10^{-21}\ \text{J}$$

In all cases,

$$\langle E \rangle = \frac{3}{2} kT$$

11. Since v_{rms} is proportional to $\sqrt{T}$,

$$\frac{T_f}{T_i} = \left(\frac{v_{rms\ f}}{v_{rms\ i}}\right)^2 = \left(1 + \frac{\Delta v_{rms}}{v_{rms}}\right)^2$$

(a) If $\frac{\Delta v_{rms}}{v_{rms}} = 0.10,$ then $\frac{T_f}{T_i} = 1.21$

(b) If $\frac{\Delta v_{rms}}{v_{rms}} = -0.10,$ then $\frac{T_f}{T_i} = 0.81$

(c) If $\frac{\Delta v_{rms}}{v_{rms}} = 0.50,$ then $\frac{T_f}{T_i} = 2.25$

13. $\langle E \rangle = \frac{3}{2} kT$

(a) $$\langle E \rangle = \frac{\frac{3}{2}(1.38 \times 10^{-23}\ \frac{\text{J}}{\text{K}})(90\ \text{K})}{1.60 \times 10^{-19}\ \frac{\text{J}}{\text{eV}}}$$

$$= \frac{3}{2}\left(8.63 \times 10^{-5}\ \frac{\text{eV}}{\text{K}}\right)(90\ \text{K})$$

$$= 0.012\ \text{eV}$$

Note that $k = 8.63 \times 10^{-5}\ \frac{\text{eV}}{\text{K}}$ when expressed in terms of $\frac{\text{eV}}{\text{K}}$.

(b) $\langle E \rangle = \frac{3}{2}\left(8.63 \times 10^{-5}\ \frac{\text{eV}}{\text{K}}\right)(300\ \text{K})$

$$= 0.0388\ \text{eV}$$

(c) $\langle E \rangle = \frac{3}{2}\left(8.63 \times 10^{-5}\ \frac{\text{eV}}{\text{K}}\right)(6000\ \text{K})$

$$= 0.776\ \text{eV}$$

15. (a) Since the center of mass is moving at 80 km/h = 22.2 m/s in the x direction,

$$\langle v_x \rangle = 22.2\ \text{m/s}, \quad \langle v_y \rangle = 0, \quad \langle v_z \rangle = 0$$

(b) According to the engineer, the center of mass of the gas is at rest, so

$$\langle v_x \rangle = 0, \quad \langle v_y \rangle = 0, \quad \langle v_z \rangle = 0$$

(c) The expression $\langle K \rangle = \frac{3}{2}kT$ is valid only in the center of mass frame. Kinetic energy is not an invariant in a change of inertial frames but temperature is, so the relation between them can only apply in one frame of reference.

17. (a) The O_2 molecule has 5 degrees of freedom, thus $U = \frac{5}{2}\ nRT$.

$$\Delta U = \frac{5}{2} nR\Delta T = \frac{5}{2}(2.2\ \text{mol})(8.31\ \text{J mol}^{-1}\ \text{K}^{-1})(130\ \text{K}) = 5.9\ \text{kJ}$$

(b) It is not possible to determine how much heat was added or how much work was done on the system to increase the internal energy, since any combination of Q and |W| which adds up to ΔU would have the same effect.

19. (a) If $\Delta V = 0$ then $W = 0$ and $Q = \Delta U$. Thus, $Q = 5.9$ kJ.

(b) $C_p = C_v + R = \frac{7}{2}R = 29.1\ \text{J mol}^{-1}\ \text{K}^{-1}$

(c) If $\Delta p = 0$, then $Q = nC_p\Delta T = 8.3$ kJ and $W = \Delta U - Q = 2.4$ kJ.

21. (a) $\gamma = \frac{C_p}{C_v} \Rightarrow C_v = \frac{C_p}{\gamma} = \frac{31.4 \text{ J mol}^{-1}\text{ K}^{-1}}{1.36} = 23.1 \text{ J mol}^{-1}\text{ K}^{-1}$

(b) $\gamma = \frac{\nu + 2}{\nu} \Rightarrow \nu = \frac{2}{\gamma - 1} = \frac{2}{0.36} = 5.6$

A rigid two–atom molecule has 5 degrees of freedom, and a vibrating one would have 6. The value of γ for F_2 at 25° C corresponds to $\nu = 5.6$, so the vibrational mode is partially active.

23. From Figure 18–5, the values of C_v at 100 K, 600 K, and 2000 K are approximately $\frac{3}{2}$R, $\frac{5}{2}$R, and $\frac{7}{2}$R, respectively.

$$\gamma = \frac{C_v + R}{C_v} = 1 + \frac{R}{C_v}$$

(a) At 100 K, $\gamma = 1 + \frac{2}{3} = \frac{5}{3} = 1.67$

(b) At 600 K, $\gamma = 1 + \frac{2}{5} = \frac{7}{5} = 1.40$

(c) At 2000 K, $\gamma = 1 + \frac{2}{7} = \frac{9}{7} = 1.29$

25. (a) From Table 18–1, the molar heats of Ar are $C_v = 12.6 \text{ J mol}^{-1}\text{ K}^{-1}$ and $C_p = 20.8 \text{ J mol}^{-1}\text{ K}^{-1}$. In this problem, 0.45 mol of Ar is subjected to (1) an isobaric compression from V = 12.0 L to V = 6.0 L, followed by (2) an isochoric heating which raises the pressure from 180 kPa to 480 kPa. Using the ideal gas law, T_i = 578 K, T_f = 770 K, and the temperature at the corner is 289 K.

Thus,
$$\begin{aligned} Q_1 &= nC_p\,\Delta T \\ &= (0.45 \text{ mol})(20.8 \text{ J mol}^{-1}\text{ K}^{-1})(289 \text{ K} - 578 \text{ K}) \\ &= -2.7 \text{ kJ} \\ Q_2 &= nC_v\,\Delta T \\ &= (0.45 \text{ mol})(12.6 \text{ J mol}^{-1}\text{ K}^{-1})(770 \text{ K} - 289 \text{ K}) \\ &= 2.7 \text{ kJ} \end{aligned}$$

For the entire process, Q = 0.0 kJ.

(b) The work done by the gas is

$$W = W_1 = p\Delta V$$
$$= (180\text{ kPa})(6.0\text{ L} - 12.0\text{ L})$$
$$= -1.1\text{ kJ}$$

(c) $\Delta U = Q - W = 1.1$ kJ. This agrees with the value

$$\Delta U = nC_v\,\Delta T = 1.1\text{ kJ}$$

calculated in Exercise 24.

27. (a) $p_f V_f^\gamma = p_i V_i^\gamma$

$$p_f = p_i\left(\frac{V_i}{V_f}\right)^\gamma = (320\text{ kPa})\left(\frac{12\text{ L}}{18\text{ L}}\right)^{1.67}$$
$$= 160\text{ kPa}$$

(b)
$$T_i = \frac{p_i V_i}{nR} = \frac{(320\text{ kPa})(12\text{ L})}{(1.4\text{ mol})(8.31\text{ J mol}^{-1}\text{ K}^{-1})}$$
$$= 330\text{ K}$$

$$T_f = \frac{p_f V_f}{nR} = \frac{(163\text{ kPa})(18\text{ L})}{(1.4\text{ mol})(8.31\text{ J mol}^{-1}\text{ K}^{-1})}$$
$$= 250\text{ K}$$

(c)

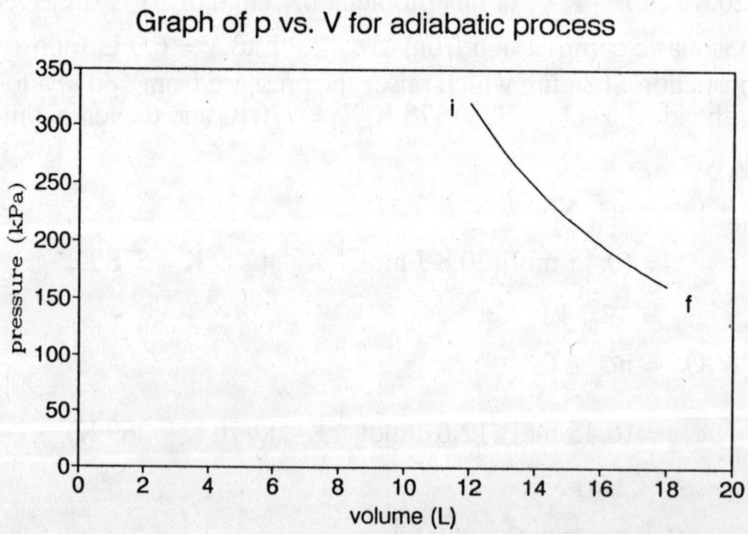

29. For an ideal gas undergoing a quasi-static adiabatic process,

$$dT = -\frac{p\,dV}{n\mathcal{C}_v} = -\frac{\left(\frac{nRT}{V}\right)dV}{n\mathcal{C}_v}$$

$$= -\frac{RT}{\mathcal{C}_v}\frac{dV}{V}$$

Now, $R = \mathcal{C}_p - \mathcal{C}_v$, so $\frac{R}{\mathcal{C}_v} = \gamma - 1$,

and $\frac{dT}{T} = -(\gamma - 1)\frac{dV}{V}$

Integrating,

$$\int_{T_i}^{T_f} \frac{dT}{T} = -(\gamma - 1)\int_{V_i}^{V_f} \frac{dV}{V}$$

$$\ln\left(\frac{T_f}{T_i}\right) = -(\gamma - 1)\ln\left(\frac{V_f}{V_i}\right)$$

or $T_i V_i^{\gamma-1} = T_f V_f^{\gamma-1}$

For an isobaric process, $T_i V_i^{-1} = T_f V_f^{-1}$

31. $n = \frac{64\text{ g}}{32\text{ g/mol}} = 2.0\text{ mol}$

(a) $Q = n\mathcal{C}_v\,\Delta T = (2.0\text{ mol})(21.1\text{ J mol}^{-1}\text{ K}^{-1})(30\text{ K}) = 1.3\text{ kJ}$

$W = 0$

$\Delta U = Q - W = 1.3\text{ kJ}$

(b and c) Since the gas is considered ideal, the change in the internal energy depends only on the change in temperature. Thus, $\Delta U = 1.3$ kJ regardless of the process involved.

33. $T_i = \frac{p_i V_i}{nR} = \frac{(200 \text{ kPa})(20 \text{ L})}{(1 \text{ mol})(8.31 \text{ J mol}^{-1} \text{ K}^{-1})} = 481 \text{ K}$

At that temperature, the values of γ given in Table 18–1 apply. They are:

Gas	γ
He	1.63
Ne	1.64
O_2	1.40
CO_2	1.30

The curves for a quasi-static adiabatic process with $V_f = 30$ L are shown below.

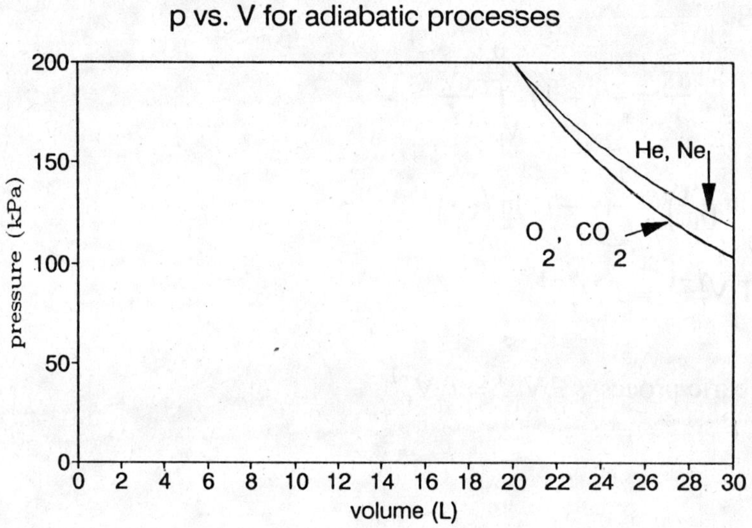

35. From equation 18–20,

$$f(v) = Av^2 e^{-\beta v^2}$$

where $A = 4\pi N \left(\frac{m}{2\pi kT}\right)^{3/2}$ and $\beta = \frac{m}{2kT}$

The most probable speed is the speed at which f(v) is a maximum, so $\frac{df}{dv} = 0$

$$\frac{df}{dv} = 2Ave^{-\beta v^2} - 2\beta v(Av^2 e^{-\beta v^2})$$

$$= 2Av(1 - \beta v^2)e^{-\beta v^2}$$

This expression is zero when $v = 0$ and when $v^2 = \frac{1}{\beta}$. The first value corresponds to a minimum, while the second corresponds to a maximum. Thus,

$$v_m^2 = \frac{1}{\beta} = \frac{2kT}{m}$$

$$v_m = \sqrt{\frac{2\,kT}{m}}$$

37. Constructing a table,

	ΔN	v m/s	$(\Delta N)(v)$ m/s	$(\Delta N)(v^2)$ m^2/s^2
	1	50	50	2500
	2	150	300	45000
	4	250	1000	250000
	3	350	1050	367500
	1	450	450	202500
Totals	11		2850	867500

(a) $\langle v \rangle = \frac{\sum (\Delta N)v}{\sum \Delta N} = \frac{2850 \text{ m/s}}{11} = 260 \text{ m/s}$

(b) $\langle v^2 \rangle = \frac{\sum (\Delta N)(v^2)}{\sum N} = \frac{867500 \text{ m}^2/\text{s}^2}{11} = 79 \times 10^3 \text{ m}^2/\text{s}^2$

$$v_{rms} = \sqrt{\langle v^2 \rangle} = 280 \text{ m/s}$$

(c) From the table, the speed for which ΔN is the largest ($\Delta N = 4$) is $v = 250$ m/s.

Thus, $v_m = 250 \frac{m}{s}$

Problems:

1. (a) For an isobaric process, $\frac{dp}{dV} = 0$

(b) For an isothermal process, $p = \frac{nRT}{V}$

$$\frac{dp}{dV} = -\frac{nRT}{V^2}$$

(c) For an adiabatic process, $p = p_i \left(\frac{V_i}{V}\right)^{\gamma}$

$$\frac{dp}{dV} = \frac{-\gamma\, p_i V_i^{\gamma}}{V^{\gamma+1}}$$

(d) The adiabatic process has the steepest slope.

3. (a) $v_{rms} = \sqrt{\frac{3\,kT}{m}} = \sqrt{\frac{3\,RT}{M}} = \sqrt{\frac{3(8.31\text{ J mol}^{-1}\text{ K}^{-1})(1000\text{ K})}{0.028\text{ kg mol}^{-1}}} = 940\text{ m/s}$

(b) $v_i = \sqrt{\frac{2\,GM_e}{r_i}} = \sqrt{\frac{2(6.67\times10^{-11}\text{ Nm}^2\text{ kg}^{-2})(5.97\times10^{24}\text{ kg})}{(6.53\times10^6\text{ m})}}$

$= 11000\text{ m/s}$

$$\frac{f(v_i)}{f(v_{rms})} = \frac{v_i^2\,e^{-\beta v_i^2}}{v_{rms}^2\,e^{-\beta v_{rms}^2}} \quad \text{where} \quad \beta = \frac{m}{2\,kT} = \frac{3}{2v_{rms}^2}$$

$$\frac{f(v_i)}{f(v_{rms})} = \left(\frac{v_i}{v_{rms}}\right)^2 e^{-3/2\,[(v_i^2/v_{rms}^2)-1]}$$

$$= (11.7)^2\,e^{-3/2[(11.7)^2-1]}$$

$$= 7.5\times10^{-88}$$

(c) For H_2, $v_{rms} = 3500$ m/s, so

$$\frac{f(v_i)}{f(v_{rms})} = 2.0\times10^{-5}$$

The ratio is much larger because the mass of the H_2 molecule is smaller, making v_{rms} larger.

5. (a) $m = \frac{0.016\text{ kg/mol}}{6.02\times10^{23}\text{ molecules/mol}} = 2.7\times10^{-26}\text{ kg}$

$I = \frac{1}{2}mL^2 = \frac{1}{2}(2.7\times10^{-26}\text{ kg})(0.2\times10^{-9}\text{ m})^2 = 5.3\times10^{-46}\text{ kg m}^2$

(b) $E_{rot} = \frac{1}{2}kT = \frac{1}{2}I\Omega^2$

$$\Omega = \sqrt{\frac{kT}{I}} = \sqrt{\frac{(1.38\times10^{-23}\text{ J/K})(400\text{ K})}{5.3\times10^{-46}\text{ kg m}^2}} = 3\times10^{12}\text{ rad/s}$$

7. (a) $U = 3N\langle E\rangle = \frac{3\,Nh\nu_o}{e^{h\nu_o/kT}-1}$

$$C_v = \frac{1}{n}\frac{dU}{dT} = 3\,N_o h\nu_o \frac{d}{dT}\left(\frac{1}{e^{h\nu_o/kT}-1}\right)$$

$$= 3\,N_o h\nu_o \left(\frac{-1}{(3^{h\nu_o/kT} - 1)^2}\right)\left(-\frac{h\nu_o}{kT^2}\,e^{h\nu_o/kT}\right)$$

$$= \frac{3\,N_o h^2 \nu_o^2}{kT^2}\,\frac{e^{h\nu_o/kT}}{(e^{h\nu_o/kT} - 1)^2}$$

$$= 3R\,\frac{(h\nu_o/kT)^2\,e^{h\nu_o/kT}}{(3^{h\nu_o/kT} - 1)^2}$$

(b)

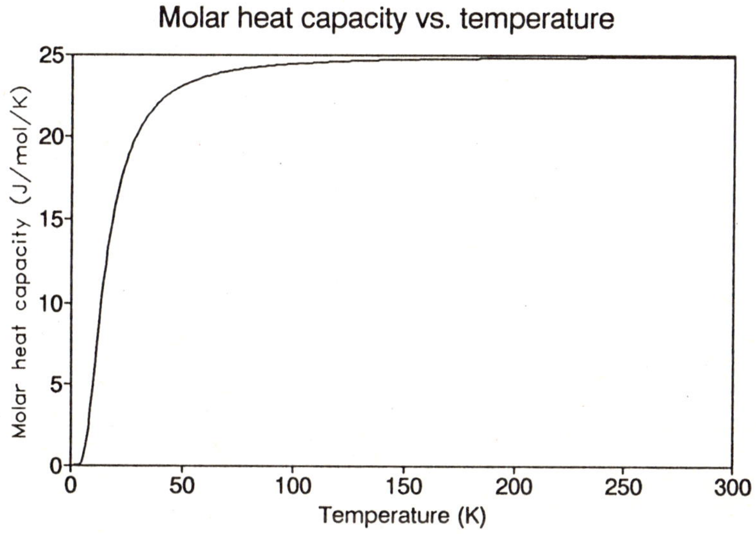

(c) As $T \to \infty$, $e^{h\nu_o/kT} \to 1 + h\nu_o/kT$

$$\text{so} \quad C_v \to \frac{3R\left(\dfrac{h\nu_o}{kT}\right)^2}{\left(\dfrac{h\nu_o}{kT}\right)^2} = 3R$$

9. Following the development in Section 18–1, the average force exerted by the molecules of gas 1 is

$$F_1 = \frac{m_1 \sum v_{1jx}^2}{L_1}, \text{ etc.}$$

so the total force is

$$F = \frac{m_1 \sum v_{1jx}^2}{L_1} + \frac{m_2 \sum v_{2jx}^2}{L_1} + \dots$$

and $p = \dfrac{F}{L_2 L_3} = \dfrac{1}{V}(m_1 \sum v_{1jx}^2 + m_2 \sum v_{2jx}^2 + \dots)$

$$= p_1 + p_2 + \ldots$$

Thus the pressure due to the mixture is equal to the sum of the pressures that the component gases would exert by themselves in the same volume.

11. From equation 18–20,

$$f(v) = Av^2 e^{-\beta v^2},$$

where $A = 4\pi N \left(\frac{m}{2\pi kT}\right)^{3/2}$ and $\beta = \frac{m}{2kT}$

The fraction of molecules having a speed less than or equal to v is

$$p = \frac{1}{N} \int_0^v f(v)\,dv$$

We are interested in finding the value of v which makes $p = \frac{1}{2}$. To simplify the calculations, let $x = v\sqrt{\beta}$. Then, $dv = \frac{dx}{\sqrt{\beta}}$ and

$$p = \frac{1}{N} \int_0^{v\sqrt{\beta}} A\left(\frac{x^2}{\beta}\right) e^{-x^2} \frac{dx}{\sqrt{\beta}}$$

$$= \frac{A}{N\beta^{3/2}} \int_0^{v\sqrt{\beta}} x^2 e^{-x^2}\,dx$$

$$= \frac{4\pi \left(\frac{m}{2\pi kT}\right)^{3/2}}{\left(\frac{m}{2kT}\right)^{3/2}} \int_0^{v\sqrt{\beta}} x^2 e^{-x^2}\,dx$$

$$= \frac{4}{\sqrt{\pi}} \int_0^{v\sqrt{\beta}} x^2 e^{-x^2}\,dx$$

To find the value of v which makes $p = \frac{1}{2}$, we set up a spreadsheet that calculates

$$F(x) = \int_0^x x^2 e^{-x^2}\,dx$$

with the following cell formulas:

```
A1: 0
B1: +A1^2*@EXP(–A1^2)
A2: A1+$D$1
B2: (copied from B1)
```

C2: +0.5*(B1+B2)*$E$1
D2: +C2*$E$2

Subsequent rows are copied from row 2. Cell E1 contains the integration step and Cell E2 contains the constant $4/\sqrt{\pi}$ (4/@ PI ^ .5).

A part of this spreadsheet is shown below.

x	f	F	Fraction
1.076	0.363755	0.217326	0.490451
1.078	0.363539	0.218053	0.492092
1.08	0.363318	0.21878	0.493733
1.082	0.363091	0.219506	0.495372
1.084	0.362859	0.220232	0.49701
1.086	0.362622	0.220958	0.498648
1.088	0.36238	0.221683	0.500284 ←
1.09	0.362133	0.222407	0.501919
1.092	0.36188	0.223131	0.503553
1.094	0.361622	0.223855	0.505186
1.096	0.36136	0.224578	0.506817
1.098	0.361092	0.2253	0.508448
1.1	0.360819	0.226022	0.510077

From this output it can be seen that the value of x corresponding to $p = \frac{1}{2}$ is $x = 1.088$, thus

$$v_{med} = 1.088\sqrt{\frac{2\,kT}{m}}$$

In units of $\sqrt{\frac{2\,kT}{m}}$, the statistical characteristics are:

$$\langle v \rangle = \frac{2}{\sqrt{\pi}} = 1.128$$

$$v_m = 1$$

$$v_{rms} = \sqrt{\frac{3}{2}} = 1.225$$

so

$$v_m < v_{med} < \langle v \rangle < v_{rms}$$

13. Following the procedure outlined in the problem statement we set up the following BASICA program:

```
10    'ch18pr13
20    DIM A(110)
30    FOR N=1 TO 100
40    RANDOMIZE (TIMER)
50    SUM=300
60    FOR I=1 TO 50
70    SUM=SUM+10*(RND-.5)
80    NEXT I
90    J=(SUM+2.5)\5
100   A(J)=A(J)+1
110   NEXT N
120   CUM=0
130   FOR J=1 TO 110
140   IF A(J)= 0 THEN 170
150   CUM=CUM+A(J)
160   PRINT J,A(J),CUM
170   NEXT J
180   END
```

Steps 40–80 evaluate v. Step 40 used the internal clock to provide a seed for the random number generator. The loop 30–110 evaluates v 100 times, and assigns the value an integer ranking J such that 5*J is at the center of the interval of width $\Delta v = 5$. Once J is calculated, an array A(J) is used to keep a running count of the number of times that v has fallen in that range.

Steps 120–180 are used to print out the frequency of occurrence A(J) and the cumulative number of occurrences (CUM). Step 140 is there simply to leave out the null entries, but can be omitted. A graph of the results of one run of this program is shown in the figure below.

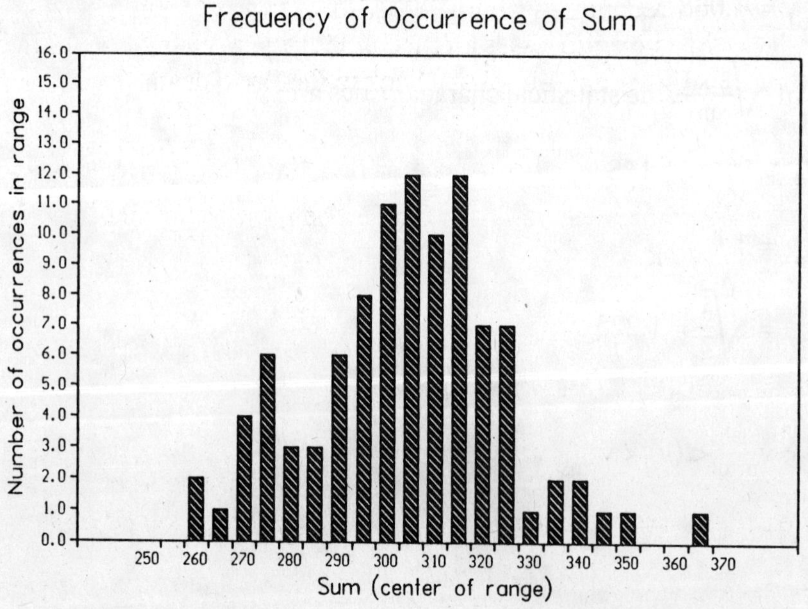

CHAPTER 19 THE SECOND LAW OF THERMODYNAMICS

Questions:

1. A cyclic process is desirable in a heat engine because it returns the working fluid to its original condition so that the process can be carried out over and over again indefinitely. Another advantage is that a cyclic engine is compact. In the single expansion process discussed in the text, the length of the cylinder increases with the amount of work to be extracted, with no limit. The disadvantage of the cyclic engine is that it requires that some heat be "dumped" in order for the working fluid to return to its original state and thus the thermodynamic efficiency is always less than 1.

3. It is not possible to cool a room by leaving the refrigerator room open. The heat extracted from the refrigerator is "dumped" into the room, along with the energy needed to run the refrigerator. The net result is that heat is added to the refrigerator-room system.

5. Some of the factors are: friction in bearings, belts, etc.; the fact that the process is not carried out precisely between two temperatures, so some heat is absorbed at temperatures that are below the highest available and "dumped" at temperatures above the coldest available. The adiabatic parts are not truly adiabatic since there are conductive and radiative heat losses.

7. Examples of nearly reversible processes are: a slow isothermal or adiabatic expansion of a gas; the slow stretching of a spring; the slow lifting of a weight under gravity. Examples of irreversible processes are: the mixing of two liquids at different temperatures; an explosion; dropping a book to the floor; death.

9. The hot gases from the combustion of the gasoline play the part of the high temperature reservoir and the ambient air cools the fluid in the radiator, which acts as the cold temperature reservoir.

11. Both processes are irreversible. The entropy change in the water is the same for both processes since the initial and final states are the same.

13. (a) The entropy of the gas increases in an isothermal compression.
    (b) The entropy remains unchanged in an adiabatic compression.

15. A liquid is more orderly than a vapor. The vapor state has the larger entropy at the boiling point since heat has to be added to the system to effect the phase change from liquid to vapor.

17. There is no conflict with the second law because the entropy of the universe increases: Living organisms are irreversible (and nearly cyclic) heat engines. In the process of becoming more ordered structures, they cause their surroundings to become less ordered.

19. The second law does not say that the entropy of a system cannot decrease. Thus the entropy of a substance can decrease upon freezing provided that there is an equal or greater increase in the entropy of the surroundings.

21. The entropy of a kg of water can be increased by heating it at constant volume and decreased by cooling it at a constant volume.

23.

Symbol	**Represents**	**Type**	**SI Unit**
Q_H	Heat gained at high temperature	scalar	J
Q_L	Heat gained at low temperature	scalar	J
T_H	Temperature of hot reservoir	scalar	K
η	Efficiency of a heat engine	scalar	dimensionless
K	Coefficient of performance	scalar	dimensionless
K_{hp}	Coefficient of performance for heat pump	scalar	dimensionless
S	Entropy	scalar	J/K

Exercises:

1. Given $Q_H = 440$ J, $\eta = 0.28$

(a) $\eta = \frac{W}{Q_H}$ so $W = \eta Q_H = 120$ J

(b) $|Q_L| = Q_H - W = 320$ J

(c) $\Delta U = 0$ because the system returns to its original state in one cycle.

3. Given $Q_H = 2200$ J, W = 620 J

(a) $\eta = \frac{W}{Q_H} = \frac{620\text{ J}}{2200\text{ J}} = 0.282\ (28.2\%)$

(b) $|Q_L| = Q_H - W = 1580$ J

(c) $\frac{dQ_H}{dt} = \frac{2200\text{ J}}{0.033\text{ s}} = 67\text{ kW}$

$\left|\frac{dQ_L}{dt}\right| = \frac{1580\text{ J}}{0.033\text{ s}} = 48\text{ kW}$

$\frac{dW}{dt} = \frac{620\text{ J}}{0.033\text{ s}} = 19\text{ kW}$

5. (a) $T_a = \frac{p_a V_a}{nR} = \frac{(101\text{ kPa})(22.4\text{ L})}{(1.00\text{ mol})(8.31\text{ J/mol/K})} = 272\text{ K}$

$T_b = \frac{p_b V_b}{nR} = \frac{2p_a V_a}{nR} = 2T_a = 544\text{ K}$

$T_c = \frac{(2p_a)(2V_a)}{nR} = 4T_a = 1090\text{ K}$

$T_d = \frac{(p_a)(2V_a)}{nR} = 2T_a = 544\text{ K}$

(b) Heat added ($a \rightarrow b \rightarrow c$)

$Q_H = n\mathcal{C}_v(T_b - T_a) + n\mathcal{C}_p(T_c - T_b)$

$\mathcal{C}_v = \frac{3}{2}R$ and $\mathcal{C}_p = \frac{5}{2}R$, so

$Q_H = (1\text{ mol})(8.31\text{ J/mol/K})\left\{\frac{3}{2}(544\text{ K} - 272\text{ K}) + \frac{5}{2}(1090\text{ K} - 544\text{ K})\right\}$

$= 14.7\text{ kJ}$

(c) W = Area abcda = $p_a V_a = 2.26$ kJ

(d) $|Q_c| = Q_H - W = 12.4$ kJ

(e) $\eta = \frac{W}{Q_L} = \frac{2.26\text{ kJ}}{14.7\text{ kJ}} = 0.153$ or 15.3%

7. $|Q_H| = 250$ J, $|W| = 80$ J

(a) $Q_L = |Q_H| - |W| = 170$ J

(b) $K = \frac{Q_L}{|W|} = \frac{170 \text{ J}}{80 \text{ J}} = 2.1$

9. $K_{hp} = 2.2, \frac{d|W|}{dt} = 3.5 \text{ kW}$

(a) $\frac{d|Q_H|}{dt} = K_{hp}\frac{d|W|}{dt} = 7.7 \text{ kW}$

(b) $Q_L = |Q_H| - |W|$

$$\frac{dQ_L}{dt} = (7.7 \text{ kW} - 3.5 \text{ kW}) = 4.2 \text{ kW}$$

(c) $\text{Cost} = \left(\frac{\$0.10}{\text{kW h}}\right)(3.5 \text{ kW})(1 \text{ h}) = \$0.35$

11.

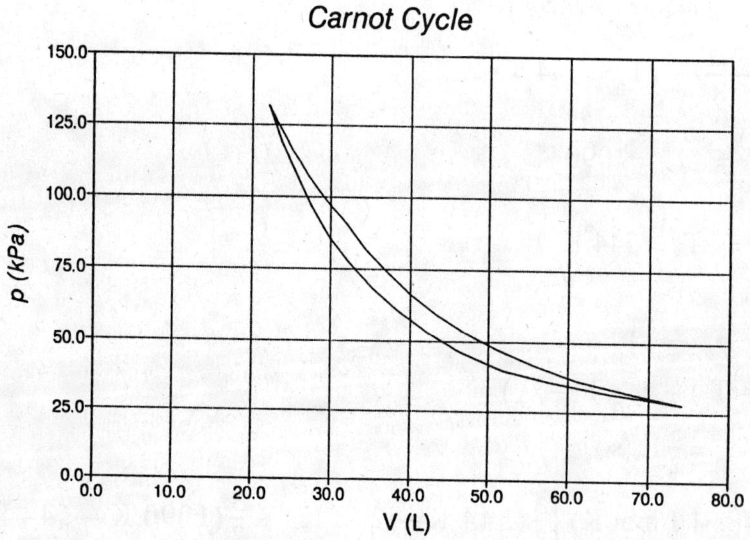

From the diagram, the work can be estimated to be about 300 J.

13. (a) $\frac{|Q_L|}{Q_H} = \frac{T_L}{T_H} = \frac{273 \text{ K}}{373 \text{ K}} = 0.732$

(b) If $Q_H = 0.0125$ J, then $|Q_L| = .732\, Q_H = 0.00915$ J

(c) The answers would be the same because $\frac{|Q_L|}{Q_H} = \frac{T_L}{T_H}$ for any Carnot cycle.

15. For a reversible heat pump,

$$\frac{|Q_H|}{T_H} = \frac{Q_L}{T_L}$$

$$\therefore\ K_{hp} = \frac{|Q_H|}{|W|} = \frac{|Q_H|}{|Q_H| - Q_L} = \frac{T_H}{T_H - T_L}$$

If $T_H = T_L$, then the room and the outdoors are at the same temperature; no work is needed to transfer heat from one to the other and the coefficient of performance becomes infinite. This is the situation that one would encounter when entering an unheated house that has reached ambient temperature. At first, no work is required to bring heat into the house, but as soon as the temperature begins to rise, the coefficient of performance drops.

17. With $T_L = 300$ K and $T_H = 400$ K, $\eta = \frac{T_H - T_L}{T_H} = \frac{100\text{ K}}{400\text{ K}} = 0.250$

(a) With $T_L = 280$ K and $T_H = 400$ K, $\eta = \frac{120\text{ K}}{400\text{ K}} = 0.300$, so $\Delta\eta = 0.050$

(b) With $T_L = 300$ K and $T_H = 420$ K, $\eta = \frac{120\text{ K}}{420\text{ K}} = 0.286$, so $\Delta\eta = 0.036$

(c) With $T_L = 270$ K and $T_H = 400$ K, $\eta = \frac{130\text{ K}}{400\text{ K}} = 0.325$, $\Delta\eta = 0.075$

With $T_L = 300$ K and $T_H = 440$ K, $\eta = \frac{140\text{ K}}{440\text{ K}} = 0.318$, $\Delta\eta = 0.068$

(d) The general conclusion is that the efficiency is improved more by a lowering of the cold temperature than by a corresponding (absolute or fractional) rise in the hot temperature.

19. $T_H = 25°\text{ C} = 298$ K

$T_L = 10°\text{ C} = 283$ K

(a) $\eta = \frac{T_H - T_L}{T_H} = \frac{15\text{ K}}{298\text{ K}} = 0.05$ (5%)

(b) If $\frac{dW}{dt} = 2$ MW, then $\frac{dQ_H}{dt} = \frac{1}{\eta}\frac{dW}{dt} = 40$ MW

(c) $\frac{dQ_H}{dt} = c_v\,\Delta T\,\frac{dm}{dt}$

$$\frac{dm}{dt} = \frac{1}{c_v \, \Delta T} \frac{dQ_H}{dt} = \frac{40 \text{ MW}}{(4180 \text{ J kg}^{-1} \text{ K}^{-1})(1\text{K})} = 9600 \frac{\text{kg}}{\text{s}}$$

$$\frac{dV}{dt} = 9600 \frac{\text{kg}}{\text{s}} \frac{\text{m}^3}{1000 \text{ kg}} = 9.6 \frac{\text{m}^3}{\text{s}}$$

21. Assuming an outside temperature of 38° C and an inside temperature of 26° C, for a Carnot engine,

$$\frac{|W|}{Q_L} = \frac{T_H - T_L}{T_L} = \frac{311 \text{ K} - 299 \text{ K}}{299 \text{ K}}$$

$$= 0.040$$

$$\frac{d|W|}{dt} = 0.040 \frac{dQ_L}{dt} = 0.040 \text{ (20 kW)}$$

$$= 800 \text{ W}$$

This is an underestimate because a real air conditioner will not perform as well as an ideal Carnot refrigerator.

23. (a) $\Delta S = \frac{Q}{T}$

(b) $\Delta S = \frac{30 \text{ J}}{300 \text{ K}} = 0.10 \frac{\text{J}}{\text{K}}$

25. $\Delta S = \frac{Q}{T} = \frac{(2.26 \times 10^6 \text{ J/kg})(1.0 \text{ kg})}{373 \text{ K}}$

$$= 6.1 \frac{\text{kJ}}{\text{K}}$$

The vapor phase has the higher entropy.

27. (a) Let T_f = final temperature.

Heat gained by ice $= m_i L_f + m_i c_p (T_f - T_i)$

$= (0.250 \text{ kg})(0.335 \times 10^6 \text{ J/kg})$

$+ (0.250 \text{ kg})(4.186 \times 10^3 \text{ J/kg K})(T_f - 0° \text{ C})$

Heat lost by water $= m_w c_p (T_w - T_f)$

$= (0.950 \text{ kg})(4.186 \times 10^3 \text{ J/kg K})(85° \text{ C} - T_f)$

Since the sytem is in an adiabatic enclosure, the heat lost by the water is gained by the ice.

$$83.7 \times 10^3\ \text{J} + \left(1.05 \times 10^3\ \frac{\text{J}}{\text{K}}\right)T_f = 338 \times 10^3\ \text{J} - \left(3.98 \times 10^3\ \frac{\text{J}}{\text{K}}\right)T_f$$

Collecting terms,

$$\left(5.02 \times 10^3\ \frac{\text{J}}{\text{K}}\right)T_f = 2.54 \times 10^5\ \text{J}$$

$$\therefore\quad T_f = 51^\circ\ \text{C or } 324\ \text{K}$$

(b) The entropy gained by the ice is

$$\Delta S_i = \frac{mL_f}{T_i} + mc_p \int_{T_i}^{T_f} \frac{dT}{T}$$

$$= \frac{(0.250\ \text{kg})\left(0.335 \times 10^6\ \frac{\text{J}}{\text{kg}}\right)}{273\ \text{K}} + (0.250\ \text{kg})\left(4.186 \times 10^3\ \frac{\text{J}}{\text{kg K}}\right)\ln\left(\frac{324\ \text{K}}{273\ \text{K}}\right)$$

$$= 484\ \frac{\text{J}}{\text{K}}$$

The entropy lost by the water is

$$|\Delta S_w| = mc_p \ln\left(\frac{T_w}{T_f}\right) = (0.950\ \text{kg})\left(4.186 \times 10^3\ \frac{\text{J}}{\text{kg K}}\right)\ln\left(\frac{358.2\ \text{K}}{323.7\ \text{K}}\right) = 402\ \frac{\text{J}}{\text{K}}$$

(c) The entire system gains 82 J/K of entropy.

(d) The process is irreversible because the water and ice cannot be unmixed. Also note that the entropy of the universe has increased by 82 J/K and it cannot be decreased.

29. (a) Entropy lost by high temperature reservoir $= \dfrac{Q_H}{T_H}$

Entropy gained by engine in this part of cycle $= \dfrac{Q_H}{T_H - \Delta T}$

Entropy gained by universe $= \dfrac{Q_H}{T_H - \Delta T} - \dfrac{Q_H}{T_H} = \dfrac{Q_H\,\Delta T}{(T_H - \Delta T)T_H} > 0$

(b) Entropy lost by engine at cooling $= \dfrac{|Q_L|}{T_L + \Delta T}$

Entropy gained by low temperature reservoir $= \dfrac{|Q_L|}{T_L}$

Entropy gained by universe $= \dfrac{|Q_L|}{T_L} - \dfrac{|Q_L|}{T_L + \Delta T} = \dfrac{|Q_L|\,\Delta T}{(T_L + \Delta T)T_L} > 0$

(c) One can conclude that for a real heat engine the entropy of the universe increases in each cycle.

31. (a) During the isochoric part of the reversible process i → a,

$$dS = \frac{nC_v dT}{T}$$

$$\Delta S_1 = nC_v \ln\left(\frac{T_a}{T_i}\right)$$

During the isobaric part,

$$\Delta S_2 = nC_p \ln\left(\frac{T_f}{T_a}\right)$$

Using the ideal gas law,

$$\frac{T_a}{T_i} = \frac{p_a V_a}{p_i V_i} = \frac{p_a}{p_i} = \frac{p_f}{p_i}$$

and $$\frac{T_f}{T_a} = \frac{p_f V_f}{p_a V_a} = \frac{V_f}{V_a} = \frac{V_f}{V_i}.$$

$$\therefore \quad \Delta S_1 = (1\text{ mol})\left(\frac{5}{2}\right)\left(8.31\ \frac{\text{J}}{\text{mol K}}\right)\ln\left(\frac{146\text{ kPa}}{210\text{ kPa}}\right)$$

$$= -7.55\ \frac{\text{J}}{\text{K}}$$

$$\Delta S_2 = (1\text{ mol})\left(\frac{7}{2}\right)\left(8.31\ \frac{\text{J}}{\text{mol K}}\right)\ln\left(\frac{31\text{ L}}{24\text{ L}}\right)$$

$$= 7.44\ \frac{\text{J}}{\text{K}}$$

$$\Delta S = \Delta S_1 + \Delta S_2 = -0.11\ \frac{\text{J}}{\text{K}}$$

(b) Since entropy is a state variable, the change in entropy of the air is –0.11 J/K.

(c) Since the process was irreversible, the universe had a positive change in entropy.

33. $$\frac{dW}{dt} = 550\text{ MW}$$

$$\frac{Q_H - |Q_L|}{Q_H} = \eta = 0.34$$

$$\therefore \quad \frac{|Q_L|}{Q_H} = 0.66; \qquad \frac{W}{Q_H} = 0.34$$

$$|Q_L| = \frac{0.66}{0.34} W = 1.94\ W$$

$$\frac{d|Q_L|}{dt} = (1.94)(550\text{ MW}) = 1070\text{ MW}$$

35. (a) $\frac{d|W|}{dt} = 1.15\text{ kW}, \frac{dQ_L}{dt} = 0.960\text{ kW}$

$$\frac{d|Q_H|}{dt} = \frac{dQ_L}{dt} + \frac{d|W|}{dt} = 2.11\text{ kW}$$

(b) $K_{hp} = \frac{|Q_H|}{|W|} = 1.83$

37. (a)
$$\Delta S = mc_p \ln\left(\frac{T_f}{T_i}\right)$$
$$= (0.50\text{ kg})\left(386\ \frac{J}{kg\ K}\right)\ln\left(\frac{300\ K}{310\ K}\right)$$
$$= -6.3\ \frac{J}{K}$$

(b)
$$\Delta S = (0.50\text{ kg})\left(386\ \frac{J}{kg\ K}\right)\ln\left(\frac{300\ K}{290\ K}\right)$$
$$= 6.5\ \frac{J}{K}$$

(c) The system comes to thermal equilibrium at 300 K. Thus the entropy change for the universe is

$$\Delta S = \left(6.5\ \frac{J}{K}\right) - \left(6.3\ \frac{J}{K}\right) = 0.2\ \frac{J}{K}$$

39. Given: $n = 0.361$ mol, $\eta = 0.500$, $T_H = 602$ K, $\frac{V_b}{V_a} = 2$.

(a) $\eta = \frac{T_H - T_L}{T_H} \Rightarrow T_L = (1 - \eta)T_H = (1 - .500)602\text{ K} = 301\text{ K}$

(b) $Q_H = \int_a^b p\,dV = nRT_H \ln\left(\frac{V_b}{V_a}\right) = (0.361 \text{ mol})\left(8.31 \frac{\text{J}}{\text{mol K}}\right)(602 \text{ K})\ln 2 = 1252 \text{ J}$

(c) $|Q_L| = (1 - \eta)Q_H = 626 \text{ J}$

(d) $W = Q_H - |Q_L| = 626 \text{ J}$

41. (a) $\eta = \frac{W}{Q_H} \Rightarrow W = Q_H\eta = (100 \text{ J})(0.60) = 60 \text{ J}$

(b) $|Q_L| = Q_H - W = 100 \text{ J} - 60 \text{ J} = 40 \text{ J}$

(c) For a Carnot cycle, $\eta = 1 - \frac{T_L}{T_H} = 1 - \frac{300 \text{ K}}{1200 \text{ K}} = 0.750$

$\therefore\ W = Q_H\eta = (100 \text{ J})(0.750) = 75 \text{ J}$

(d) Since the Carnot cycle could have produced 75 J while the real engine only produces 60 J, 15 J have become unavailable.

(e) In a complete cycle, the entropy of the working substance is unchanged, $\Delta S = 0$.

(f) For the universe,

$$\Delta S = -\frac{Q_H}{T_H} + \frac{|Q_L|}{T_L} = -\frac{100 \text{ J}}{1200 \text{ K}} + \frac{40 \text{ J}}{300 \text{ K}} = 0.50 \frac{\text{J}}{\text{K}}$$

Problems:

1. For the isothermal expansion at T_H, $\Delta S = \int \frac{dQ}{T_H} = \frac{Q_H}{T_H}$

For the adiabatic cooling to T_L, $\Delta S = 0$ because $\Delta Q = 0$

For the isothermal compression at T_L, $\Delta S = \int \frac{dQ}{T_L} = \frac{Q_L}{T_L}$

For the adiabatic compression that returns the working fluid to the initial state, $\Delta S = 0$.

3. Suppose that the "superengine" (the one that is more efficient than a reversible engine) operates between T_H and T_L and that it has an efficiency η_S. If it extracts Q_H^S from the high temperature reservoir, then $W^S = \eta_S Q_H^S$ and $|Q_L| = (1 - \eta_S)Q_H^S$. Now let us take a reversible engine with efficiency η and use it as a refrigerator to extract an amount $|Q_L|$ of heat from the low temperature reservoir and relaease $|Q_H|$ to the high temperature one. Now, $Q_H = \frac{|Q_L|}{1 - \eta}$ and the work done on the engine is $|W| = \eta Q_H$. For the entire system, the following are the net results:

(1) The low temperature reservoir is unchanged: it neither loses nor gains heat.

(2) The high temperature reservoir loses heat in the amount

$$Q_H^S - |Q_H| = \frac{|Q_L|}{1 - \eta_S} - \frac{|Q_L|}{1 - \eta} = \frac{(\eta_S - \eta)|Q_L|}{(1 - \eta_S)(1 - \eta)} > 0$$

(3) The engine does work in the amount

$$W^S - |W| = \eta_S Q_H^S - \eta Q_H = \frac{[\eta_S(1 - \eta) - \eta(1 - \eta_S)]\,|Q_L|}{(1 - \eta_S)(1 - \eta)} = \frac{(\eta_S - \eta)\,|Q_L|}{(1 - \eta_S)(1 - \eta)};$$

that is, an amount equal to the heat transferred from the high temperature reservoir. This would be a violation of the Kelvin–Planck statement of the second law.

5. (a) For the cycle a → b → c → d → a shown in Figure 19–21, we can calculate Q and W as follows:

During the adiabatic compression a → b,

$$W_{ab} = \int_a^b p\,dV = \int_a^b p_a V_a^\gamma \frac{dV}{V^\gamma} = \frac{p_a V_a^\gamma}{\gamma - 1}\left(\frac{1}{V_a^{\gamma-1}} - \frac{1}{V_b^{\gamma-1}}\right)$$

$$= \frac{1}{\gamma - 1}(p_a V_a - p_b V_b) = \frac{nR}{\gamma - 1}(T_a - T_b)$$

and $Q_{ab} = 0$.

During the isochoric process b → c, $W_{bc} = 0$ and $Q_{bc} = nC_V(T_c - T_b)$.

During the expansion c → d, $W_{cd} = \frac{nR}{\gamma - 1}(T_c - T_d)$ and $Q_{cd} = 0$.

In the isochoric process d → a, $W_{da} = 0$ and $Q_{da} = nC_V(T_a - T_d)$.

For the entire cycle,

$$Q_H = Q_{bc} = n\mathcal{C}_v(T_c - T_b)$$

$$W = \frac{nR}{\gamma - 1}(T_a - T_b + T_c - T_d)$$

$$\eta = \frac{W}{Q_H} = \frac{R}{\mathcal{C}_p - \mathcal{C}_v}\frac{(T_a - T_b + T_c - T_d)}{T_c - T_b}$$

$$= 1 - \frac{T_d - T_a}{T_c - T_b} \text{ (since } R = \mathcal{C}_p - \mathcal{C}_v)$$

Now, $T_dV_d^{\gamma-1} = T_cV_c^{\gamma-1}$ and $T_aV_a^{\gamma-1} = T_bV_b^{\gamma-1}$

so
$$\eta = 1 - \frac{T_c\dfrac{V_c^{\gamma-1}}{V_d^{\gamma-1}} - T_a}{T_c - T_a\dfrac{V_a^{\gamma-1}}{V_c^{\gamma-1}}} = 1 - \frac{\dfrac{T_c}{r^{\gamma-1}} - T_a}{T_c - T_a r^{\gamma-1}}$$

$$= 1 - \frac{1}{r^{\gamma-1}}$$

(b) For $r = 8$ and $\gamma = 1.4$, $\eta = 1 - \frac{1}{8^{.4}} = 0.56$

7. $Q_H = 1200\text{ J}, \eta = 0.25$

$$W = \eta Q_H = (0.25)(1200\text{ J}) = 300\text{ J}$$

$$|Q_L| = Q_H - W = 900\text{ J}$$

(a) $$\Delta S_{univ} = -\frac{Q_H}{T_H} + \frac{|Q_L|}{T_L}$$

$$= -\frac{1200\text{ J}}{550\text{ K}} + \frac{900\text{ J}}{350\text{ K}} = 0.390\ \frac{\text{J}}{\text{K}}$$

(b) $$\eta_{rev} = 1 - \frac{T_L}{T_H} = 1 - \frac{350\text{ K}}{550\text{ K}} = 0.364$$

$W_{rev} = \eta_{rev}Q_H = 436\text{ J}$, so the reversible engine could perform an additional 136 J of work per cycle.

(c) $$\Delta S_{univ} = -\frac{Q_H}{T_H} + (1-\eta)\frac{Q_H}{T_L}$$

$$W_{rev} - W = (\eta_{rev} - \eta)Q_H$$

$$= \left(1 - \frac{T_L}{T_H} - \eta\right)Q_H$$

$$= T_L \Delta S_{univ}$$

For this example, $T_L \Delta S_{univ} = (350\ K)\left(0.390\ \frac{J}{K}\right) = 136\ J$, verifying the relation in this particular instance.

9. To calculate the entropy change, we effect the change reversibly in two steps: (1) expand isothermally at $T_1 = 280$ K from $V_1 = 22$ L to $V_2 = 44$ L; (2) heat at constant volume to raise the temperature to $T_2 = 1120$ K.

$$\Delta S = nR \ln\left(\frac{V_2}{V_1}\right) + nC_V \ln\left(\frac{T_2}{T_1}\right)$$

$$= (1.0\ mol)\left(8.31\ \frac{J}{mol\ K}\right) \ln\left(\frac{44\ L}{22\ L}\right) + (1.0\ mol)\left(\frac{3}{2}\right)\left(8.31\ \frac{J}{mol\ K}\right) \ln\left(\frac{1120\ K}{280\ K}\right)$$

$$= 23\ \frac{J}{K}$$

11. $$\Delta S = n\int_{T_i}^{T_f} \frac{C_V\, dT}{T}$$

$$= nR\int_{T_i}^{T_f}\left[\frac{5}{2} + \frac{\left(\frac{\theta}{T}\right)^2 e^{\theta/T}}{(e^{\theta/T} - 1)^2}\right]\frac{dT}{T}$$

Let $x = \frac{\theta}{T}, \frac{dx}{x} = -\frac{dT}{T}$

$$\Delta S = nR\int_{x_f}^{x_i}\left[\frac{5}{2} + \frac{x^2 e^x}{(e^x - 1)^2}\right]\frac{dx}{x}$$

Here $\theta = 85$ K, $T_i = 10$ K, $T_f = 200$ K, so that $x_f = \frac{85\ K}{200\ K} = 0.425$, $x_i = 8.5$,

so $$\frac{\Delta S}{nR} = \int_{0.425}^{8.5}\left[\frac{5}{2} + \frac{x^2 e^x}{(e^x - 1)^2}\right]\frac{dx}{x}$$

This integral is evaluated in the spreadsheet shown below.

x	f(x)	F(x)
8.0	0.315186	9.198918
8.025	0.314154	9.206785
8.05	0.313129	9.214626
8.075	0.312112	9.222442
8.1	0.311102	9.230232
8.125	0.310099	9.237997
8.15	0.309103	9.245737
8.175	0.308114	9.253452
8.2	0.307131	9.261143
8.225	0.306156	9.268809
8.25	0.305187	9.27645
8.275	0.304224	9.284068
8.3	0.303269	9.291662
8.325	0.302319	9.299232
8.35	0.301376	9.306778
8.375	0.300439	9.3143
8.4	0.299509	9.3218
8.425	0.298584	9.329276
8.45	0.297666	9.336729
8.475	0.296754	9.344159
8.5	0.295848	9.351567

The value of the integral is 9.35, so $\Delta S = (1\text{ mol})\left(8.31\ \frac{\text{J}}{\text{K mol}}\right)(9.35) = 77\ \frac{\text{J}}{\text{K}}$